S0-EDB-726

COLONIZATION CONTROL OF HUMAN BACTERIAL ENTEROPATHOGENS IN POULTRY

FOOD SCIENCE AND TECHNOLOGY

A Series of Monographs

A complete list of the books in this series appears at the end of the volume.

COLONIZATION CONTROL OF HUMAN BACTERIAL ENTEROPATHOGENS IN POULTRY

Edited by

Leroy C. Blankenship

Coedited by

J. Stan Bailey
Nelson A. Cox
Stephen E. Craven
Richard J. Meinersmann
Norman J. Stern

Richard B. Russell Agricultural Research Center
Agricultural Research Service
United States Department of Agricultural
Athens, Georgia

ACADEMIC PRESS, INC.
Harcourt Brace Jovanovich, Publishers
San Diego New York Boston London Sydney Tokyo Toronto

Academic Press Rapid Manuscript Reproduction

This book is printed on acid-free paper. ♾

Index prepared by Alan M. Greenberg.

Academic Press, Inc.
San Diego, California 92101

United Kingdom Edition published by
Academic Press Limited
24–28 Oval Road, London NW1 7DX

Library of Congress Cataloging-in-Publication Data

Colonization control of human bacterial enteropathogens in poultry /
edited by Leroy C. Blankenship.
p. cm.
Includes index.
ISBN 0-12-104280-4
1. Salmonellosis in poultry--Prevention. 2. Campylobacter infections in poultry--Prevention. 3. Bacterial diseases in poultry--Prevention. 4. Meat--Contamination. I. Blankenship, Leroy C
[DNLM: 1. Campylobacter--isolation & purification. 2. Food Contamination--prevention & control. 3. Poultry--microbiology. 4. Salmonella--isolation & purification. QW 138.5.S2 C718]
SF995.6.S3C65 1991
664'.93--dc20
DNLM/DLC
for library of Congress 91-4577
CIP

PRINTED IN THE UNITED STATES OF AMERICA
91 92 93 94 9 8 7 6 5 4 3 2 1

CONTENTS

CONTRIBUTORS

Numbers in parentheses indicate the pages on which the authors' contributions begin.

T. S. Agin (267), Department of Microbiology, Miami University, Oxford, Ohio 45056

R. C. Axtell (3), North Carolina Agricultural Extension Service, North Carolina State University, Raleigh, North Carolina 27695

J. S. Bailey (1, 105, 347, 371) Poultry Microbiological Safety Research Unit, United States Department of Agriculture, Agricultural Research Service, Richard B. Russell Agricultural Research Center, Athens, Georgia 30613

P. A. Barrow (199) Department of Microbiology, Institute for Animal Health, Houghton Laboratory, Houghton, Huntingdon, Cambridgeshire, PE17 2DA, United Kingdom

C. W. Beard (161) Southeast Poultry Research Laboratory, United States Department of Agriculture, Agricultural Research Service, Athens, Georgia 30605

R. C. Beier (299) Veterinary Toxicology and Entomology Research Laboratory, United States Department of Agriculture, Agricultural Research Service, College Station, Texas 77840

W. H. Benjamin, Jr. (365) Department of Microbiology, University of Alabama, Birmingham, Alabama 35294

C. E. Benson (149) Department of Clinical Studies, University of Pennsylvania, New Bolton Center, Kennett Square, Pennsylvania 19348

B. Blanchfield (323) Bureau of Microbial Hazards, Health Protection Branch, Health and Welfare Canada, Ottawa, Ontario K1A 0L2

L. C. Blankenship (105, 347, 371, 375) Poultry Microbiological Safety Research Unit, United States Department of Agriculture, Agricultural Research Service, Richard B. Russell Agricultural Research Center, Athens, Georgia 30613

N. M. Bolder (77, 353, 359) Spelderholt Centre for Poultry Research and Information Services, Agricultural Research Service, 7361 DA, Beekbergen, The Netherlands

D. E. Briles (365) Department of Microbiology, University of Alabama, Birmingham, Alabama 35294

W. E. Briles (365) Department of Biologic Sciences, Northern Illinois University, DeKalb, Illinois 60115

W. J. Cameron (287) Willmar Poultry Company Inc., Willmar, Minnesota 56201

K. S. Chung (219) Protozoan Diseases Laboratory, United States Department of Agriculture, Agricultural Research Service, Beltsville Agricultural Research Center, Beltsville, Maryland 20705

R. C. Clarke (133) Department of Veterinary Microbiology and Immunology, Veterinary College, University of Guelph, Guelph, Ontario, Canada N1G 2W1

D. E. Corrier (299) Veterinary Toxicology and Entomology Research Laboratory, United States Department of Agriculture, Agricultural Research Service, College Station, Texas 77840

P. F. Cotter (335) Biology Department, Framingham State College, Framingham, Massachusetts 01701

N. A. Cox (105, 347, 371) Poultry Microbiological Safety Research Unit, United States Department of Agriculture, Agricultural Research Service, Richard B. Russell Agricultural Research Center, Athens, Georgia 30613

S. E. Craven (371) Poultry Microbiological Safety Research Unit, United States Department of Agriculture, Agricultural Research Service, Richard B. Russell Agricultural Research Center, Athens, Georgia 30613

R. Curtiss III (169) Department of Biology, Washington University, St. Louis, Missouri 63130

J. R. DeLoach (299) Veterinary Toxicology and Entomology Research Laboratory, United States Department of Agriculture, Agricultural Research Service, College Station, Texas 77840

H. W. Dickerson (343) Department of Medical Microbiology, College of Veterinary Medicine, University of Georgia, Athens, Georgia 30602

M. P. Doyle (121) Department of Food Microbiology and Toxicology, Food Research Institute, University of Wisconsin-Madison, Madison, Wisconsin 53706

R. J. Eckroade (149) Department of Clinical Studies, University of Pennsylvania, New Bolton Center, Kennett Square, Pennsylvania 19348

H. E. Ekperigin (287) Poultry Consultant, Davis, California 95616

D. A. Gabis (23) Silliker Laboratories, Inc., Chicago Heights, Illinois 60411

R. K. Gast (161) Southeast Poultry Research Laboratory, United States Department of Agriculture, Agricultural Research Service, Athens, Georgia 30605

C. Gentry-Weeks (169) Department of Biology, Washington University, St. Louis, Missouri 63130

T. M. Gleeson (323) Bureau of Microbial Hazards, Health Protection Branch, Health and Welfare Canada, Ottawa, Ontario K1A 0L2

C. L. Gyles (133) Department of Veterinary Microbiology and Immunology, Veterinary College, University of Guelph, Guelph, Ontario, Canada N1G 2W1

T. L. Hanner (309) Department of Animal Science, North Carolina A&T State University, Greensboro, North Carolina 27411

J. O. Hassan (169, 277) Department of Biology, Washington University, St. Louis, Missouri 63130

A. Hinton, Jr. (299) Veterinary Toxicology and Entomology Research Laboratory, United States Department of Agriculture, Agricultural Research Service, College Station, Texas 77840

M. Hinton (259) Department of Veterinary Medicine, University of Bristol, Langford House, Langford, Avon BS18 7DU, United Kingdom

A. M. Horne (315) Georgia Poultry Laboratory, Oakwood, Georgia 30566

W. T. Hubbert (xxix) United States Department of Agriculture, FSIS, Washington, D.C. 20250

F. Jones (3) North Carolina Agriculture Extension Service, North Carolina State University, Raleigh, North Carolina 27695

S. M. Kelly (169) Department of Biology, Washington University, St. Louis, Missouri 63130

C. J. Kim (243) Department of Veterinary Pathobiology, College of Veterinary Medicine, University of Minnesota, St. Paul, Minnesota 55108

J. D. Klinger (335) GENE-TRAK Systems, Inc., Framingham, Massachusetts 01701

M. C. Kumar (243) E. B. Olson Farms, Division of Jennie-o Foods, Inc., Willmar, Minnesota 56201

H. S. Lillehoj (219) Protozoan Diseases Laboratory, United States Department of Agriculture, Agricultural Research Service, Beltsville Agricultural Research Center, Beltsville, Maryland 20705

R. H. McCapes (287) University of California, Davis, California 95616

G. C. Mead (91) Agricultural and Food Research Council, Institute of Food Research, Bristol Laboratory, Langford, Bristol BS18 7DY, United Kingdom

R. J. Meinersmann (167, 331, 343) Poultry Microbiological Safety Research Unit, United States Department of Agriculture, Agricultural Research Service, Richard B. Russell Agricultural Research Center, Athens, Georgia 30613

A. Morris-Hooke (267) Department of Microbiology, Miami University, Oxford, Ohio 45056

R. W. A. W. Mulder (77, 353, 359) Spelderholt Centre for Poultry Research and Information Services, Agricultural Research Service, 7361 DA, Beekbergen, The Netherlands

M. Munson (169) Department of Biology, Washington University, St. Louis, Missouri 63130

J. E. Murphy (335) GENE-TRAK Systems, Inc., Framingham, Massachusetts 01701

C. Murray (309) Department of Animal Science, North Carolina A&T State University, Greensboro, North Carolina 27411

K. V. Nagaraja (243, 287) Department of Veterinary Pathobiology, College of Veterinary Medicine, University of Minnesota, St. Paul, Minnesota 55108

E. Nurmi (57) National Veterinary Institute, Helsinki, Finland

B. S. Pomeroy (243) College of Veterinary Medicine, University of Minnesota, St. Paul, Minnesota 55108

C. Poppe (133) Department of Veterinary Microbiology and Immunology, Veterinary College, University of Guelph, Guelph, Ontario, Canada N1G 2W1

S. B. Porter (169) Department of Biology, Washington University, St. Louis, Missouri 63130

R. Redus (287) VE Corporation, Arlington, Texas 76011

W. E. Rigsby (331) Electron Microscopy Laboratory, Poultry Microbiological Safety Research Unit, United States Department of Agriculture, Agricultural Research Service, Richard B. Russell Agricultural Research Center, Athens, Georgia 30613

W. L. Ritchie (287) Farm Service Elevator Company, Willmar, Minnesota 56201

D. V. Rives (3) North Carolina Agriculture Extension Service, North Carolina State University, Raleigh, North Carolina 27695

R. D. Rolfe (59) Department of Microbiology, School of Medicine, Texas Tech University Health Sciences Center, Lubbock, Texas 79430

D. M. Rollins (47) Infectious Diseases Department, Naval Medical Research Institute, Bethesda, Maryland 20814

S. E. Scheideler (3) North Carolina Agriculture Extension Service, North Carolina State University, Raleigh, North Carolina 27695

S. M. Shane (29) Department of Epidemiology and Community Health, School of Veterinary Medicine, Louisiana State University, Baton Rouge, Louisiana 70803-8404

G. H. Snoeyenbos (xxxi) Department of Veterinary and Animal Sciences, Paige Laboratory, University of Massachusetts, Amherst, Massachusetts 01003

S. Stavric (323) Bureau of Microbial Hazards, Health Protection Branch, Health and Welfare Canada, Ottawa, Ontario K1A 0L2

N. J. Stern (119, 331, 371) Poultry Microbiological Safety Research Unit, United States Department of Agriculture, Agricultural Research Service, Richard B. Russell Agricultural Research Center, Athens, Georgia 30613

F. R. Tarver (3) North Carolina Agriculture Extension Service, North Carolina State University, Raleigh, North Carolina 27695

Robert V. Tauxe (xv) Division of Bacterial Diseases, Center for Infectious Diseases, Center for Disease Control, Atlanta, Georgia 30333

R. L. Taylor, Jr. (335) Department of Animal and Nutritional Sciences, University of New Hampshire, Durham, New Hampshire 03824

S. A. Tinge (169) Department of Biology, Washington University, St. Louis, Missouri 63130

W. D. Waltman II (315, 365) Georgia Poultry Laboratory, Oakwood, Georgia 30566

W. L Willis (309) Department of Animal Science, North Carolina A&T State University, Greensboro, North Carolina 27411

M. J. Wineland (3) North Carolina Agriculture Extension Service, North Carolina State University, Raleigh, North Carolina 27695

R. L. Ziprin (299) Veterinary Toxicology and Entomology Research Laboratory, United States Department of Agriculture, Agricultural Research Service, College Station, Texas 77840

FOREWORD: TRANSMISSION OF HUMAN BACTERIAL PATHOGENS THROUGH POULTRY (BANQUET ADDRESS)

Robert V. Tauxe

Centers for Disease Control
Atlanta, Georgia

My thanks to the organizers of the symposium and particularly to Norman Stern for that introduction. Of all the things he could have said, mentioning my passion for Belgian beer was pretty mild. To introduce myself: I came to the CDC after completing my residency in internal medicine. I joined the Epidemic Intelligence Service, a two year training program for physicians and veterinarians and others on conducting the work of public health epidemiology. After that, I joined the staff in the Enteric Diseases Branch. That Branch is active in many different areas, from botulism and cholera to *Yersinia* infections and includes both laboratory and epidemiologic components. I am an epidemiologist, and I have little expertise in the lab.

The ultimate objective of our group, like that of the entire CDC, is disease control. There are many methods available to us for use in disease control. The main ones have been to encourage people to stop smoking, to fasten their seat belts, and whatever they do, to use condoms! It doesn't seem that any of these are going to make any great deal of difference with the illnesses that we are addressing in this symposium. This means we are still trying to understand the transmission of bacterial enteropathogens well enough to be able to control them.

When I first came to CDC, I was interested in the moment of contact when a pathogen meets a host. For instance, what happens when, in an African country, someone dies of cholera? The victim's family holds a large funeral, and a number of people contract cholera from the meal that is served. What happens during the equally interesting rites of university fraternity parties where foodborne *Campylobacter* can be transmitted? The basic tools of epidemiology are simple and include, for example, the carefully conducted case control study. We also have the basic outbreak investigation, the collection of routine surveillance data, and from time to time, the conducting of special surveys.

I want to talk tonight about the transmission of human disease pathogens through poultry. There may be many possible pathogens to consider, but *Salmonella* and *Campylobacter* lead the list. I will provide a quick overview of the magnitude of the public health problem as we perceive it and include the results of some of our investigations into the transmission of these organisms. In organizing this talk, I had to make the difficult decision about whether to talk about the chicken or the egg first. I chose the chicken, and we will get to the egg a little later.

Salmonella infections are now relatively common in the United States, but they weren't always, or at least they were not recognized to be. It was not until after World War II that we recognized strains of *Salmonella* other than *Salmonella typhi* (which causes typhoid fever) as the cause of a public health problem in the United States. State Health Departments began reporting non-typhoid causes of human salmonellosis about 1943. In 1963, state health departments began serotyping the *Salmonella* referred to them from clinical laboratories and forwarding that information to the CDC. That system of reporting began as a series of outbreaks of salmonellosis in hospitals up and down the East coast, usually due to *Salmonella derby* from contaminated eggs. The eggs were unwashed and fecally contaminated, and an enormous number of outbreaks occurred. Even in those times (which were primitive by comparison to our modern epidemiological standards), the findings eventually pointed to eggs as the problem. This was the genesis of modern *Salmonella* surveillance as we know it. The problem of egg-associated salmonellosis virtually disappeared.

However, *Salmonella* continued to increase steadily over the years, and in recent years, 40,000 to 50,000 isolates per year from humans have been reported in this country. This number has strained our data handling capacities to the limit. We estimate that in one percent of reported cases, the person from whom the *Salmonella* was isolated died in the setting of that infection. We estimate the cost of the reported cases to be at least 50 million dollars in medical expenses and lost wages. I should stress that the reported cases represent a statistically improbable sequence of events including: 1) individuals who were sick enough to visit physicians (who in turn decided to take stool cultures), 2) microbiologists who succeeded in isolating *Salmonella* from the stool culture, and then 3) remembered to send the cultures to the state health department for serotyping. There are a lot of reasons why a case would not be reported or diagnosed. Clinical laboratories in this country are growing ever more expensive and the state health departments, reeling under the impact of the AIDS epidemic, are trying to stretch smaller budgets for such things as serotyping to cover the larger number of isolates referred. So, we estimate that a small fraction, between one and five percent of all cases are reported, and the actual number of infections is much much greater than 40,000, falling roughly between one and four million.

For *Campylobacter*, the story began more recently. We only began collecting nationwide surveillance data in 1982. Many states have not yet made this a reportable infection and so our reporting for *Campylobacter* is less complete than for *Salmonella*. However, several studies have shown that *Campylobacter* infections are even more common than *Salmonella*. We conducted a study with about eight clinical laboratories around the country back in 1979, when *Campylobacter* was first recognized as a human pathogen. We asked them to use Blaser's diarrheal medium for *Campylobacter* whenever they were going to culture a stool for anything. In this study, the laboratories found *Campylobacter* in stools with twice the frequency of *Salmonella*. We think that if *Campylobacter* reporting were as good as even the imperfect *Salmonella* reporting system, we would be overwhelmed by reports of close to 100,000 cases a year out of an actual number of infections that must also be in the millions.

A question that we are often asked is: What proportion of these infections come from one or another food vehicle that has been reported in the press? We have had some refreshing questions, such as: What proportion of salmonellosis cases come from quail eggs served in sushi bars? In the last two years since a report on a popular television program, a frequently asked question has been: What proportion comes from poultry? A problem we have is that patients do not arrive at the doctor's office with a label on their forehead saying, "I got this from poultry." It is a bit of a challenge to figure out which illness they had in the first place. The problem is that an individual's case could come from any number of possible reservoirs in their home, such as infected foods, the pets, or some other source.

The potential sources of contamination are very broad. For instance, we had a very unusual case of *Plesiomonas* infection, which is not on the agenda tonight and has nothing to do with poultry as far as I know. It is hard to catch *Plesiomonas* in this country. In this case, a young child was being cared for in the home of a baby-sitter. The baby-sitter kept piranhas and once a week she dumped out the fish tank water into the bathtub. The baby-sitter would also bathe the baby in the tub without cleaning it. That is what it took to get a *Plesiomonas* infection. Patients do not come in labeled, "I just took a bath in old piranha water." You have to ask or you never find out.

How we go about deciding what proportion of the cases come from piranha water is difficult. It is almost impossible to assemble 50 cases of *Plesiomonas* in the first place. We tried that once, but unfortunately we did not know to ask victims about piranhas, though they were asked if they had been fishing. To find out, we select about 50 cases and ask them standard questions. Then we search for someone else in their neighborhood of similar age, but not infected, and ask them all the same questions. If we are lucky, a pattern begins to emerge. It may turn out that the selected cases didn't come from piranha water. Our cases might turn out to have a real fondness for something the healthy control people never ate, like beef jerky or rattlesnake steaks. Or the cases may report that they ate a common food much more frequently than controls did, or that they prepared it in a different way than the controls. If these differences come up more often than we expect by chance alone, then we associate a specific vehicle with that infection, be it the piranha water or the sandwich from the vending machine. This simple but powerful technique, the case-control study, is our epidemiologic bread and butter. We can use it in the outbreak setting if we have 50 cases of *Plesiomonas* infection at the day care center, or we can use it with individual sporadic cases.

Let me review some of our data about *Campylobacter* infections. We started collecting surveillance data in 1982 and continued through 1986. A summary of the data was published as a supplement in the *Morbidity Mortality Weekly Report* in 1988. Who gets *Campylobacter* infections in this country? *Campylobacter* infections peak among infants in their first year of life, then they decline. Then there is a very surprising second peak among adults in the age group of 20-30 years. This Brontosaurus-shaped curve is unique to *Campylobacter*. When you are first exposed to food of any sort in the home, weaning is when you get your first chance to learn what to eat and what not to eat. Then there comes a critical time in everyone's life when they leave of their own free will and set up homes for themselves. All of a sudden, they have to cook! That's when you know they didn't learn how. They weren't paying attention in Home Economics class, and people don't pay much attention to how teenagers learn to cook these days. I have a strong hunch that a lot of them never really learn to cook. So I call this the "second weaning effect," when you are weaned from your good home cooking and all of a sudden you must make do for yourself. A clue to what I think is going on is that men more than women experience campylobacteriosis. Having been in some of the small dormitory rooms with a hot-plate and five guys trying to stretch a budget, it is easy to see how.

I want to talk about a couple of the unique features of *Campylobacter*; 1982 was not a good year. Very few states were reporting, but the reports suggested a slight increase in cases. Actually, that's because a few more states signed on every year, and it is hard for us to use these data to show any increase at all. The increase we observed was due to more and more states reporting to us. About 98 percent of cases reported to us were caused by *Campylobacter jejuni*. There are several interesting points. One is that every year the winter low is brief; but, in May of every year it swoops up to a very sharp peak, and never really falls off again. It continues until about October or November and then falls again. In 1984 it went up, then went down and came up again in November, and then fell again in December. If I showed you the *Salmonella* incidence curve, it would have a smooth characteristic. To me, the *Campylobacter* incidence curve has almost a square wave, on/off characteristic, and it puzzles me that it can drop so sharply from November to January. Maybe there is an end of year clean-out, and in January things are base line again. The majority of cases result when one individual who doesn't know anybody else gets sick and doesn't seem to be part of an outbreak. Sometimes, we do get fairly large outbreaks. About 60 of them have been reported in the last ten years. A very small number of these outbreaks, maybe three, have been traced to poultry, but the great majority are from contaminated fluid vehicles, mainly raw milk or drinking water.

Let us consider the seasonal distribution of reported *Campylobacter* isolations (1982-1986) and outbreaks (1978-1986) by month and vehicle in the United States. Waterborne incidences occurred in the months of March, May, June, and September. I have no explanation for this. Raw milk outbreaks were predominant in April, May, June, October, and December. Interestingly though, the greatest number of isolations were reported in June, July, August, and September, and were attributed to foods other than raw milk and water. The fact that the differences occur is an important clue to us, and I think it is telling us that the outbreaks caused by raw milk and contaminated water are something different from what is going on with the cases occurring in the middle of the summer.

Fortunately, sporadic *Campylobacter* infections have been pretty well studied. When *Campylobacter* came along and the CDC became concerned in 1978-79, tremendous interest in this remarkable pathogen developed. A great deal had to be learned, and there was a great interest in doing sporadic case control studies. Where were those sporadic cases coming from? Three large studies were organized and conducted fairly early. The FDA funded a big study in Seattle, Washington; Dr. Deming and myself conducted a study at the University of Georgia in Athens; and another CDC group conducted a study in Dubuque, Iowa.

The technique was basically the same as the technique I outlined to you before. The results were a little different in each case, but they compliment each other. In Seattle, the basic finding was that at least 50 percent of the cases could be attributed to poultry. At the same time, raw milk accounted for five percent. Having pets in the home, particularly ones with diarrhea, accounted for about six percent. Nine percent were related to foreign travel. Another eight percent were linked to hiking in the Cascade Mountains and drinking mountain water or exposure to surface water.

We got involved at the University of Georgia when a mysterious ailment affected the Bulldogs and their performance started to slip. The Student Health Service identified the reason for their poor performance as *Campylobacter*. At that point, it was time for a Federal Investigation. We were able to do a very nice collaborative study with the university and the county health department. The results were good. The university had an extremely aggressive Student Health Service microbiologist. Basically, we learned that the students in the University of Georgia study were particularly unusual as far as *Campylobacter* goes. We attribute 70 percent of those cases to poultry consumption.

Another 30 percent were attributed to having some kind of contact with a cat, and nobody had been traveling or drinking mountain water or raw milk.

In Dubuque, Iowa, the situation was different. Generally rural, the area surrounding Dubuque was inhabited by many dairy farmers. Drinking raw milk was very popular there, and a full 47 percent of the cases of campylobacteriosis were attributed to raw milk consumption.

Exactly what is involved? The bottom line for us in these studies is that a large proportion, easily half, of *Campylobacter* infections are associated with poultry. What do you have to do to eliminate *Campylobacter* from poultry? The case control study in Seattle identified an important co-factor, which was simply that people reported using a cutting board for cleaning poultry. However, they didn't routinely wash the board with soap. Some smaller case control studies in Colorado had similar findings. Sometimes it was the person who was handling the raw chicken meat, and sometimes it was the person who was eating under-cooked chicken.

Now, in Georgia, we found that there were three things the students were facing. This is part of the reason why I became convinced that someone needed to teach students how to properly cook poultry. We found a very small number of students who were actually eating chicken raw! It wasn't a prank, either. Incredible! A much greater number of the students were eating chicken that was under-cooked. A typical scenario goes like this: individuals under the influence of mind-altering substances would attend a fraternity party where chicken was being cooked on a grill; The grill was too hot and the outside of the chicken cooked faster than the inside. Result: raw chicken rapidly eaten, and a *Campylobacter* epidemic occurred that affected everyone in attendance! A third factor was that the students reused a knife or cutting board after cutting raw chicken without washing it first.

The educational message should be fairly straight forward. I've brought this up three times before to groups of university students, and I haven't figured out yet how to get them interested. They think that this has got to be the most boring thing to talk about in the world and there are a whole lot more interesting things to talk about.

I'm going to leave you with a puzzle. It is plausible that simple food handling errors are going to transmit infections because there is a lot of information on what the infectious dose is for *Campylobacter*. Unfortunately, it is very low. Robinson produced illness in himself with a single gulp of milk containing 500 cells. There have been volunteer studies which show that an infectious dose can be as low as 800 cells. This means that *Campylobacter* doesn't need to multiply on food in order to cause illness. Moreover, the quantities present on a raw market chicken are sufficient to cause an infection. The puzzle is that low infectious dose diseases usually transmit easily from one person to the next, and people working with that organism in the laboratory are likely to get it. Neither seems to be the case with *Campylobacter*. Thus, why is it so efficient at causing individual sporadic cases among persons who are handling or eating chicken, but so inefficient at causing large outbreaks due to consumption of chicken transmitted from person to person? Some of it has to do with the delicate nature of the organism. It is sensitive to destruction by drying, exposure to oxygen, and is readily killed during proper cooking. The closer you are to the raw chicken as it comes out of the door of the refrigerator, the more likely you are to become infected.

I want to turn to the problem of *Salmonella*. Part of our job at CDC is answering questions, and one of the most common questions asked recently is how many *Salmonella* infections come from poultry. It always surprises the reporters, though it probably won't surprise you, that there is no easy answer to that question. Unlike

Campylobacter, the best epidemiological investigation of the sporadic cases that would answer that question has simply not been done. In the case of *Salmonella*, attention is constantly diverted by the large disruptive and complicated outbreaks it causes. There are currently 80 or 90 outbreaks a year of salmonellosis that are reported to us. We certainly don't investigate them all; but we try to keep track of them, and that is more than ten times the number due to *Campylobacter*! *Salmonella* is clearly a much more efficient cause of big outbreaks than *Campylobacter* and no doubt because of its ability to survive and multiply under a variety of conditions. Nonetheless, just as for *Campylobacter*, sporadic cases do occur. Those outnumber the outbreak-associated cases. So how do we estimate the proportion of these infections that are due to poultry without having that good case-control study?

Two other methods are available, and I don't think either of them are satisfactory. What are the vehicles that we identify in investigations of *Salmonella* outbreaks? How much is that going to tell us about the sources of all salmonellosis? When we evaluate the foodborne disease outbreak data from 1983 to 1987, for which we can identify the vehicle based on the reports of outbreaks that are funneled into the CDC every year from State Health Departments, we find that they account for about 40 percent of all outbreaks. This does not include the one large single outbreak which had 16,600 cases of culture confirmed *Salmonella typhimurium*. That was the Chicago milkborne outbreak in 1985. It was a giant outbreak and if we included it in our analyses, dairy products would rise to the top, much above all other outbreaks. You do need to know it's there. If we look at the percentages of known vehicles, 7.5 percent of those were chicken. If we look at the number of cases that are associated with those outbreaks, 16.7 percent were associated with chicken. Egg-associated outbreaks, which were not a problem throughout the seventies, are a growing problem now and we'll talk about that later. In second position is beef with 9.5 percent of outbreaks and 8.7 percent of cases. Thus, the percent of outbreaks due to beef is more than chicken. Turkey is in fourth position, then pork, dairy, seafood, and lastly, the most dangerous food of all, is "other." And as long as you stay away from "the other," you will probably be all right. "Other" includes potato salad, homemade ice cream, and a variety of other foods in that category. I feel that this list probably defines the spectrum pretty well.

If we listed all the "other" vehicles for salmonellosis, it would go on down to the second floor. I have a lot of trouble saying that this actually reflects the total number of sporadic cases and that this is where salmonellosis comes from. For one thing, large outbreaks are much more likely to be investigated than small ones. If two or three persons are ill after a private meal, that is not likely to be recognized as an outbreak, but when 50 persons become ill after the firehall benefit, the outbreak would be studied quickly. A considerable quantity of food must be mishandled to cause an outbreak that large, and some vehicles such as stuffed turkey are more likely to cause that large outbreak than others. But the "others" may be more important because they may cause many more sporadic cases. I feel that in the absence of case control data these have to be taken with a grain of salt. It's the only thing we've had for some time now, and I don't ever feel comfortable saying that these are the percentages. In addition, foodborne outbreaks are investigated and reported to CDC with varying degrees of skill and interest by local health authorities. In general, they tend to overrate severe or easily diagnosed illnesses and underrepresent the common, milder, less easily diagnosed illnesses. For these and other reasons it's difficult to use these proportions to estimate sources of sporadic salmonellosis.

We can also consider data showing the distribution of isolated serotypes in relation to food source, but it probably has just as many problems. This data is compiled on *Salmonella* that are isolated and serotyped from a broad variety of food animal sources as a result of routine surveillance of clinical illness in animals. The serotyping is done by the National Veterinary Services Lab and is included in our annual *Salmonella* surveillance data. What happens if we estimate the contribution of each of the food animal reservoirs, say chicken, turkey, pig, and the cow for each serotype using these data? In 1986, for instance, there were 5,595 isolates of *S. heidelberg* from humans reported and a similar number of *S. enteritidis*. If we look at where the *S.heidelberg* isolates from animals came from, a real predominance in chickens appears with a small number in turkeys, pigs, and cows. If we weight that percentage for each of those, we have to multiply that by the proportions of all the human isolates that are *S. heidelberg*. If we do that for each serotype we come up with relative weights for the ten most common serotypes: 36.8 percent attributable to chicken, 17.7 percent to turkey, 6.5 percent to pork, and 39 percent to the bovine reservoir, driven largely by the high number of *S. typhimurium* isolations.

This whole process of weighting is a statistical nightmare and very inappropriate. Let me be the first to say that it is a very flawed way of coming up with an estimate, because these isolates were not collected for this purpose, but for a variety of different reasons. It would be a mistake to relate these in any way to estimate the proportion of salmonellosis caused in these animals. I'm not much happier with using our outbreak data. One point that we can draw from this information is that there are some important serotypes for humans. *S. typhimurium* ranked one, *S. enteritidis* - two, *S. heidelberg* - three, and *S. newport* - four. *S. newport* is hard to find in pigs; apparently, it is not as hard to find it in cattle. *S. hadar* was number five. It is likely that the National Veterinary Service is going to get isolates from turkeys and chickens and not from the bovine reservoir or from swine. This may be telling us that the serotypes you really want to pay attention to when talking about poultry-associated salmonellosis infections are going to be *S. enteritidis*, *S. heidelberg* , *S. hadar*, *S. infantis,* and *S. agona*. The first four strike me as being the target organisms if we're going to use this to suggest which reservoir they are in.

I want to turn to a different topic. For those of you who think that the egg should precede the chicken, it's time to wake up! One of the reasons we haven't been able to organize the case control study of sporadic salmonellosis is that an ever increasing proportion of our time is eaten up by the egg-associated *S. enteritidis*. We first became aware of this epidemic in 1986, when a large outbreak of salmonellosis was caused by contaminated commercial pasta and was traced to the raw eggs used to make the pasta. The dramatic increases in this serotype actually began well before then, as early as 1979, in the Northeast. If we go back to 1978 or 1977, the incidence was fairly constant in the Piedmont region, followed by a few small outbreaks in 1979, followed by significant increases beginning to build in New England.

I'm going to jump forward now to 1987. The incidence of *S. enteritidis* is beginning to move down from New England into the Mid-Atlantic and South-Atlantic area. A funny two to three year cycle appears to be occurring. I've never really understood why it would cycle. The Mid-Atlantic area came up much later and extended into Pennsylvania and New York. New Jersey started in 1984-1985 and now has reached the same isolation rate as New England. The good news for New England is that between 1987 and 1988 the number of outbreaks actually decreased, although it has done that before. Unfortunately, *S. enteritidis* outbreaks have continued to increase in the Mid-Atlantic area and extended into the South-Atlantic area as far south as North Carolina. The really bad news is that it isn't just sporadic cases, but also a lot of outbreaks. Between 1985 and 1988, we had reports of 140 outbreaks of *S. enteritidis* alone. Outbreak incidence in 1988 was a little bit lower than 1987, but higher than 1986, and I'm afraid that 1989 is probably going to crack the record of 1987 and be more than 50 for the year by the time the year is out. We have tried to learn as much as we can about the outbreaks. We've investigated a very small number of these, but we have County Health Departments doing these investigations by phone, and we have gathered as much information as we can. Those 140 outbreaks occurred in 12 states, involved 4,976 cases, 30 of whom died. The deaths associated with *Salmonella* are almost all restricted to outbreaks occurring in hospitals and nursing homes, and unfortunately there have been far too many of these. Twenty-four outbreaks during that four year period were associated with 27 of the 30 deaths. Thus, with an outbreak in a hospital or nursing home, about 3.6 percent of the people infected may die. That is our main problem as public health officials. Our first goal is to prevent deaths.

We must prevent outbreaks in hospitals and nursing homes. The food vehicle implicated in 89 of the 140 outbreaks was the Grade A shelled egg (73 percent). That percentage is holding pretty steady from the first time we looked at it. Recent outbreaks continue to be egg-associated. The eggs that were involved in these outbreaks were traced to egg producers and distributors in 12 different states including five outside of the Northeast. The eggs have almost always been Grade A, and there has been a growing experience of evaluating the suspect farms that includes enough cultures of ovaries of hens to identify source farms and to convince us that the ovaries of those hens on suspect farms are contaminated with *S. enteritidis* far too often. At the end of 1986 and early 1987, we began to suspect that contamination of the yolk occurred before shell formation and was likely to play an important role in this epidemic. The question of how much of this is accounted for by transovarian transmission has fascinated us from the beginning. As I said, our primary job as public health officials is to prevent death and then to prevent morbidity. So, in December 1988, we issued some guidelines. Before that we made some recommendations about reducing the risks of hospital and nursing home outbreaks, saying that in all affected parts of the United States the safest alternative was to use pasteurized egg instead of bulk pooled egg in recipes that call for bulk pooled egg. This has not been uniformly or totally adopted, because outbreaks are still occurring. We also suggested that in all parts of the country, in expectation that this problem would spread, it was important to review food handling in hospitals and nursing homes to make sure that nobody was being exposed to raw eggs. We also recommended minimizing any error in handling bulk pooled egg, and particularly, to think about the possibility of cross-contamination in blenders. All of you are familiar with the model state program for detecting and eliminating infected flocks based on the cross-reactivity between *S. enteritidis* and *S. pullorum*. Let's stop for a minute and ask how we are doing -- certainly not as well as we need to! We continue to see outbreaks in nursing homes this year and clearly have further to go in educating nursing homes about the problem and getting them to switch to the use of pasteurized eggs. In the first half of this year, 20 more outbreaks were reported. The total is now over 40, and more outbreaks will be reported in 1989.

Another problematic issue and one that in a way is really particularly interesting and challenging is that this appears to be a global problem. About 40 countries report the result of *Salmonella* serotyping to a reference center in London at the Public Health Laboratory Service. In the past, *S. typhimurium* has almost universally been the most frequent serotype. This is changing rapidly. In 1979, fewer than 10 percent of countries reported that *S. enteritidis* was their most common serotype, and in 1987, 43 percent of reporting countries, mostly in Europe, reported that *S. enteritidis* was their most common serotype. When we consider all countries of the world from 1979 through 1986, there appears to have been at least a 25 percent increase in *S. enteritidis* reported, including countries actually on the North American continent, South American continent, Europe, and Tunisia in Africa. In addition, there are a number of countries that do not report to WHO, but for which we also know that there is a *S. enteritidis* problem, including some central African countries such as Zaire, Rowanda, Burundi, and many countries in the eastern block of Europe. Outbreaks of infection caused by *S. enteritidis* have been linked to eggs in the United Kingdom, Spain, France, Switzerland, Hungary, Norway, as well as in the United States. The United Kingdom has witnessed a dramatic increase in isolates of *S. enteritidis* and attributes this to infected broiler flocks in addition to the problem in eggs. The phage type of comparisons across the different countries remains incomplete. Our phage type in this country used to be 13 A. Now our predominant phage type is type 8. The predominant phage type in Europe is type 4. Our abilities to control this pandemic at the moment are really quite limited.

Many aspects of the experimental model state control program really need systematic evaluation, including evaluating the reliability of our serologic approach for screening tests. How reliable and useful is it compared to other more efficient methods? The sanitization of infected farms, before reintroduction of new *S. enteritidis*-free birds, is inadequate, and the resources that are currently devoted to these evaluations are minimal. However, if we do not succeed in constructing an effective and useful strategy for limiting the contamination of the Grade A shell egg for *Salmonella*, public health concern decisions, which 80 years ago rocked our country and led to the uniform pasteurization of milk, may soon begin for eggs. *S. enteritidis* currently accounts for 17 percent of all *Salmonella* isolated in the United States. I cannot predict how much further that will grow before the public demands the safety of pasteurized products. In addition to the immediate research needs of the control program, many aspects of the biology of ovary-associated salmonellosis are unexplored. What is the level of contamination in the egg when it is laid? Where precisely does the contamination occur? What is the natural history of the infection in the hen? Can the infected fertile egg hatch and develop into a healthy hen with ovarian infections? What are the microbial genetic factors that permit this deep tissue invasion, and can manipulation of gut flora prevent deep tissue invasion?

In summary, the major poultry-associated human pathogens include various serotypes of *Salmonella*, *Campylobacter jejuni*, and the specific challenge of egg-associated *S. enteritidis*. The recurring theme for all three of these is that foodborne illness is the result of silent carriage in the food animal, either in the gut or the food tissue, and this brings the concern of the public health official to agricultural specialists in the search for an understanding of the biology of this silent carriage, i.e., commensal colonization, that will lead to effective control technologies. Thank you very much for your attention.

PREFACE

Leroy C. Blankenship

USDA, ARS, Russell Research Center
Athens, Georgia

Fresh poultry and meat are recognized as important sources of *Salmonella* as well as the more recently recognized human bacterial enteropathogens, including *Campylobacter, Listeria*, and *Escherichia coli* 0157:H7. Foodborne enteric human disease continues to plague society on a worldwide scale and appears to be increasing despite research and management efforts to remedy the problem. A great deal of criticism and publicity have been directed toward poultry processing plants as that part of the integrated poultry industry which is responsible for the unacceptably high incidence of *Salmonella* contamination of fresh poultry carcasses. It is important to point out that current processing technology and practice is capable of converting enteropathogen-free birds into enteropathogen-free carcasses and products. But, if chickens that are intestinally enteropathogen-colonized and/or surface (feathers, feet, and skin)-contaminated are delivered to processing plants, it is obvious and well documented that the enteropathogens will become widely distributed by cross-contamination among a majority of the processed carcasses as a consequence of the mechanics of processing. Thus, it is obvious that our first line of defense to control enteropathogen contamination of poultry resides in the production phase of integrated poultry industry operations. Processing phase intervention certainly continues to be our second line of defense, while consumer and institutional food handler education is our last line of defense.

In order to emphasize the importance of enteropathogen control in poultry during production, it is important to note that research efforts of the past thirty years that were directed toward improving the microbiological quality of poultry during processing have failed to make any significant improvement in the product. Indeed, product contamination incidence and disease incidence among consumers have slowly, but steadily, increased.

The recognition of the necessity to control enteropathogen commensal colonization of chickens during production served as the catalyst for organizing this symposium. Our knowledge of the factors controlling commensal intestinal colonization of chickens and animals in general is sadly lacking. It is obvious from what we do know that commensal colonization is an extremely complex process. Thus, much research is needed on many fronts to quickly and efficiently develop the required control technologies. We can also look forward to important spin-off benefits for human and animal health as a consequence of our greater understanding of commensal intestinal colonization. Our research efforts must be focused on production oriented intervention projects with the objective of developing the technologies and management practices to prevent commensal intestinal colonization of poultry by human bacterial enteropathogens, so that enteropathogen-free chickens (poultry) can be delivered to slaughter plants for processing. This research will, of necessity, have to be directed toward breeder/multiplier flock intervention, hatchery operations, and grow-out flock intervention.

Our invited speakers and poster presenters are the leaders in research concerning commensal intestinal colonization in poultry and other animals. It is our hope that the results of this International Symposium on Colonization Control of Human Bacterial Enteropathogens in Poultry will stimulate and focus worldwide research that will accelerate our progress toward the knowledge and technologies with which microbiologically safer, more wholesome poultry products can be made available to consumers.

ACKNOWLEDGMENTS

The editors acknowledge with great appreciation the contributions of Mr. Harold E. Ford, Executive Vice President, Southeastern Poultry and Egg Association, Decatur, Georgia, and Mr. Larry Brown and Ms. Pat Dyer, also with the Southeastern Poultry and Egg Association. Their cooperation and help were invaluable in successfully conducting the international symposium upon which this volume is based.

Also, appreciation is directed to Ms. Julia Bandler for help in making symposium arrangements, taking care of correspondence, and for providing camera-ready copy of this book to the publisher.

OPENING REMARKS

William T. Hubbert

USDA, FSIS
Washington, D.C.

It's a pleasure to be here for the opening session of this International Symposium. Food safety is in the news, and foodborne enteric pathogens are receiving a major part of the attention. Although the weather makes news too, we can do little to alter it. I commend the Agricultural Research Service and the Southeastern Poultry and Egg Association for bringing together this outstanding international panel of experts to examine what we can do about controlling enteric pathogens in poultry.

Although we may bring different perspectives to the issues at hand, a common bond unites us as well: we are all consumers. Not only do we care about food safety from a professional standpoint, but from our own personal standpoint as well. Today's consumers are vitally interested in food safety. It has become the issue of the 80's and 90's, just as nutrition was the issue of the 70's. Just look at the number of newspaper articles and television broadcasts that have explored subjects such as pesticide contamination and *Salmonella*. The average consumer couldn't tell you what *Salmonella* was ten years ago. Today it's a different story.

The job ahead of us is to keep food safe, and to help the public understand why it's safe. It's a large responsibility, and one that is shared by all of us. In a regulatory environment, we believe the most effective control is accomplished by preventing problems, not simply by condemning products of a failed process. Despite our comprehensive regulatory programs, we can't do the job ourselves. We must depend on the dedication of professionals such as yourselves in the agricultural and food industries to help us pursue these activities.

Public health professionals share with us the role of insuring a safe food supply and educating the public. From dieticians, who oversee food preparation in health care institutions, to veterinarians, who insure that foreign animal diseases are kept out of the country; all have an opportunity to guard the public health and the health of the U.S. agriculture.

Risk communication is an area in which public health professionals can play a critical role. An entertaining as well as informative article in the *EPA Journal* defined risk as the sum of hazard and outrage. The public pays too little attention to hazard, and the experts pay absolutely no attention to outrage.

The point is well taken. We must try to reconcile the differences. Public health professionals can help by educating the public about risk and by giving them the information they need to make their own decisions.

The food industry, of course, also has a role in safety. Farmers and ranchers have a responsibility to practice good animal husbandry. Food processors have a responsibility to maintain high-quality operations. Without the industry's commitment, our job of insuring the quality and safety of the food supply becomes very difficult. Not only does this make sense from our point of view, but from the industry's as well. Unless the industry moves out in front of emerging safety issues, it could find itself in a precarious position. One has only to look as far as the Alar issue to see what can happen, and how fast it can happen.

Although science must play the major role in food safety, we must recognize that food safety is not just a scientific issue. Food safety is also a political issue, an economic issue, and a trade issue. All of us, no matter what our profession, must recognize this as we struggle with the issues of the day. Nowhere is this more evident than in the international arena. As food safety issues surface in this country, they reverberate around the world and are frequently magnified.

Foodborne enteric pathogens are significant worldwide and those of animal origin play a major role. Therefore, I wish the participants of this symposium success in presenting a model to effectively reduce this important source of human disease.

INTRODUCTION

Glen H. Snoeyenbos

University of Massachusetts
Amherst, Massachusetts

This is the third International Symposium in North America in the past twelve years with salmonellosis as the central topic. The first of these was held in 1977 at the University of Guelph, Guelph, Ontario, Canada. The second was held in 1984 in New Orleans, Louisiana, USA. Primary sponsorship of these symposia has been the University of Guelph, The American Association of Avian Pathologists, and the United States Department of Agriculture, Agricultural Research Service, respectively. In each instance, substantial financial assistance was provided by various government agencies and private industry, and in some cases, by a number of professional and producer organizations. This breadth of support has not only been essential for developing the symposia, but clearly demonstrates the breadth of recognition of the importance of gatherings such as this. Organizers of this symposium have, for the first time among these symposia, recognized in the program content the great significance of campylobacteriosis as a major foodborne disease. The fact that it rivals or exceeds the incidence of salmonellosis in humans demands the most careful consideration, even though effective and practical methods for control of contamination by *Campylobacter jejuni* in fresh poultry are almost nonexistent. Although campylobacteriosis in humans is undoubtedly an old disease, it is of interest that only a few diagnoses were made prior to the development of an efficient selective growth medium by Skirrow in 1977.

At the 1984 symposium, the following conclusions and recommendations were made by consensus:

1. The eradication of *Salmonella* in domestic animals is not attainable at this time except for specific infections such as *S. pullorum* and *S. gallinarum*. Serious efforts should be made to control and reduce the incidence of *Salmonella* commensal colonization and infections in domestic animals.
2. Methods are available to reduce *Salmonella* contamination of processed feeds to safe levels, but feed contamination remains a widespread and serious problem. Further improvement is necessary.
3. It should be candidly recognized that raw foods of animal origin are frequently contaminated by *Salmonella* and that such contamination levels cannot be expected to change greatly in the near future.

4. Trade barriers should not be erected and import requirements with respect to *Salmonella* should not exceed standards attained in the importing country.
5. There should be continuing programs to provide education and information on proper food handling to the consumer and to all those handling foods.

There is no doubt that these points are as applicable now as they were five years ago and likely will remain applicable for many years in the future.

Given the great number of variables that can influence data on the incidence of *Salmonella* infections in man and other animals, it is not clear that there has been any pronounced change in total incidence in poultry in many years. We are all acutely aware that the apparent incidence is intimately related to detection methods and breadth of search, and that our best published data usually represents only a small sample of the poultry population.

Since the 1984 symposium, two noteworthy problems have emerged. The first was the recognition by epidemiologists of CDC that the incidence of *Salmonella enteritidis* in humans in our Northeastern states had been steadily increasing since the early 1970's, and that many of these cases were associated with use of whole-shell chicken eggs. Subsequent work by several investigators in the past two years has documented the presence of *S. enteritidis* at or near the yolk level in fresh eggs from some infected flocks. The infection in a small but probably very important percentage of hens is deep-seated and in organs such as the ovary, which could potentially lead to transovarian transmission. It is significant that bacteria in this position are not effectively inhibited by ovotransferrin and other protective compounds in the albumen. We are working on the assumption that only some strains of *S. enteritidis* are dangerous in this respect, but we do not now have clear markers for their differentiation. Phage typing (PT) of isolates by the Colindale scheme indicated that they are primarily PT 8 and 13 A in the United States.

The second somewhat related problem, also *S. enteritidis* in humans traceable to whole eggs, has been most widely publicized as a problem in England during the last year. There the problem is caused by PT 4 which has not been found in the United States. At least some strains of PT 4 posses substantial virulence in chicks compared to most of our strains of PT 8 and 13 A. Actually, the problem with PT 4 was recognized at least eight years ago in Spain by avian pathologists, but received little publicity. Typical of foodborne infections, food handling abuses have been a prerequisite of these egg-associated infections in man.

It is likely that the *Salmonella*, as resourceful and formidable opponents, will construct additional nasty surprises in the future. There is scant hope of having fresh, processed poultry carcasses which are dependably free of *Salmonella*, other than by developing and maintaing *Salmonella*-free poultry at all levels of production. Utilizing promising new vaccines, including killed whole cells for specific serotypes in breeding stock, outer membrane proteins or live attenuated mutant strains, as well as competitive exclusion may be very valuable aids in the control of *Salmonella*. But, in my opinion, these cannot be expected in themselves to lead to production and maintenance of *Salmonella*-free chickens as has been demonstrated in Sweden and production of *Salmonella*-free turkeys as has been demonstrated in Minnesota.

Both of these approaches include use of *Salmonella*-free breeding stock, rigid sanitary management during all stages of production to preclude adventitious introduction of *Salmonella*, and periodic monitoring to assure freedom from *Salmonella*. The question of economic feasibility is critical and highly important because of the added production costs which will likely be very substantial and which will require major restructuring of the poultry industries in most countries. If market demands become sufficiently coercive, I am confident that over time our U.S. industries could produce a majority of flocks for processing that are *Salmonella*-free.

In the view of many, the cost effectiveness of *Salmonella* control should be of primary importance. Even if it were possible to accurately define costs and benefits, there have been numerous examples of dangers perceived by the public, chiefly in respect to possible chemical contamination of foods, which forced involved food producing industries to make major and costly changes in production methods which were not justified by clear scientific evidence. One can hope that no such public emotionalism overtakes the poultry industries that we work for or with. Difficult as the formulation may be, it is necessary to keep cost/benefit relations clearly in mind as we continue to attempt improvement in methods to control salmonellae. It is probably overly optimistic to expect that the large number of *Salmonella* serotypes which have little or no virulence for poultry, that some consider to be commensals, can be ignored by the industries. Although there is evidence of significant difference in genetic resistance of chickens to *Salmonella*, it remains to be demonstrated that genetic resistance can be developed to the point that it could be utilized as a major factor in control.

I look forward to hearing the important part of the program which immediately follows. Selected world leaders in their respective areas of expertise will report many significant advancements and bring us up to date in their fields. The immediate benefit of this symposium will be to help all of us to better appreciate what we can and should be done now in commercial poultry production, processing and marketing, and in basic and applied research.

Of equal or greater importance are the long-term benefits that can be secured by extending personal communications among those working on related problems in various parts of the world. The problems to be addressed here are of international scope which require close interpersonal cooperation and assistance among the international community for most rapid progress.

Lastly, Dr. Blankenship and his colleagues should bethanked for their leadership and large expenditure of effort required to organize this symposium. I am not alone in hoping that some other group, perhaps another continent, will develop the next similar symposium by the middle of the next decade.

SESSION I: ENVIRONMENTAL FACTORS AND SOURCES

Convener: *J. Stan Bailey*

USDA, ARS, Russell Research Center
Athens, Georgia

There are numerous potential areas of salmonellae contamination in an integrated poultry operation. These include: the hatchery, chicks, feed, rodents, birds, insects, transportation, and processing. In fact, exposure to salmonellae can occur anywhere during grow-out or processing and breaking this chain of passing the salmonellae from one generation of chickens to the next has been the goal of researchers for 25 or more years.

Preventing live chickens from becoming colonized with salmonellae during grow-out is extremely difficult because, for the most part, salmonellae do not adversely affect the chickens' health. In fact, what we have is a commensal relationship between the colonizing salmonellae and the host. Since the bird is not being impaired by the salmonellae, it does little to prevent the presence of, or once established, try to exclude the salmonellae. For this reason, we say that chickens are colonized, not infected, with salmonellae with the distinction being that an infection would cause a pathological problem for the bird.

Numerous factors affect the susceptibility of chickens to *Salmonella* and *Campylobacter* colonization. These include age, level of stress on the bird, health of the bird, type and amount of feed additives, and the genetics of the bird. Different intestinal colonization patterns in chickens are also seen with *Salmonella* and *Campylobacter*. Because of contamination in the hatchery and feed, *Salmonella* can be found in a small number of chickens as they are placed in the grow-out house. Intestinal colonization rates in flocks will slowly increase until three to four weeks of age when they will begin to decrease until chickens go to the processing plant. In contrast, because *Campylobacter* is not usually found in the hatchery or feed, *Campylobacter* colonization is not usually found in young chicks. However, as soon as *Campylobacter* is introduced into the flock by some environmental source such as a fly or rodent, the subsequent intestinal colonization rate rapidly increases to close to 100% of the chickens in the flock and remains at this high rate until the chickens go to the processing plant.

These differences in colonization patterns and the different environmental sources of *Salmonella* and *Campylobacter* which contribute to the intestinal colonization of chickens will be examined in the following session.

ENVIRONMENTAL FACTORS CONTRIBUTING TO *SALMONELLA* COLONIZATION OF CHICKENS

F. Jones
R. C. Axtell
F. R. Tarver
D. V. Rives
S. E. Scheideler
M. J. Wineland

North Carolina State University
Raleigh, North Carolina 27695

ABSTRACT

A project was conducted in order to determine the degree of *Salmonella* contamination and the specific farm management practices which were related to *Salmonella* contamination. Samples were screened using an enzyme immunoasssay and confirmed with biochemical and serological tests. Samples collected from the feed mill were contaminated most frequently, followed by samples from the breeder/multiplier house, the processing plant, the broiler house, and the hatchery. Two breeder/multiplier flocks whose ovaries were contaminated with *S. typhimurium* were discovered, and progeny of these flocks were followed in the field. While *S. typhimurium* was found in 40-60% of yolk sacks of the day-old chicks tested, the percentage of birds in which *Salmonella* (any serotype) was detected decreased to about 3% at market age. *Salmonella* was found in about 35% of the ceca collected at the plant, while slightly over 75% of the processed birds were positive. Dominant *Salmonella* serotypes shifted as the birds were processed, with *S. hadar* and *S. muenster*, rather than *S. typhimurium*, dominating in the finished product. Only reduced litter moisture and water chlorination appeared related to *Salmonella* contamination rates in broilers on the farm. While feeds frequently contain *Salmonella*, these data suggest that breeder/multiplier flocks can determine whether or not broilers contain *Salmonella*. However, these data also suggest that *Salmonella* contamination of ready-to-cook broilers probably occurs mainly during catching and loading, live-haul, or processing.

Colonization Control of Human Bacterial Enteropathogens in Poultry

INTRODUCTION

The costs associated with human salmonellosis in the Unites States have been estimated at $4 billion annually (Todd, 1989). While poultry products are not the vehicle with which salmonellosis cases are most often associated in the United States (Bryan, 1985), recent news reports have apparently gone a long way toward linking salmonellae with poultry products in the mind of the consumer (Carey, 1987; McAuliffe, 1987). Yet, paradoxically, the consumption of poultry products has continued to increase (Bates and Hall, 1989). Nonetheless, the poultry industry finds itself in a very precarious position, and I believe it is fair to say that everyone would like to see a solution to the salmonellae problem. Yet the situation with respect to salmonellae in poultry and poultry products seems to become increasingly more complex.

The primary objectives of this study, funded by the Southeast Poultry and Egg Association, were: 1) to estimate salmonellae contamination rates at various points in the broiler production and processing system, and 2) to determine what management factors are related to salmonellae contamination.

MATERIALS AND METHODS

Laboratory Procedures. The laboratory methods by which samples were examined for the presence of salmonellae are outlined in Figure 1. The method is an AOAC-approved method for nonfat dry milk, milk chocolate, meat and bone meal, dry whole egg, and ground pepper (Flowers *et al.*, 1987).

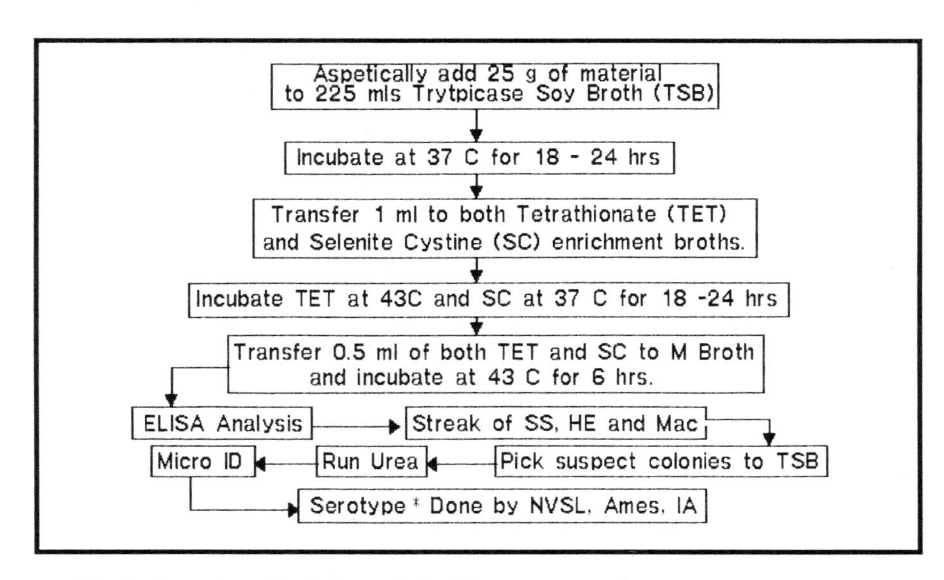

Fig. 1. Procedures for *Salmonella* isolation.

Sampling Methods. Solid samples such as litter, feed, and dust were collected by using sterile tablespoons to transfer the material directly to tryptic soy broth (TSB) for incubation. Bird rinses, both those on the farms and those in the plant, were done using the method described by Cox *et al.*, 1983. Egg rinses were done in an analogous manner. Insects and mice were trapped and transported to the laboratory for analysis. The entire gastrointestinal tract of captured mice was aseptically removed and placed in TSB for incubation. Insects were rinsed in sterile distilled water prior to analysis and, either sanitized with a 20 ppm chlorine solution or placed directly into TSB for analysis. Setters, chick boxes, and environmental swabs were collected by wetting either a sterile cotton-tipped swab or a sterile gauze pad with TSB and rubbing it over an area of about 1 sq. ft. before returning it to the medium for incubation. A new pair of disposable latex gloves was always worn during the collection of swab samples with a fresh pair being used to collect each sample. Yolk sac samples were collected from chicks following cervical dislocation by aseptically opening the chick with scissors and excising about a 1 g portion of yolk material which was placed in TSB for incubation. Samples of yolk material were collected from eggs following aseptic opening using sterile disposable tuberculin syringes which were discarded after one use. Egg content samples were collected following soaking the egg in ethanol by cracking the egg on the lip of the media container and allowing the egg contents to fall into the medium. Ceca were collected at the plant by hand, aseptically opened and the contents expressed directly into the TSB for analysis. Egg folicle samples were collected at the plant by hand, sanitized in ethanol and aseptically opened using a sterile scalpel. A sterile disposable syringe was used to remove 1 ml of yolk material which was then placed in TSB for incubation. Vaccinator samples were collected by allowing the vaccinating apparatus to deliver one dose of vaccine directly into TSB for incubation. Litter moistures were estimated when birds were between three and five weeks of age, using a Delmhorst Hay Moisture Detector Model F-4, and the methods described by Koelliker and Dieker (1978). Water pH was estimated using Fisher brand pH paper.

Production and Processing Records. Production and processing records were obtained from the participating companies who maintained the records.

RESULTS AND DISCUSSION

The points at which samples were taken during the first phase of this project are shown in Figure 2 along with the total number of samples collected at each location, the number of samples positive, and the percentage of samples positive. *Salmonella* was most frequently isolated (20%) from samples collected at the feed mill. *Salmonella* was found in slightly over 16% of the samples collected at the processing plant, while nearly 13% of the samples collected at the multiplier farm were contaminated. Slightly over 7% of the samples collected at the hatchery contained *Salmonella*, while only about 4.5% of the samples collected in the broiler house contained the organism.

Figure 3 compares the data we collected with those collected by Morris *et al.* (1969) over two years ago. The data collected from the feed mill and the multiplier house appear to be comparable. The isolation rates in our study would appear to be higher than those of Morris *et al.* (1969) in the processing plant and the hatchery, while the reverse would appear to be true in the broiler house. The data collected from the feed mill and the multiplier house appear to be comparable.

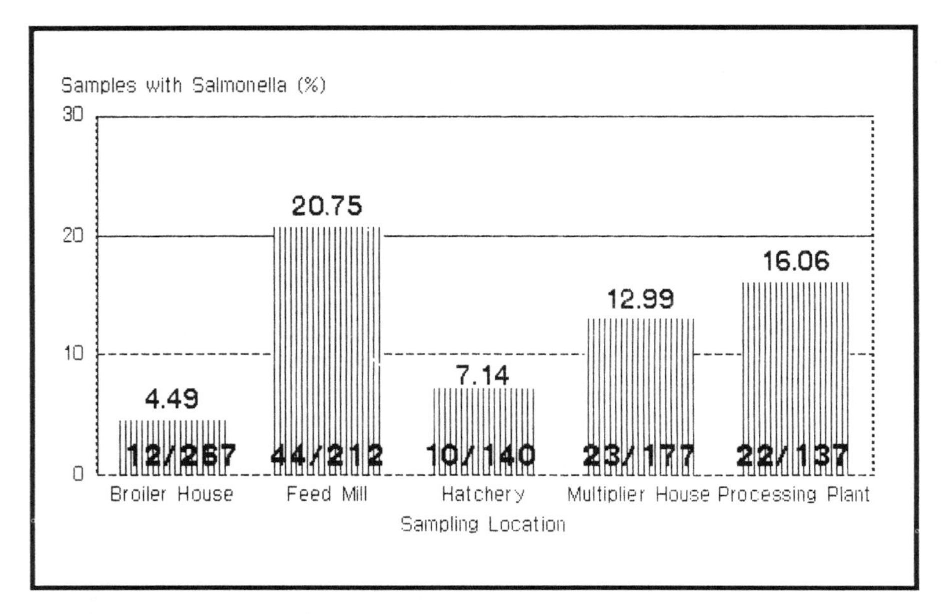

Fig. 2. *Salmonella* results overview.

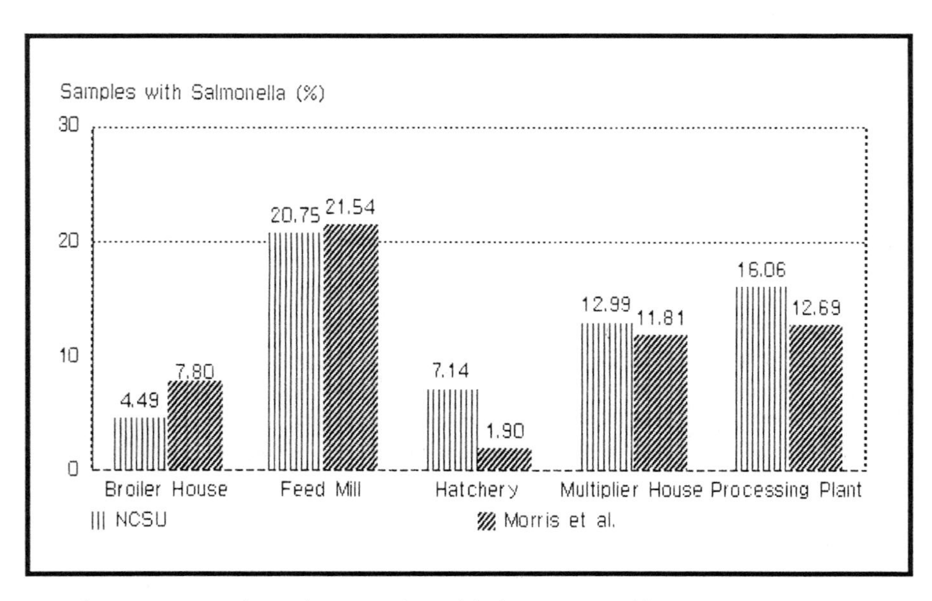

Fig. 3. A comparison of current data with those reported in 1969.

Salmonella was found in two of the 14 dead bird rinses (14.3%) collected in the broiler house (Figure 4), while only one positive was found when live birds were rinsed. About 5% of the fecal samples tested were found to contain *Salmonella*, while only one of the 42 environmental swabs collected was positive. *Salmonella* was not found in feed, water, or litter samples collected on the broiler farm. The isolation frequencies reported here for feces are considerably lower than those reported by Lahellec and Colin (1985) (13.2%) and Lahellec *et al.* (1986) (22.9%), but are higher than those reported by Doughtery (1976) (2.5%). Since Soerjadi-Leim and Cumming (1984) have shown that *Salmonella* are isolated much less frequently from flocks reared on old litter than those reared on new litter, perhaps these differences can be explained, at least in part, by differences in litter type. Our study, and that of Dougherty, were conducted all or in part on old litter, while the other two studies were conducted on new litter.

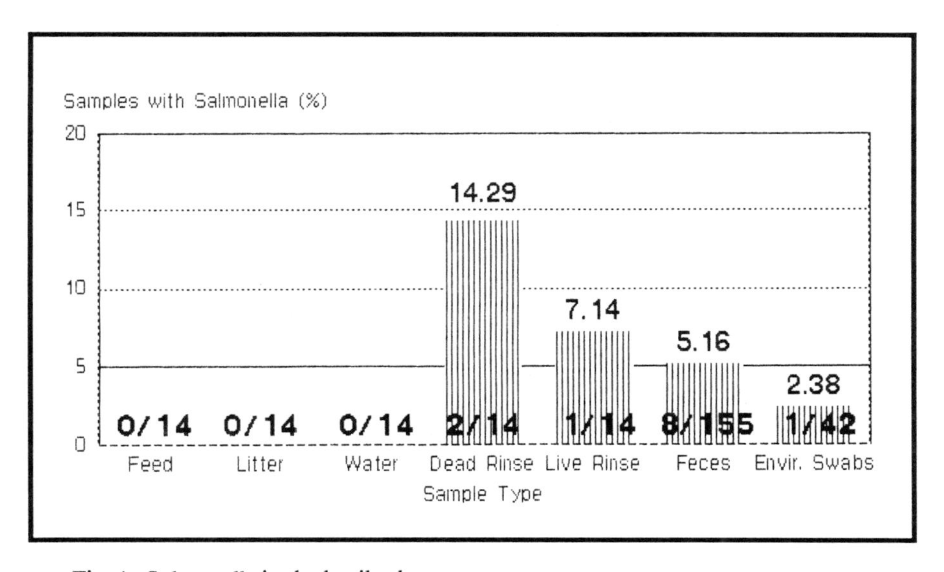

Fig. 4. *Salmonella* in the broiler house.

From samples collected at the feed mill, *Salmonella* was found most frequently in meat and bone meal (Figure 5), while fish meal and poultry meal samples were contaminated at a slightly lower rate. *Salmonella* was found in nearly 21% of the other feed ingredient samples collected, while the pelleting process reduced *Salmonella* contamination rates by about 80%. *Salmonella* was not isolated from environmental swabs, while one of eight dust samples collected was found contaminated. The contamination rates reported here for meat and bone meal are lower than those reported by Morris *et al.* (1969) (87.2%) and MacKenzie and Bains (1976) (72%), but are comparable to those recently reported for animal protein products by Shrimpton (1989) (67%). Shrimpton's data suggest that pelleting reduced *Salmonella* isolation rates by about 90% in feeds.

However, his data suggest that feeds are recontaminated following pelleting; so, if recontamination rates are considered, Shrimpton's data suggest that pelleting reduces *Salmonella* isolation rates by 81.1%.

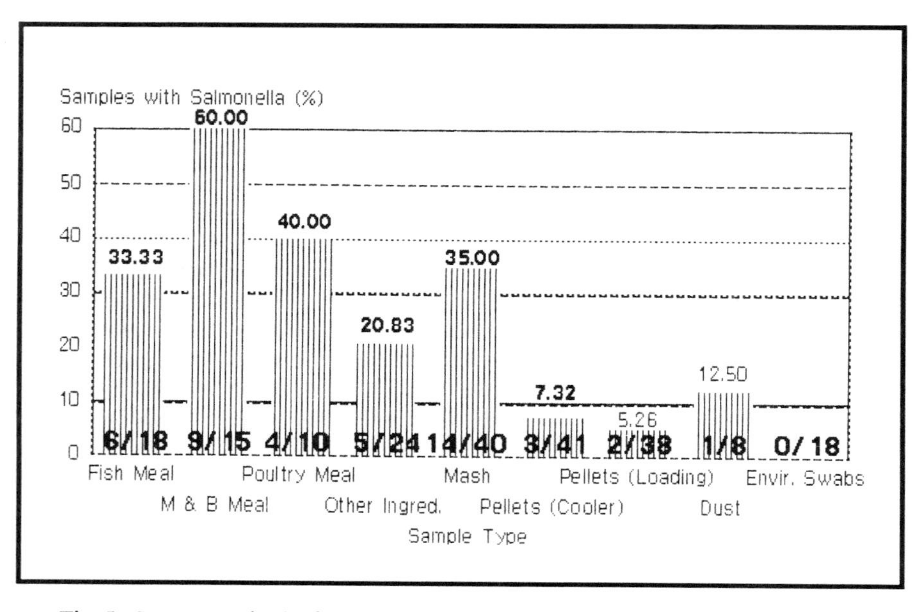

Fig. 5. *Salmonella* in the feed mill.

Salmonella was found in the yolk sac of slightly over 9% of the chicks sampled at the hatchery (Figure 6), while only one of the 14 swab samples collected in the setters was contaminated. *Salmonella* was not found in dead embryos, chick boxes, or in the vaccinator. The results obtained here in the setters are comparable to those obtained by Lahellec and Colon (1985) who found a 6.3% contamination rate. While both these isolation rates are considerably lower than those reported by Cox and Bailey (1989) (88%), these data can, in reality, not be compared since Cox and Bailey collected samples from chick belts while the other two studies collected swabs from setters. The contamination rate reported here for chicks sampled at the hatchery (9%) is higher than that reported by Bhatia *et al.* (1979), who reported a 6.6% isolation rate.

Salmonella isolation rates were nearly the same for feed, nest litter, and nest egg rinse samples collected in the multiplier house (Figure 7). Furthermore, floor egg rinse and floor litter samples had similar contamination rates. These data suggest that feed was perhaps the ultimate source of cantamination at the multiplier farm and that the surface of floor eggs was contaminated as it contacted the litter. The *Salmonella* isolation rate in cecal droppings was only slightly higher than the rate seen earlier in day-old broiler chicks (12.3% vs. 9.4%), suggesting that perhaps the two are connected.

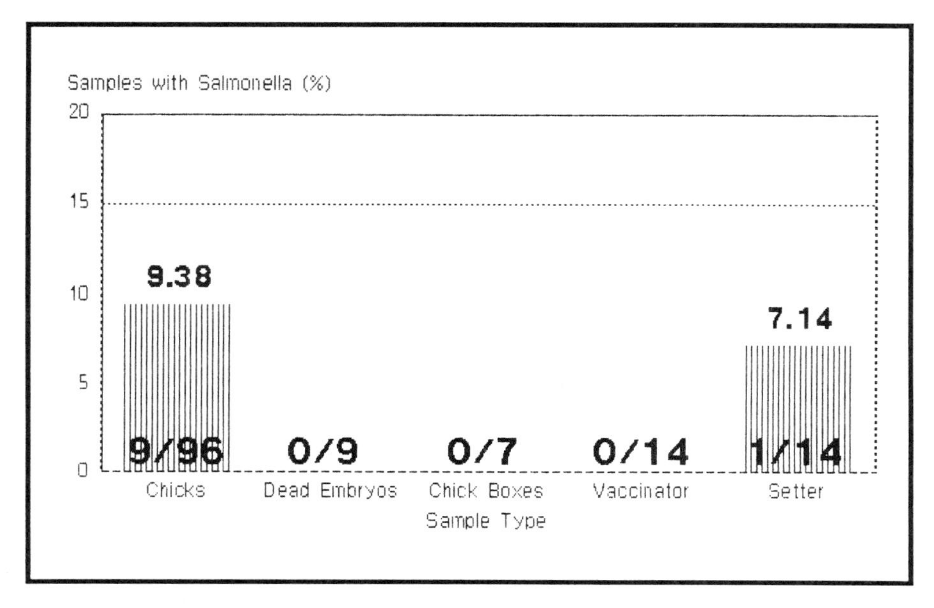

Fig. 6. *Salmonella* in the hatchery.

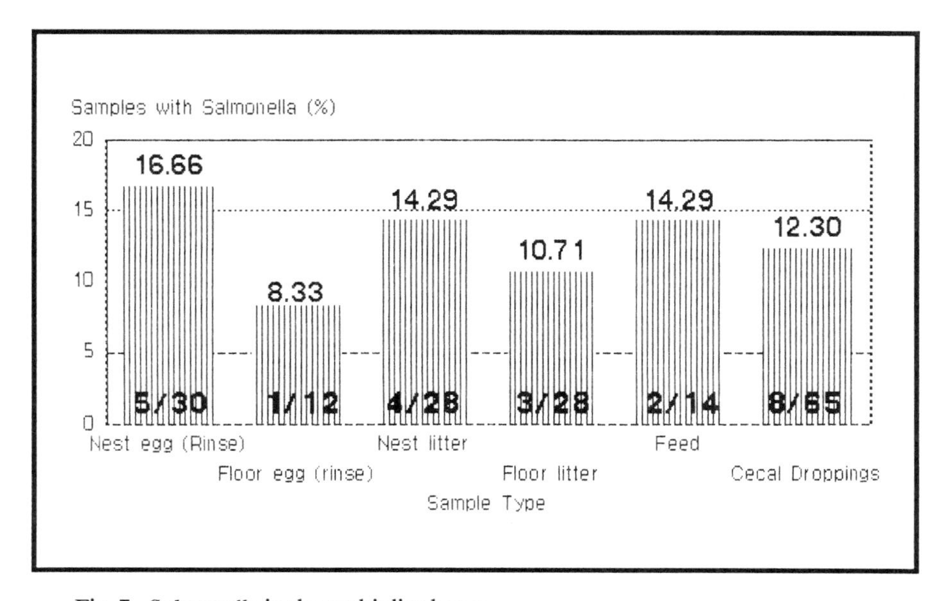

Fig. 7. *Salmonella* in the multiplier house.

Salmonella was found in one-third of the samples collected from live haul trucks at the plant (Figure 8), but was not found in cloacal swabs collected from birds just prior to kill. Our data on live haul trucks are similar to those reported by Goren *et al.* (1988) (29.6%), but are lower than those reported by Rigby *et al.* (1980) (86.6%). Our data on cloacal swabs are similar to those collected by Rigby and Pettit (1979) and Bailey and Cox (1989), who both concluded that more recoveries were made from cecal droppings than from cloacal swabs. *Salmonella* was found about half as frequently in birds exiting the chiller as in birds at packaging. Since birds were rehung for transportation to packaging, these data suggest that processing plant workers may be a significant source of cross contamination in the plant. The 21% contamination rate reported here is lower than the 37% reported by Green *et al.* (1982). These differences are probably due to sampling methods and sample size (i.e., number of carcasses examined).

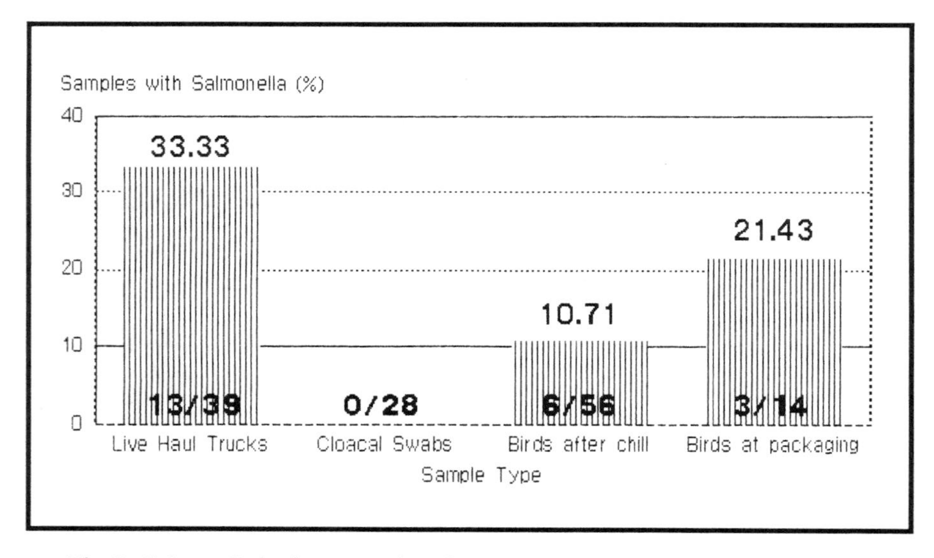

Fig. 8. *Salmonella* in the processing plant.

Salmonella was isolated from only one of the 19 mice captured, while about 13% of the insect samples were positive. However, *Salmonella* was isolated from insects which were not disinfected with 20 ppm chlorine over five times more frequently than from disinfected insects. These limited data suggest that insects are probably mechanical carriers of *Salmonella* and seldom carry the organism in their intestinal tracts.

The *Salmonella* serotypes isolated during phase 1 are detailed in Table 1. Twenty of the 23 serotypes isolated were found in feed. While feed is certainly not the only source of *Salmonella*, these data suggest that feed could be a great repository of *Salmonella* serotypes.

In the second phase of the project we looked at the management factors which are are related to *Salmonella* contamination. Two important questions had to be answered prior to beginning work: 1) What method of sample collection should be employed for *Salmonella* isolation from poultry flocks?, and 2) How can we be sure the birds we are following will carry *Salmonella*?

Table 1. *Salmonella* serotypes isolated in Phase 1.

Serotypes	Broiler House	Feed Mill	Hatchery	Multiplier House	Processing Plant	Totals
Adelaide		13				13
Agona		2				2
Anatum	2			5		7
Binza		3				3
Broughton		5				5
Cerro		3				3
Hadar	1	1			5	7
Havana		4				4
Heidelberg	2	3	6	5	4	20
Indiana		2		7		9
Infantis		3				3
Lexington		1				1
Mbandaka		4				4
Muenster					4	4
Neumuenster		3				3
Newington		4				4
Ohio		1				1
Orangienburg		1				1
Othmarchen		2	2			4
Thomasville		3				3
Typhimurium	8	30	2	10	7	57
Untypable			2	5		7
Worthington		3			3	6
TOTAL	**13**	**91**	**12**	**32**	**23**	**171**

Since the ceca are generally where *Salmonella* is most commonly found in the bird (Bailey and Cox, 1989), the only two practical methods of following flock contamination levels were either cloacal swabs or cecal droppings. To answer our questions about sampling methods we collected samples from breeder hens on our own research farm and examined them using both sampling methods. Only one of 50 cloacal samples were positive for *salmonellae*, while six of 50 cecal droppings were positive, which suggests that cecal droppings is the best of the two methods for detecting *Salmonella* contamination in breeders, and we have assumed the same is true for broilers.

In answer to the second question, we decided that perhaps the best method of ensuring that the broilers we were following were carrying *Salmonella* was to find a multiplier flock which carried a heavy load of cecal *Salmonella* and follow the chicks from that multiplier flock. Therefore, we began to examine breeder flocks for cecal-*Salmonella* contamination. The results of our survey are shown in Figure 9.

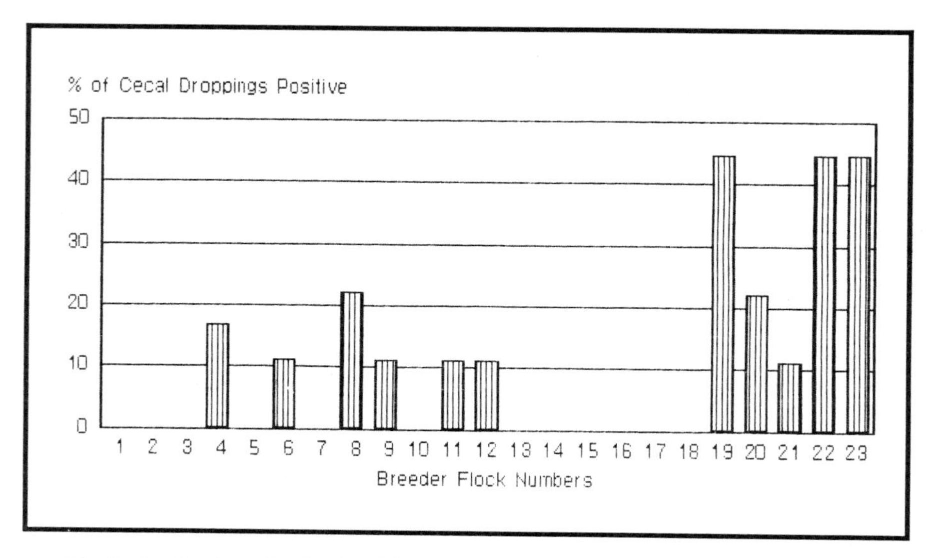

Fig. 9. Broiler breeder flocks with *Salmonella*.

While contamination levels were low in most multiplier flocks examined, several flocks were heavily contaminated with *Salmonella*. We classified breeder flocks as either problem flocks (flocks which, according to the farm manager, have been associated with some type of problem) or normal flocks, and the results are shown in Table 2. Problem flocks carried significantly more *Salmonella* than did normal flocks. The data in Table 3 were obtained from six of these problem breeder flocks. *Salmonella* was found in all egg samples examined from the breeder flocks whose problem was associated with low level cholera, while none of the egg samples from breeder flocks whose problem was associated with water system were found positive. We chose what seemed to us to be the most heavily contaminated flock and resampled the flock to ensure our contamination figures were accurate.

On subsequent sampling, we examined chicks and dead embryos from the flock at the hatchery and cecal droppings from the flock at the multiplier farm. In addition, we were able to obtain 15 breeder birds so that we could sample the egg folicles in their ovaries (Figure 10). Eight of the 15 ovary samples contained *Salmonella* and almost all of the serotypes we found were *S. typhimurium*. I was not convinced we had found a multiplier flock with contaminated ovaries. Therefore, we decided to sample some hatching eggs from the flock in order to determine whether the flock had contaminated ovaries. The results obtained from the 300 hatching egg samples examined are shown in

Figure 11. Our laboratory could only handle 100 samples per week so we had to store 200 samples in the freezer. Apparently, this storage period reduced the number of *Salmonella* isolations from these eggs, since the longer we stored the samples the fewer the number of positives we found. Nonetheless, we found *Salmonella* in 33% of the samples, and all of the nearly 200 isolates obtained were *S. typhimurium*.

Table 2. Problem breeder farms and *Salmonella.*

Description	Number of Farms	Positive/Total	% Positive
Normal Flocks	10	13/168	7.74
Problem Flocks	13	52/177	29.38

Table 3. A brief examination of problem breeder flocks.

Flock No.	Sample Type	Positives/Total	Problem
1	Egg Contents Cecal Droppings	0/7 6/10	Water Problem
2	Egg Contents Cecal Droppings	0/6 5/10	Water Problem
3	Egg Contents Cecal Droppings	7/7 6/10	Low Level Cholera
4	Egg Contents Cecal Droppings	7/7 8/10	Low Level Cholera
5	Egg Contents Cecal Droppings	7/7 5/10	Low Level Cholera
6	Egg Contents Cecal Droppings	7/7 5/10	Low Level Cholera

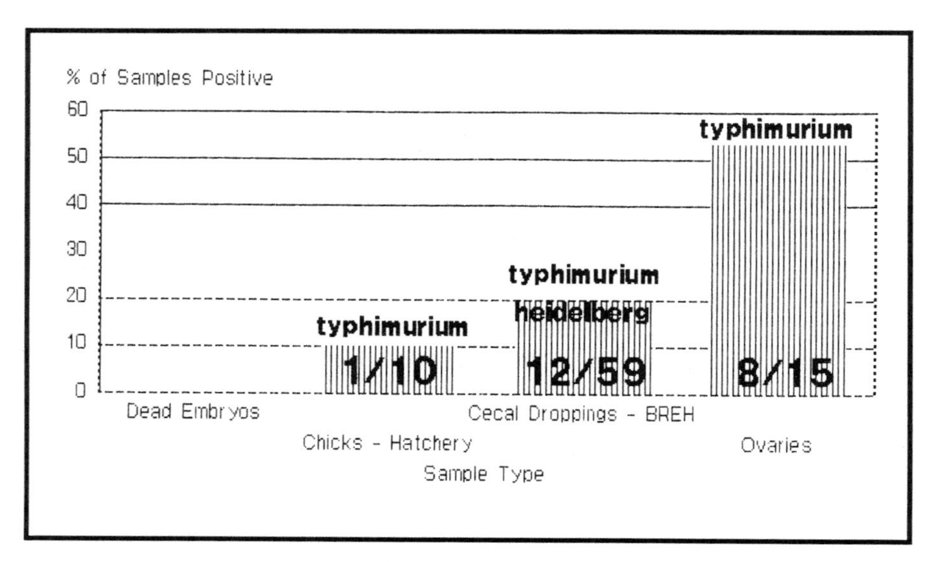

Fig. 10. Recheck of contaminated breeder birds.

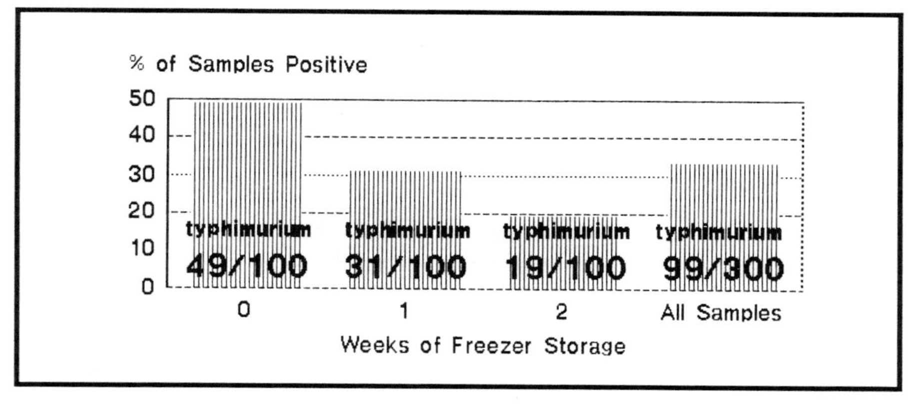

Fig. 11. *Salmonella* in hatching eggs from problem breeder birds.

Unfortunately, the breeder flock we were working with was sold. We began examining other breeder flocks when we learned of a sister flock to the first one. We sampled eggs, chicks, and cecal droppings from this second flock and found about one-third of the chicks, about 10% of the eggs, and over 40% of the cecal droppings contaminated. As with the previous flock, we found only one serotype, *S. typhimurium*. We also collected samples from the breeder house environment and found *S. typhimurium* to predominate in all samples except water where *S. agona* was most prevalent. We later confirmed, when the birds were processed, that 69 of 101 ovaries examined were contaminated with *S. typhimurium*. Before we go any further, let us examine this situation with respect to *Salmonella* contamination of ovaries. The data generated in this study give some clues as to what factors may be involved. Table 4 shows the egg production and hatchability data from two flocks with contaminated ovaries. Since one flock did very well at egg production and hatchability, while the other flock did not produce as well, it seems from these data that production and hatchability are unrelated to whether or not a bird has contaminated ovaries. The only consistent factor for both flocks are mouth lesions, consistent with those found during exposure to T_2 or another *Fusarium* mycotoxin. *Salmonella* was found in these birds after their apparent exposure to the *Fusarium* mycotoxins. This finding is consistent with the findings of Taylor (1989), Corrier *et al.* (1989) and Boonchuvit *et al.* (1975), all of whom found that exposure to T_2 toxin significantly reduced host resistance to colonization with *Salmonella*. These data suggest that at least four factors are associated with contamination of ovaries by *Salmonella*: management problems, a contaminated environment, mycotoxin exposure (particularly *Fusarium* mycotoxins)) and disease. A reduction of the severity or impact of factors would appear to reduce the liklihood of birds acquiring contaminated ovaries.

Table 4. Egg production and hatchability data from two *Salmonella* contaminated multiplier flocks.

Flock No.	Eggs/Hen Housed	Hatchability (%)
1	171.07	82.81
2	159.95	75.82

We were able to separate birds from the contaminated breeder flock, place them in houses, follow the birds in the field and then through the processing plant. The data in Figure 12 allow us a quick overview of the situation with respect to *Salmonella* in these birds. As previously mentioned, the multiplier house environnment was heavily contaminated, and eggs and chicks from this flock were also contaminated. However, the contamination level drops to slightly over 3% by the time birds are ready for processing. Contamination levels increased as the birds reached the plant and several other *Salmonella* serotypes appeared. The serotypes *S. hadar* and *S. muenster* were isolated only twice in the broiler house. Thus, these data suggest that birds may have been contaminated during catch and/or transportation to the plant. Ten of 14 samples taken from shipping crates were positive for *Salmonella*. *S. senftenberg*, *S. heidelberg*, and *S. hadar* were the serotypes isolated from the coops. The coop was the only location where *S. senftenberg* was found. Since the contamination rate nearly doubled, it seems apparent from these data that cross-contamination was occurring in the plant. However, we will come to that subject in a moment.

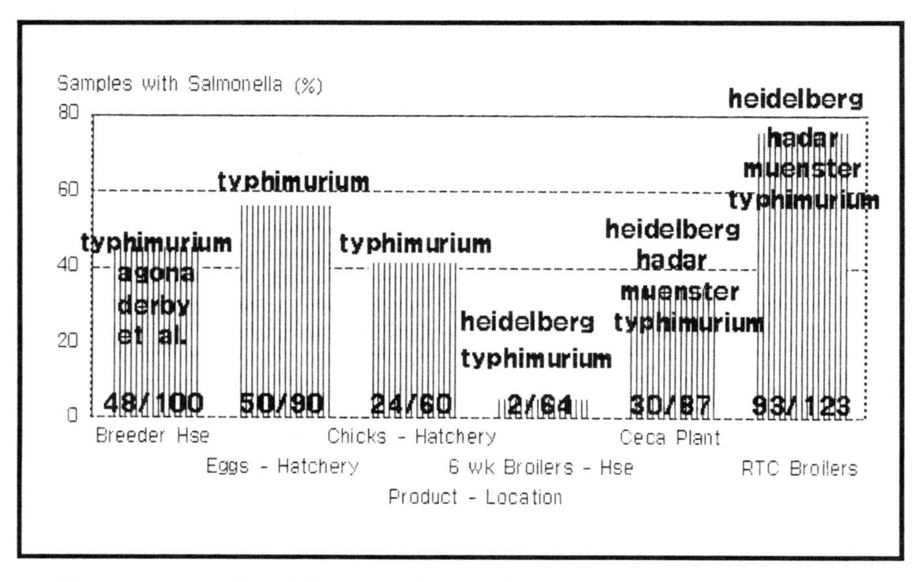

Fig. 12. An overview of *Salmonella* from an integrated broiler operation.

The data in Figure 13 show a steady decrease in *Salmonella* positives as birds age. This is consistent with the data of other researchers (Morris *et al.*, 1969; Sadler *et al.*, (1969). Two explanations may be hypothesized for this reduction in levels of *Salmonella*. First of all, that the samples collected from chicks in the hatchery were collected from the yolk sac, not from the ceca. We had simply assumed that if it is in the yolk sac it will be in the ceca of the chick. However, in their book on the development of the avian embryo, Freeman and Vince (1974) state, "Although the yolk sac membrane is essentially an extension of the gut, little or no yolk material passes directly into it." Thus, our assumption that all of the chicks we were following had *Salmonella* in their ceca at the hatchery may not have been correct. The second explanation involves immune response. Schat and Myers (1989) stated at a recent meeting, "While IgG is passively acquired serumimmunoglobulin, IgA can also be passively transferred from the hen to the intestine of the chick through its secretions into albumen by the oviduct. It has been suggested that antigen specific IgA prevents the adhesion of bacteria or virus to receptors on the surface of epithelial cells." Since the multiplier house was an environment which was heavily contaminated with *Salmonella*, it seems reasonable to assume that the exposed hens mounted an immune response to the organisms and passively transferred this response to their chicks as described by Schat and Myers (1989). A further examination of the data in Figure 12 suggests that two serotypes (*S. typhimurium* and *S. heidelberg*) were competing in the broiler house. Other serotypes appeared briefly, but did not persist.

We attempted to correlate the overall *Salmonella* contamination rate found in broiler houses with the following factors: market weight, feed conversion, livability, condemnations, average daily gain, bird age, waterer type, house size, water chlorination, house age, grower rank, management rating, general house condition, litter moisture,

and water pH. Only two management factors were correlated with *Salmonella* contamination; water chlorination and litter moisture. Flocks that were given chlorinated water at least once a week had less *Salmonella* contamination than did birds not exposed to the chlorine, and birds maintained on litter of between 20% and 25% moisture had less *Salmonella* contamination than did other birds. Unfortunately, the two factors would appear to be confounded with each other, since flocks raised on litter lower in moisture also apparently received chlorine in their water. Our data conflict with those of Poppe *et al.* (1986) and Al-Chalaby *et al.* (1985), both of whom showed that cleaning or sanitation of water does not reduce *Salmonella* isolations in birds.
However, both of these studies were conducted under controlled conditions, while ours was a field study. Since there have been several scattered field reports of *Salmonella* contamination of well water, perhaps these observations can at least partially resolve the discrepancies between these data. The correlation of litter moisture with *Salmonella* contamination rates agrees with the data of Fanelli *et al.* (1970) and Turnbull and Snoeyenbos (1973). In fact, Fanelli *et al.* (1970) states, "The shed rate of birds on built up litter suggests an inhibitory effect for *Salmonella* in litter. This effect may reduce salmonellae in either the litter or the intestinal tract. Cycling between litter and the intestinal tract appears of significance in maintaining intestinal infection."

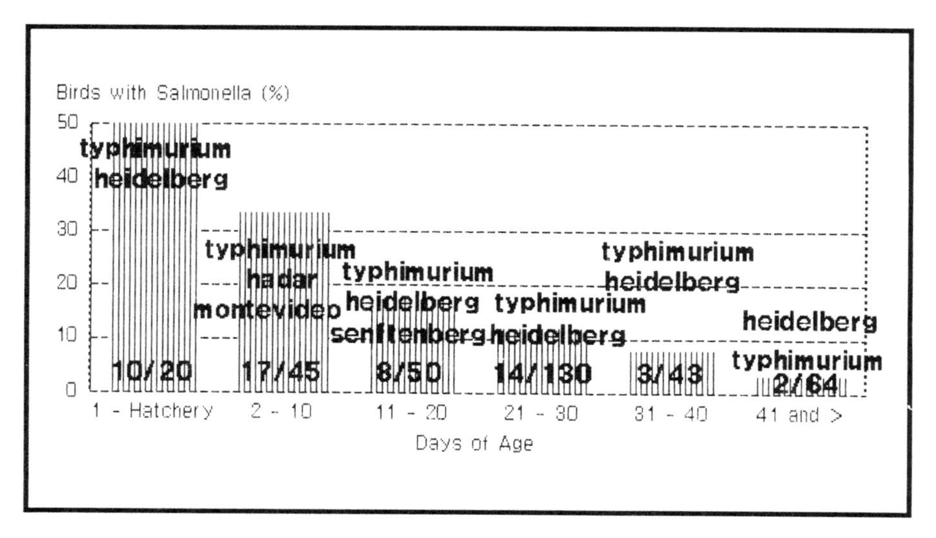

Fig. 13. Contamination of broilers from *Salmonella*-positive breeders.

Figure 14 lists the 18 factors which we attempted to correlate with *Salmonella* contamination rates in the processing plant. The presence of extraneous material on carcasses was most strongly correlated with the increase in *Salmonella* isolation rates. Since most of this extraneous material is probably fecal and we already knew these flocks were carrying *Salmonella*, this result was not surprising. The strong correlation betweeen run order (i.e., the order in which the flock was run in relation to other flocks) and cecal contamination rates suggests that the longer birds remain in shipment coops the greater the

Salmonella isolation rates. The weak, but significant correlation between catching crew and cecal contamination rates, would appear to underline the fact that humans can also be a source of *Salmonella* contamination. While line speed was not correlated with the increase in *Salmonella* isolations, it does appear to be correlated with extraneous material contamination of carcasses which in turn is correlated with the increase in *Salmonella* isolations. Line speed also appears to be negatively correlated with Grade A' s (data not shown); that is as line speed increases Grade A' s decrease.

Factors	
Cecal Positives	Shrink
Carcass Positives	Moisture
Increase in Positives	Carcass Extraneous Matter
Catching Crew	Oil Gland on Carcass
Bird per Coop	Lungs on Carcass
Total Kill	Bursa on Carcass
Line Speed	Crop on Carcass
Grade A's	Trachea on Carcass
Run Order	Feathers on Carcass

Fig. 14. Plant management factors examined.

In closing, I believe the data presented here support the following conclusions:

1. Feed remains as a great repository of *Salmonella* serotypes.
2. Multiplier flocks with contaminated ovaries do occur. While apparently not related to production parameters, the following factors would appear to be involved:
    - Disease status of the bird
    - Exposure to mycotoxins (particularly *Fusarium* toxins)
    - Poor manangement
    - Contamination in the multiplier house
3. While young birds are susceptible to *Salmonella* colonization, contamination rates tend to decrease as birds age.
4. Reduced litter moisture and water chlorination appear to be correlated with reduced *Salmonella* contamination rates in field situations.
5. Catching, loading, and live haul procedures appear to be major contributors to the contamination rates seen in ready-to-cook broilers.
6. The presence of extraneous material on carcasses is most closely associated with increased *Salmonella* isolation rates in the plant.

REFERENCES

1. **Al-Chalaby, Z.A.M., M.H. Hinton, and A.H. Linton.** 1985. Failure of drinking water sanitisation to reduce the incidence of natural *salmonella* in broiler chickens. Vet. Rec. **116**:364-365.
2. **Bailey, J.S., and N.A. Cox.** 1989. Effect of age of chick and challenge dose level on *Salmonella typhimurium* colonization of broilers. Poultry Sci. **68(suppl. 1)**:168.
3. **Bates, E., and B. Hall.** 1989. Ruling the roost. Southern Exposure **17 (2)**:11.
4. **Bahatia, T.R.S., G.D. McNabb, H. Wyman, and G.P.S. Nayar.** 1979. *Salmonella* isolation from litter as an indicator of flock infection and carcass contamination. Avian Dis. **23**:838-847.
5. **Boonchuvit, B., P.B. Hamilton, and H.P. Burmeister.** 1975. Interaction of T_2 toxins with *Salmonella* infections of chickens. Poultry Sci. **54**:1693-1696.
6. **Bryan, F.L.** 1988. Risks associated with vehicles of foodborne pathogens and toxins. J. Food Prot. **51**:498-508.
7. **Carey, J.** 1987. Serving up *Salmonella* for dinner. U.S. News & World Report. March 9, 1987, pp. 60-61.
8. **Corrier, D.E., J.R. DeLoach, and R.L. Ziprin.** 1989. Effect of T_2 toxin, cyclophosphamide or dexamethasone on the resistance of young chickens to *Salmonella* colonization. Poultry Sci.**68(suppl. 1)**:34.
9. **Cox, N.A., and J.S. Bailey.** 1989. The role of the hatchery and hatchery environment in the colonization of baby chicks with salmonellae. Poultry Sci. **68 (Suppl. 1)**:178.
10. **Cox, N.A., J.E. Thomson, and J.S. Bailey.** 1983. Procedure for isolation and identification of *Salmonella* from poultry carcasses. USDA ARS Agric. Handbook No. 603.
11. **Dougherty, T.J.** 1976. A study of *Salmonella* contamination in broiler flocks. Poultry Sci. **55**:1811-1815.
12. **Fanelli, M.J., W.W. Sadler, and J.P. Brownell.** 1970. Preliminary studies on persistence of salmonellae in poultry litter. Avian Dis. **14**:131-141.
13. **Flowers, R.S., M.J. Klatt, B.J. Robison, J.A. Mattingly, D.A. Gabis, and J.H. Silliker.** 1987. Enzyme immunoassay for detection of*Salmonella* in low-moisture foods: Collaborative study. J.A.O.A.C. **70**:530-535.
14. **Freeman, B.M., and M.A. Vince.** 1974. Development of the avian embryo-A behavioural and physiological study. Chapman and Hall Ltd. London, England, p. 172.
15. **Goren, E., W.A. de Jong, P. Doornenbal, N.M. Bolder, R.W.A.W. Mulder, and A. Jansen.** 1988. Reduction of *Salmonella* infection of broilers by spray application of intestinal microflora: a longitudinal study. Vet. Quart. **10(4)**:249-255.
16. **Green, S.S., A.B. Moran, R.W. Johnston, P. Uhler, and J. Chiu.** 1982. The incidence of *Salmonella* species and serotypes in young whole chicken carcasses in 1979 as compared with 1967. Poultry Sci. **61**:288-293.
17. **Lahellec, C., and P. Colin.** 1985. Relationship between serotypes of salmonellae from hatcheries and rearing farms and those from processed poultry carcasses. Brit. Poultry Sci. **26**:179-186.
18. **Lahellec, C., P. Colin, G. Pennejean, P. Paquin, A. Guillerm, and J.C. Debois.** 1986. Influence of resident *Salmonella* on contamination of broiler flocks. Poultry Sci. **65**:2034-2039.

19. **Koelliker, J.K., and R. Dieker.** 1978. Comparison of probe hay moisture detector and conventional method for determining poultry litter moisture. Poultry Sci. **57**:858-859.
20. **McAuliffe, K. 1987.** Will your dinner make you sick? U.S. News & World Report. November 16, 1987, p. 73.
21 **MacKenzie, M. A., and B.S. Bains.** 1976. Dissemination of *Salmonella* serotypes from raw feed ingredients to chicken carcasses. Poultry Sci. **55**:957-960.
22. **Morris, G.K., B.L. McMurray, M.M. Galton, and J.G. Wells.** 1969. A study of the dissemination of salmonellosis in a commercial broiler chicken operation. Am. J. Vet. Res. **30**:1413-1421.
23. **Poppe, C., D.A. Barnum, and W.R. Mitchell.** 1986. Effect of chlorination of drinking water on experimental *Salmonella* infection in poultry. Avian Dis. **30**:362-369.
24. **Rigby, C.E., and J.R. Pettit.** 1979. Some factors affecting *Salmonella typhimurium* infection and shedding in chickens raised on litter. Avian Dis. **23**:442-455.
25. **Rigby, C.E., J.R. Pettit, M.F. Baker, A.H. Bentley, M.O. Salomons, and H. Loir.** 1980. Flock infection and transport as sources of salmonellae in broiler chickens and carcasses. Canad. J. Comp. Med. **44**:328-337.
26. **Sadler, W.W., J.R. Brownell, and M.J. Fanelli.** 1969. Influence of age and inoculum level on shed pattern of *Salmonella typhimurium* in chickens. Avian Dis. **13**:793-803.
27. **Schat, K.A., and T.J. Myers.** 1989. Avian intestinal immunity. Proc. Avian Enter. Dis. Symp. July 16, 1989. AAAP/AVMA Annual Meeting.
28. **Shrimpton, D.H.** 1989. The *Salmonella* problem of Britain. Milling Flour and Feed. January 1989, pp. 16-17.
29. **Soerjadi-Liem, A.S., and R.B. Cumming.** 1984. Studies on the incidence of *Salmonella* carriers in broiler flocks entering a poultry processing plant in Australia. Poultry Sci. **63**:892-895.
30. **Todd, E.C.D.** 1989. Preliminary estimates of costs of foodborne disease in the United States. J. Food Prot. **52**:595-601.
31. **Taylor, M.J.** 1989. The immunomodulatory actions of trichothecene mycotoxins. Presented at: Third Pan-American Biodeterioration Society Meeting, George Washington University, Washington, DC., August 3-6, 1989.
32. **Turnbull, P.C.P., and G.H. Snoeyenbos.** 1973. The roles of ammonia, water activity and pH in the salmonellacidal effect of long-used poultry litter. Avian Dis. **17**:72-86.

DISCUSSION

R. ROLFE: Did you do any *Salmonella* serology in particular on the birds heavily contaminated with *typhimurium*?

F. JONES: No, unfortunately we didn't. It would have been very useful to do blood testing on the contaminated birds. We had all we could handle.

R. ROLFE: What sample size did you use for your feed and how many different types of feeds and samples were monitored.

F. JONES: The sample size was 25 grams. We looked at samples of feed and the feed ingredient in seven different feed mills.

L. SHIPMAN: When you were following the chicks over time up to 41 or more days what was your methodology?

F. JONES: The method is not an easy one and not one that is too popular because cecal dropping are dropped either early in the morning or roughly mid-afternoon most of the time. Basically, what we did was to walk the house itself and obtain cecal droppings. Ones were as best as we could, but were not continuous with any other fecal material so that one cecal dropping represented one bird.

L. SHIPMAN: How did you know old cecal dropping from today's?

F. JONES: Cecal droppings dry out pretty quickly. These were fresh droppings.

ENVIRONMENTAL FACTORS AFFECTING ENTEROPATHOGENS IN FEED AND FEED MILLS

D. A. Gabis

Silliker Laboratories, Inc.
Chicago Heights, Illinois 60411

ABSTRACT

Poultry feeds containing *Salmonella*-contaminated meat and bone meal are sources of infection for live poultry. The incidence of *Salmonella* in meat and bone meal is on the order of 50% of the samples tested. Salmonellae in raw offal are destroyed by the high temperatures of rendering processes, but the product is recontaminated during handling in the dry state after cooking. The factory environment where dry meat and bone meal is handled and stored is the primary direct source of salmonellae. The post-cooking product environment is expected to be dry, and the presence of water is a hazard for microbial growth. When water mixes with accumulated, dry, finished product residue, niches are established where salmonellae are encouraged to multiply. Growth niches of salmonellae in the plant are the main sources of contamination for meat and bone meal. Water may be present as a result of water vapor condensation on the interiors of conveyors, leaking roofs, unsealed buildings, leaking water and steam lines, water remaining after wet cleaning, and unsanitary design, fabrication, and repair of equipment. Control of *Salmonella* growth in dry processing environments requries that the design, fabrication, construction, installation, maintenance, and repair of buildings and equipment be carried out to prevent the development of sites where water and product residue are allowed to accumulate. *Salmonella* cannot be controlled by cleaning and sanitizing practices alone. Unsanitary maintenance and repair practices are the predominant causes of the high incidences of *Salmonella* contamination in rendered animal by-products. The incidence of *Salmonella* in meat and bone meal is reduced when the processing environment is dry.

Colonization Control of Human Bacterial Enteropathogens in Poultry

INTRODUCTION

Rendered animal by-products have high incidence of contamination by non-host adapted salmonllae. Rendered animal by-products such as meat and bone meal, feather meal, and other products are used as ingredients of animal feeds (Bisplinghoff, 1984).

Epidemiological evidence exists which suggests that salmonellae-controlled animal feeds are associated with infection of food animals including poultry and hogs, and, ultimately with human foodborne salmonellosis. For example, *Salmonella agona* was initially detected in Peruvian fishmeal imported into the southeastern United States in 1970 (Clark *et al.*, 1973). After this initial isolation, *S. agona* was reported with increasing frequency from nonhuman sources, and by 1976 this serotype ranked third among the most frequently isolated serotypes from humans and fourth from nonhumans (Gangarosa, 1978). In England, *Salmonella hadar* was first identified in 1973 in poultry meal imported from Israel. By 1976, this serotype had become the fourth most frequently isolated *Salmonella* from humans in England and Wales (Watson and Kirby, 1984).

Samonellae in raw materials are destroyed by the heat treatment applied during the rendering process. Rendered animal by-products may become contaminated at many points after cooking, and by the primary sources of salmonellae are microbial growth niches in the post-cooking environment.

Past efforts by government and industry groups have failed to eliminate or reduce the incidence rate of salmonellae contamination in rendered products; nor has the addition of organic acids and other bacterial compounds to contaminated feed ingredients been successful. Historically, control efforts have failed to address the need to correct specific defects which are the causes of contamination of finished products from individual rendering plants. The key to reducing the incidence rate of salmonellae in rendered animal by-products is prevention of recontamination after cooking.

MICROBIAL GROWTH NICHES

Much of the effort placed on *Salmonella* control has been through the development of general guidelines and procedures for sanitation (USDA, 1964). General good manufacturing and sanitation practices are insufficiently specific to bring about a significant reduction in the incidence of salmonellae in rendered animal by-products. To be effective, *Salmonella* control programs must be directed towards eliminating specific conditions in individual plants which lead to post-cooking contamination. Control programs based on destruction of salmonellae which enter the product after processing are akin to closing the barn door after the horses have escaped. The most reliable, consistent, and long-term economic salmonellae control programs must require prevention of post-cooking contamination.

Without a doubt, adequate cooking destroys any salmonellae in raw materials, but there are many opportunities in the post-cooking rendering system for recontamination of the cooked product. For the most part, rendered animal by-products become contaminated by salmonellae from sites which support the growth of large populations of microorganisms (microbial growth niches) in the post-cooking process environment.

A distinction should be made between direct contamination of product from environmental growth niches on or near the product flow stream and cross-contamination from raw materials. Microbial growth niches are permanently ensconced in the processing environment and are ongoing sources of high numbers of microorganisms. On the other hand, the probability of cross-contamination from raw materials is comparatively less because vectors and fomites are required to transfer the salmonellae, and the vector must come in direct contact with a tremendously large amount of product in order to account for the high incidence rate in finished material.

Segregation of raw material areas from finished product areas is an important component of good manufacturing practices, but microbial growth niches in the finished product environment are responsible for the bulk of high incidence rate. Microbial growth niches are established when salmonellae are introduced by raw materials into growth-promoting environments in the plants. If the salmonellae are not provided with a suitable environment in which to grow, the risk of finished product contamination is greatly reduced because the original populations do not proliferate.

Along with water, sufficient levels of nutrients from the product, adequate incubation temperatures, enough time to allow growth, and, of course, microorganisms must be present in order to establish a microbial growth niche. However, water in the dry finished product environment is by far the most important requirement for the establishment of microbial growth niches. In all practicality, water is the only one of the elements of the growth niche which can be controlled. The other three elements cannot be controlled because finished product residue is omnipresent due to the very nature of the process; the ambient temperatures of most rendering plants are ideal for microbial growth; salmonellae will always be present at some level in raw materials; and, plenty of time is available to achieve high populations. If water cannot be eliminated from the dry product environment, then significant reduction is unlikely to occur in the incidence rate of salmonellae in rendered animal by-products. Experience shows that those rendering plants with the driest environment are the most likely to have the lowest incidence rate of salmonellae in finished product.

SOURCES OF WATER IN THE DRY PROCESSING ENVIRONMENT

Some of the most significant sources of uncontrolled water in dry processing environments of rendering plants and feed mills are: water from wet cleaning procedures used on equipment and plant areas which are not designed for wet cleaning and which cannot be dried completely and promptly; water vapor condensation at hot/cold zone interfaces in equipment, such as transfer conveyors and inadequately vented storage bins; direct addition of water to product to adjust the moisture content; unsealed buildings (including roofs, walls, windows, and doors) which allow water to enter the plant; leaking water and steam valves and pipelines; processing equipment installed on the exteriors of plants without adequate protection against entry of rain water; and finished product loading and unloading facilities which are not adequately protected against entry of water from rain and snow; the use of transportation vehicles which are not designed to keep out water.

UNDERLYING CAUSES OF WATER IN THE DRY PROCESSING ENVIRONMENT

On an industry-wide basis, the three most important causes of uncontrolled water in the dry processing environment are: (a) unsanitary design and installation of buildings, equipment, and storage/loading facilities, (b) failure to hygienically maintain and make timely repairs to buildings and processing equipment, (c) use of wet cleaning in dry environments.

In practice, the economic costs to replace existing plants and equipment with new, hygienically designed and installed units are prohibitive in light of the short term benefits to be derived. Nevertheless, great reductions in incidence rates of salmonellae contamination could be obtained with reasonable expense by adhering to sanitary maintenance and repair practices, and curtailing the use of water for cleaning.

PROGRAMS TO REDUCE THE INCIDENCE OF SALMONELLAE IN RENDERING PLANTS AND FEED MILLS

There are no magic bullets in the arsenal available to control salmonellae in rendered animal by-products. No new and expensive technology is required to solve the problem. To reduce the incidence of salmonellae in rendered by-products, the basic principles of microbial ecology used to control contamination in the environments of other dry food processing plants must be followed. The most effective and cost beneficial options for salmonellae control in the processing environment must be identified for individual plants.

Management and workers must be trained and motivated to identify environmental microbial growth niches and be taught specific procedures to eliminate them. For example, the Animal Protein Producing Industry's (APPI) *Salmonella* Control Education Program (1989), if followed, will result in removal of many niches.

The development of a successful, model demonstration, environmental salmonellae control program in one or more rendering plants will provide effective leadership and encouragement for the industry. Existing plants with low incidence rates should stand as examples for the rest of the industry to follow. Governmental regulation and sanctions mandating salmonellae control programs would undoubtedly provide strong impetus for the industry to take action.

In the end, the primary driving force required for salmonellae control must come from the owners and managers of the rendering industry. The owners and managers must provide the motivation and commitment to make the industry-wide changes required to reduce the incidence of salmonellae.

REFERENCES

1. **Bisplinghoff, F.D.** 1985. Bacterial contamination of animal proteins. *In* Proceedings of the International Symposium on *Salmonella*, pp. 232-238. New Orleans, Louisiana, July 19-20, 1984. G.H. Snoeyenbos, (ed.), American Association of Avian Pathologists, Inc. Kennett Square, PA.
2. **Clark, G.M., A.F. Kaufman, E.J. Gangarosa, and M.A. Thompson.** 1973. Epidemiology of an international outbreak of *Salmonella agona*. Lancet. **11**:490-493.
3. **Gangarosa, E.J.** 1978. History and surveillance programs for salmonellosis. *In* Proceedings National Salmonellosis Seminar. Washington, DC.
4. **Watson, W.A., and F.D. Kirby.** 1985. The *Salmonella* problem and its control in Great Britain. *In* Proceedings of the International Symposium on *Salmonella*, pp. 35-47. New Orleans, Louisiana, July 19-20, 1984. G.H. Snoeyenbos,(ed.), American Association of Avian Pathologists, Inc. Kennett Square, PA.
5. **U.S. Department of Agriculture.** 1964. Recommended sanitation guidelines for processors of poultry and animal by-products. ARS 91-47. Animal Disease Eradication Division, Agricultural Research Service. Hyattsville, MD.

DISCUSSION

F. JONES: Damien, I noticed you mentioned moisture. I would summarize your talk by saying that if you can keep moisture down in the plant you have almost got the battle won. Have you looked at any correlation in the samples between moisture and Salmonella?

D. GABIS: We didn"t measure it, Frank. Therefore, I can't answer that.

F. JONES: Do you think it might be there?

D. GABIS: Not in the finished product itself. The moisture in the niches is important. By the time you innoculate a few grams in there you have millions and millions of organisms. By the time it gets blended out through the system, moisture is dissipated equally throughout, so the water activity is dropped and that's the end of the story.

K. NAGARAJA: In an analysis of samples, how much *Salmonella enteritidis* is involved?

D.GABIS: In the 1985 study, we saw none. In this study, I don't know how many group D's we have because I haven't gone back to look at that. However, I've instructed the people that whenever they get a group D to come yelling, and they haven't done that yet.

K. NAGARAJA: The second question I have about the warranted testing program the FDA has, is: Do you see a trend in this decrease?

D. GABIS: I'm not ready to say that yet because the interpretation of the data looked at on an industry-wide basis, when we composite all of the results together in the soup and take the average, is still close to 50% positive. However, we need to go back and compare results on a plant-to- plant basis to see if that plant had an improvement because they may have had a 60-80% rate and they nailed it down to 20%, but in the total conglomerate of all the data that improvement may be lost. We're going to have that later in our program.

back and compare results on a plant-to- plant basis to see if that plant had an improvement because they may have had a 60-80% rate and they nailed it down to 20%, but in the total conglomerate of all the data that improvement may be lost. We're going to have that later in our program.

H. OOSTEROM: I'm very much delighted with your presentation. I was in charge of the bacteriological culturing in rendering plants in the Netherlands for about eight years, and I know exactly what you are seeing and what you have felt and what you have smelled especially. There are a few things I want to discuss with you. I completely agree with you that water, feed, and time is the combination that is very much important for cross-contamination. I do not agree with you that there is no new technology for solving the problems. We think that one very important thing in rendering and in making a *Salmonella*-free product is the separation between clean and unclean areas. The borderline, of course, is the cooking or the sterilizing as we do it. I think that is very important. We even have one rendering plant in which the resuction hole is physically and completely separated from the other rooms.

There is a second point. We used only closed pipelines. Once the material is dumped and crushed, you do not see it again before it is milled and put inside them again. These are very important things: separation, clean, unclean, and using closed pipelines. Furthermore, I agree with you that we try to prevent all condensation of water because water is the main factor. We also had dustcleaners and very huge vacuum cleaners to take away all the meal that, in spite of the closed pipelines, could spread all over the premises. Thank you.

D. GABIS: In my use of the word new technology I don't mean to imply that the ability to put in new equipment and so on would vastly improve the situation. I'm thinking of the tendency that I have heard from the rendering industry to try to find chemical additives to try to kill the *Salmonella* once they have gotten back into the feed. By using the words "new technology", I'm trying to discourage that as the solution as opposed to trying to do what our friends from the Netherlands have suggested which is to make these commitments for mechanical improvements in separation of raw and something else.

L. BLANKENSHIP: I would like to emphasize the point that you made earlier that the rendering plant and the feed processing plant can do the best job possible and deliver top quality enteropathogen-free feed to the farm. However, it is critical to look at the farm handling of the feed, the silos, the conveying equipment, and the feed once it gets into the house. There are opportunities for contamination to occur once the feed manufacturer has finished his job.

ENVIRONMENTAL FACTORS ASSOCIATED WITH *CAMPYLOBACTER JEJUNI* COLONIZATION OF POULTRY

S. M. Shane

Department of Epidemiology and Community Health
School of Veterinary Medicine
Louisiana State University
Baton Rouge, Louisiana 70803-8404

ABSTRACT

Commercial broiler and turkey products are recognized as a significant source of *Campylobacter jejuni* enterocolitis in humans. Prevalence studies have confirmed a high rate of intestinal carriage in mature breeder hens and pre-slaughter broiler and turkey flocks. Infection can generally be identified two weeks after commencement of the brooding cycle. The proportion of carriers rises rapidly by horizontal dissemination through water receptacles and ingestion of contaminated litter to levels approaching 100% in infected flocks. Vertical transmission is not involved in the field occurrence of *C. jejuni* infection.

Because of the sensitivity of the organism to desiccation and commercial disinfectants, thorough decontamination of houses followed by a 5- to 7-day intercrop rest interval will eliminate residual *C. jejuni* in the environment. Flies, fomites, wild birds, and vermin can serve as vehicles to introduce infection into flocks. Survival of *C. jejuni* in damp litter, prolonged periods of shedding by colonized birds, and coprophagy contribute to persistence of infection in flocks.

Procedures to maintain acceptable standards of biosecurity limit introduction of *C. jejuni*. Operation of single-age farms and thorough decontamination and resting between successive flocks are recommended control measures. Competitive exclusion has not demonstrated any potential for control of infection under commercial conditions.

From the standpoint of human health, post-processing treatment of poultry products with pasteurization levels (3-7 kGy) of ionizing radiation, or alternatively, decontamination with an approved chemical compound, offer the greatest potential for neutralizing the impact of human *C. jejuni* enterocolitis contracted from poultry.

Colonization Control of Human Bacterial Enteropathogens in Poultry

INTRODUCTION

Campylobacteriosis is now regarded as a significant foodborne disease (6) attributed to infection with *Campylobacter jejuni*. Surveys have shown that the pathogen is frequently associated with broilers (28), ducks (38), and turkeys (2), which serve as reservoirs. In addition to poultry, other domestic food animal species, including cattle and goats (64), sheep, and swine (91), function as intestinal hosts of *C. jejuni*. Prevalence studies in companion animals have shown that dogs (82) and cats (17) are carriers of *C. jejuni*, and are significant with respect to juvenile contacts. The organism has also been isolated from a wide range of laboratory (19) and captive animals in zoological parks (45), although these species are not generally involved in human intestinal infection.

Campylobacteriosis in humans is caused predominantly by *C. jejuni* (94). The 1984 isolation rate in the United States was reported as 4.9/100,000 population, but this value understates the prevalence of the disease. It is estimated that 2.1 million cases occur annually in the United States (52). The cost of human campylobacteriosis in terms of lowered productivity, death, and medical treatment ranges from $700 million to $1400 million per annum.

AVIAN VIBRIONIC HEPATITIS

Chronic, degenerative hepatopathy ascribed to infection with vibrio-like organisms has been described in commercial egg-producing flocks (60). The description of the organism, retrospectively identified as a campylobacter, was consistent with isolates made from cases of a field syndrome termed "avian vibrionic hepatitis" (30). Although vibrio-like organisms isolated from cases of human enterocolitis and the avian isolates derived from hens with hepatitis showed some morphological similarities, obvious biochemical differences existed (18). Epidemiologic studies conducted in New York State suggested an association between the hepatitis syndrome and the presence of the vibrio-like isolate (27). The organisms derived from the cecum and bile of affected birds were shown to be similar, but the isolates obtained from different cases varied in both pathogenicity and colonizing ability. Despite the intensity of research on this problem during the late 1960s, the etiology of the condition was not determined. The disease complex could not be reproduced under controlled conditions using a field isolate cultured on an artificial medium (96). During the period 1967-1973, the disease disappeared in both the United States and western Europe. Characteristic hepatopathy cannot be induced in either chicks or mature hens by inoculation of *C. jejuni* (71). Only low-grade enteritis manifested by transitory diarrhea occurs following infection of day-old chicks (71). Definitive evidence that the vibrio-like organisms were in fact *Campylobacter* spp. is lacking. It is unfortunate that an attempt to culture the suspected pathogen from lyophilized yolk stored for 25 years was unsuccessful (9).

CHARACTERISTICS AND ISOLATION OF *C. JEJUNI*

Campylobacter jejuni (86) is the predominant species of clinical significance associated with commercial poultry, although *C. coli* may occasionally be isolated from the intestinal tract of chickens and turkeys (70). *C. jejuni* is a Gram-negative (75), motile (85), spirally-curved rod. The organism is thermophilic, showing optimum growth on artificial media at 43 ° C (83). Campylobacters are microaerophilic, requiring an atmosphere comprising 5% O_2, 10% CO_2, and 85% N_2 (65). In view of the sensitivity of the organism to desiccation, liquid transport media have been developed for holding specimens prior to isolation (47). Enrichment in broth medium can be carried out to enhance the rate of isolation from meat and fecal samples (68). Selective culture media, such as Butzler's or Skirrow's, are generally used to isolate *C. jejuni* (59). Most commercial media contain brucella agar and ovine blood as a base, and a mixture of antibiotics to inhibit the growth of competitive contaminating organisms (53).

Campylobacter jejuni is oxidase and catalase positive, and reduces selenite (86). This species can be differentiated from other thermophilic *Campylobacter* spp. on the basis of nalidixic acid sensitivity and hydrolysis of hippurate (84). Confirmed isolates of *C. jejuni* may be biotyped according to hydrogen sulfide reaction (42) and DNA hydrolysis (43). The Penner scheme is widely used to serotype isolates (61). Differentiation is based on the presence of heat-stable somatic antigens, which are detected using a passive hemagglutination test.

PREVALENCE OF *C. JEJUNI* INFECTION

The prevalence of *C. jejuni* in domestic poultry can be regarded as a function of surveillance intensity (76). With the advent of commercially available selective media and gas systems for incubation under microaerobic conditions, the intensity and sensitivity of flock sampling has increased (64).

During the mid-1970s the extent of infection of broiler and turkey flocks became apparent (87). Fecal contamination with *C. jejuni* was confirmed in 82% of a batch of 46 broilers sold in a live-bird market in New York City in 1980 (26). In a 1982 survey conducted in 14 broiler houses in England, *C. jejuni* was isolated from three units. The proportion of affected birds in each infected batch was relatively high, and in some cases all specimens within a batch were positive (12). A survey conducted in Chile showed that 90% of a sample of 50 mature live broilers were infected with *C. jejuni* (16). A study conducted in Israel during 1982 showed that 38% of a sample of 124 poultry flocks were infected with *C. jejuni* (62). The prevalence of infection was similar among broilers (10%), replacement laying pullets (50%), and growing turkeys (40%). Mature layer flocks showed a high (89%) rate of infection. Both the probability of infection and the number of *C. jejuni* carriers in a flock increased in proportion to age in this study.

These observations were confirmed by a detailed series of surveys conducted on broilers in California during 1983 (88). Fecal swabs yielded a 15% recovery of *C. jejuni* from two of four farms investigated. In specific flocks infection was detected during the second week, but in some cases carriers were only diagnosed at seven weeks. The prevalence rate within affected flocks increased from 2% at ten days to 80% at the end of the growout period (21). The proportion of birds infected with *C. jejuni*, as determined by fecal excretion, increased rapidly, with practically all broilers within a sample positive within seven days of initial isolation. Infection spread readily among houses on a specific farm, and may have persisted between successive cycles.

A cross-sectional survey in a multi-age egg production unit in Sweden showed the absence of *C. jejuni* in newly hatched chicks, but a recovery rate of 5% and 72%, respectively, in five-week and sixteen-week-old pullets reared on wire floors. Mature hens in cages showed a 20%-30% rate of infection (41).

Campylobacter jejuni/coli was isolated from eight of 11 batches of broilers delivered to a processing plant in Israel in 1985 (62). In seven of the flocks positive isolations ranged from 60% to 100% of the birds sampled. In longitudinal studies conducted by these authors, four out of five broiler flocks were positive by the eighth week. Infection was detected in one flock at seven days of age, and in an additional two flocks at 28 days.

In a Canadian study, 28 of 60 broiler flocks yielded *C. jejuni* from cloacal swabs collected at slaughter. A weighted average rate of infection of 90% was determined within the sampled population (66). A study in Ireland confirmed the presence of *C. jejuni* in 11 of 12 broiler flocks ranging in age from two to ten weeks, with more than 75% of individuals affected in seven flocks examined after the fourth week. The organism was also isolated from 12 of 21 broilers submitted to a diagnostic laboratory. Affected flocks ranged in age from 15 to 50 days. *Campylobacter jejuni* could only be obtained from one one-day-old chick out of 310 sampled by the cloacal route (54).

A comprehensive vertical study conducted in an integrated broiler operation in Switzerland during 1985-1986 (31) showed that both breeder and broiler flocks were infected. Three out of four layer units yielded *C. jejuni*, with 11 of 17 flocks showing infection after 25 weeks of age. In broilers, six of 153 flocks in four of 31 units were shown to be positive between the 34th and 42nd days of the growing cycle. An Australian study (80) confirmed cloacal *C. jejuni* carriage in 74% of 240 mature broiler breeders in six commercial flocks in an integrated operation, suggesting that hens housed on litter are frequently affected.

Evaluation of cage-housed commercial laying flocks in Louisiana showed a prevalence rate in fecal samples ranging from 13% in a 50,000 hen unit to 62% in a farm housing 200,000 birds (78). In Chile, 55% of a sample of 200 caged hens yielded *C. jejuni* from the cloaca (16).

Commercial turkeys show a relatively high carriage rate compared to broilers. A survey conducted over 12 months in a processing plant in California showed that all of 600 samples of cecal contents, and 30 cloacal swabs were infected with *C. jejuni* (46). As with broilers, growing turkeys demonstrate onset of fecal shedding at approximately two weeks of age. A survey in Texas on brooded turkeys yielded *C. jejuni* in cloacal swabs obtained from 16% of 25 fifteen-day-old poults and 76% of 25 nineteen-day-old birds (1).

Ducklings reared for meat production in California during 1985 were shown in a six month survey to be carriers of *C. jejuni*. In contrast to broilers, which generally show onset of shedding at two weeks, 40% of a sample of ducklings reared on wire yielded positive cloacal swabs at four days of age, with all birds becoming infected during the 7- to ll-day period. Two out of three mature duck breeders (12-months-old) yielded *C. jejuni* (38). Infection rates in commercial ducks generally exceed the cecal carriage rate of 35% reported in a sample of 445 hunter-killed ducks representing seven common species (44).

PATHOGENICITY OF *C. JEJUNI* FOR CHICKENS

Oral infection of one-day-old chicks with 10^8 CFU *C. jejuni* of a pathogenic strain induces clinical depression and diarrhea, which may persist for eight days and result in up to 30% mortality (71). Transient diarrhea and bacteremia have been described in chicks to which 10^6 organisms were administered either at day-old (89) or two days of age (74). *Campylobacter jejuni* isolates from humans with severe enterocolitis were responsible for enteritis in these three trials.

The virulence of an avian isolate shown to be invasive was enhanced by a series of six passages in chicks. In contrast, neither a feline- nor a human-origin strain was affected by passage. Pathogenicity of *C. jejuni* strains in day-old chicks was denoted by habitus, the severity of growth depression, and the extent and duration of diarrhea (73).

The literature suggests that clinical changes can be caused only by oral infection during the first 48 hours after hatch. Neither mucoid nor watery diarrhea can be induced in chicks older than three days, although colonization without development of clinical signs occurs following oral infection (98). The observation that chicks do not show clinical signs when infected after four days of age may relate to the change in ventricular pH from 5.5 to 4.0 after initiation of normal feeding.

The predominant gross lesion associated with campylobacteriosis in newly hatched chicks comprises distention of the jejunum by mucoid or watery fluid (74). Enterotoxigenic strains of *C. jejuni* may produce hemorrhagic changes (98). Corresponding histological lesions include mucosal edema, destruction of enterocytes, and villous atrophy. Inflammatory changes comprise mononuclear infiltration of the lamina propria and accumulation of mucus and erythrocytes in the lumen of the jejunum and ileum (98).

EPIDEMIOLOGY OF CAMPYLOBACTERIOSIS

Longitudinal surveys have shown that broiler, breeder, and replacement pullet flocks are infected with *C. jejuni* at two to three weeks of age. Following introduction of infection, horizontal transmission occurs rapidly, resulting in prevalence rates of 70% to 100% of the birds sampled. Colonization of the intestinal tract leads to chronic fecal shedding of the organism.

Congenital Infection. The roles of vertical transmission and environmental contamination in the occurrence of *C. jejuni* in commercial poultry flocks have been the subject of considerable research (76). The weight of experimental and epidemiologic evidence suggests that vertical transmission of *C. jejuni* does not occur under commercial conditions.

A field survey of eggs derived from commercial laying hens demonstrated to be fecal excretors of *C. jejuni* failed to yield the organism from either the shell surface or from homogenates of yolk and albumen. In a parallel study inoculation of egg shells with a fecal suspension containing 1.4×10^8 CFU/g, *C. jejuni* resulted in shell penetration in three of 70 eggs and recovery of the organism in only one of 70 eggs. Because of the sensitivity of *C. jejuni* to desiccation, viability of the inoculum at 20 ° C was restricted to a maximum of 16 hours. A low rate of egg shell penetration was detected despite the high concentration of organisms in the inoculum (78).

Shell membranes are regarded as a significant barrier to passage of *C. jejuni* to the albumen (13). A study to investigate the ability of the organism to penetrate the egg shell and membranes and survive in the egg contents demonstrated the improbability of vertical transmission. Although three avian isolates of *C. jejuni* could penetrate the shell, survival under aerobic conditions within the egg was limited to six hours. It was not possible to recover *C. jejuni* from egg contents, dead-in-shell embryos, or from the intestinal contents of chicks hatched from specific-pathogen-free (SPF) eggs which had been immersed in a suspension of viable organisms prior to incubation (55).

A 1982 study conducted in an integrated turkey production unit in Texas failed to demonstrate *C. jejuni* in either a sample of 20 fertile hatching eggs or in 20 one-day-old poults. In addition, 25 poults reared in a campylobacter-free environment, represented by an inflatable incubator with filtered air and sterilized feed and water, did not excrete *C. jejuni* during a 21-day observation period. This is contrasted to litter-reared poults from the same hatch, which commenced shedding the organism at 15 days and showed a 76% prevalence rate in fecal samples collected on the 19th day (1).

The possibility of vertical transmission of *C. jejuni* in an integrated broiler unit was investigated by a combined survey and egg-inoculation trial conducted in Australia during 1985 (80). Of 187 eggs derived from a breeder flock with a 74% prevalence of fecal shedders,*C. jejuni* could only be isolated from two floor eggs (1%). All of the 162 chicks hatched (63%) from 257 eggs which were surface-inoculated with *C. jejuni* were free of the organism.

Injection of *C. jejuni* (10^1 to 10^4 CFU/g) into the albumen reduced hatchability to 7%, compared to saline-injected controls (56%) (81). Two of 16 chicks hatched (13%) were colonized, and all unhatched eggs yielded *C. jejuni*. An 8% rate of intestinal infection in newly-hatched chicks was achieved by exposing eggs to a suspension (10^8 CFU/ml) of mixed strains of *C. jejuni*, using a temperature differential technique. When eggs were stored for eight days prior to incubation, *C. jejuni* could not be isolated from either viable chicks or dead-in- shell embryos. After pre-incubation storage for five and one-half days, 20% of inoculated eggs yielded *C. jejuni*, but the organism could not be isolated from egg contents after two days of incubation (10). The destruction of *C. jejuni* in eggs is attributed to lysozyme, which is active at the post-inoculation pH of 9, and to the bacteriocidal action of conalbumin.

Decontamination of fertile eggs with formalin gas or a suitable contact disinfectant within two to three hours of collection will destroy *C. jejuni*. Rejection of fertile eggs with fecally soiled or defective shells further reduces the extremely low probability of egg transmission.

The horizontal dissemination of *C. jejuni* among newly-hatched chicks was strikingly demonstrated by a study involving direct-contact transmission in a hatcher (11). Single, first-hatched, dry chicks were inoculated with an oral suspension of 10^7 *C. jejuni* (either avian or human strain) and were returned to the hatcher trays containing pipping and unhatched eggs. Chicks which had hatched by the 22nd day were transferred from incubator trays to cardboard transport boxes. These boxes were modified to permit free movement among the 100 chicks in each box. Examination of sacrificed chicks after a 24 hour holding period showed that 13 of 14 of the strain-dose combinations yielded *C. jejuni* from the intestinal tract. The range extended from no recovery to 70%, although there was no consistent correlation between the origin and toxigenicity of the isolate and rate of transmission. Four strains of human origin were shown to cause intestinal distention and hepatic necrosis. Two poultry strains evaluated yielded high recovery rates without producing gross lesions. In some cases, although lesions were present, *C. jejuni* could not be isolated or was recovered from only 3% to 4% of the chicks. Horizontal infection among the batches of chicks was attributed to direct contact in the hatcher and holding boxes. Cooling water was eliminated as a possible route of transmission, since in a number of replicates chicks were contaminated but the hatcher reservoirs, originally filled with sterile water, remained free of *C. jejuni*.

Environmental Contamination. The fact that broiler chicks and poults demonstrate shedding of *C. jejuni* approximately two weeks after initiation of brooding presumes that the infection is acquired from the environment. The presence of *C. jejuni* in litter has been demonstrated in epidemiologic studies in broiler growout units (12, 20, 63) and turkey farms (1).

A structured series of experiments confirmed the role of litter in the transmission of campylobacter in broilers subjected to controlled exposure (51). Fourteen-day-old SPF chicks housed in Horsfall isolator units were placed on either sterilized or untreated rice hulls previously used to rear broilers. Chicks exposed to litter inoculated with *C. jejuni* (10^6 CFU/g) commenced fecal shedding within five days, and yielded the pathogen through to the termination of the experiment at 46 days. Chicks infected by the oral route (10^4 CFU/bird) continued to excrete *C. jejuni* in feces for 63 days after transfer to a Horsfall isolator fitted with a floor of stainless steel rods which prevented coprophagy.

Since *C. jejuni* is intolerant to desiccation (47), the organism may not survive in litter for periods exceeding ten days at 20 ° C. A 5.4 decimal reduction in numbers of *C. jejuni* was determined in inoculated litter after four days at either 17 ° C or 30 ° C (22). No enrichment stage was incorporated into the isolation procedure, and it is possible that *C. jejuni* may have survived beyond this eleven-day limit of detection. Under laboratory conditions, viability of the organism is dependent upon strain, temperature, pH, and the presence of organic matter (14). Campylobacters can retain viability for three weeks in feces stored at 4 ° C (5). These results have implications for broiler and turkey growers who reuse litter for a number of successive cycles.

Publications relating to the efficiency of decontamination have involved superficial sampling of both litter and the environment in brooder houses. Comprehensive disinfection of a turkey house apparently eliminated *C. jejuni*, although the organism was isolated from cloacal swabs obtained from the subsequent flock at 22 days of age. In a specific Texas survey, assay of litter was carried out on a 400 g pooled sample "collected from various areas in each brooder house" (1). Persistent infection in a specific broiler growing house was documented in California. The problem was attributable in part to retention of litter and ineffective decontamination (88). The moisture content of litter in the subject building, which was apparently flooded at frequent intervals, may have promoted prolonged survival of *C. jejuni*. Failure to isolate *C. jejuni* from litter in units housing flocks with a fecal excretion rate of 80% can be attributed to the low number of samples examined.

Water. Contaminated water is regarded as a significant reservoir of *C. jejuni* for human populations (49). An outbreak of campylobacteriosis in Sweden in 1981 involved 2000 patients. The non-chlorinated water supply for the community was obtained from a stream contaminated by the effluent from a commercial laying farm. A cluster of *C. jejuni* enteritis cases in a small rural community in Florida in 1983 was attributed to a defective chlorination system. The source of infection was an open reservoir to which free-living birds infected with *C. jejuni* had access (72). *Campylobacter jejuni* can survive for four days at 25 ° C in sterilized stream water, but at 4 ° C viability was extended to 32 days (5).

In a field survey involving turkey flocks in Texas in 1982, *C. jejuni* was demonstrated in the drinkers used by the flock (1). Since no attempt was made to isolate the organism from the water supply, it is not known whether infection was primary in origin or introduced into drinkers by the flock. An epidemiologic study conducted in Sweden in 1986 showed the presence of *C. jejuni* in drinkers and feeders in houses containing infected broilers. Neither the water supply nor delivered feed was contaminated (41). A second study conducted in Sweden suggested that *C. jejuni* might be associated with a contaminated source of water (15). Eight campylobacter-negative broiler flocks and six positive flocks obtained water from either deep wells or a municipal supply. Two flocks with fecal shedders of *C. jejuni* used untreated water from a lake.

The 1983 California survey of broiler growing units (88), in which a specific problem house was identified, confirmed the correlation between fecal shedders of *C. jejuni* (26/30) and the detection of the organism in water receptacles (3/20). This observation contrasted with the absence of *C. jejuni* in waterers in two adjoining units housing flocks free of fecal *C. jejuni*.

The structured trial involving exposure of chicks to contaminated litter in Horsfall isolators demonstrated *C. jejuni* contamination of the water receptacles five days after initiation of the experiment (77). Isolation of the organism from water continued for up to 39 days, consistent with the presence of *C. jejuni* in litter. Swabs obtained from the palatine cleft of birds infected both by the oral route and by exposure to contaminated litter yielded *C. jejuni* on the 11th day of the trial. This was in all probability due to either coprophagy or consumption of contaminated water.

Neither experimental nor epidemiologic evidence is available to determine the precise role of contaminated water in either introducing or spreading infection. It is possible that non-chlorinated surface water containing viable *C. jejuni* may infect flocks. It is also apparent that drinkers, which become contaminated with feed and litter, and hatcher cooling systems (11) are capable of intra-flock dissemination.

Insect Transmission. Infected houseflies (*Musca domestica*) have been suggested since 1983 as possible vectors of *C. jejuni* (69). This is based on the isolation of *Campylobacter* spp. from 50% of 146 flies captured on or in the vicinity of chicken farms in southeastern Norway. Of the 74 flies shown to be contaminated, the relative percentages for *C. jejuni*, *C. coli*, and *C. laridis* were 13%, 80%, and 7%, respectively. Since *C. jejuni* is principally associated with poultry and *C. coli* with hogs, the flies yielding *C. coli* may have acquired the infection in swine barns. It is noted that *C. coli* represented 99% of the isolates from flies obtained from piggeries in the Norwegian study. No attempts were made to correlate isolation of *Campylobacter* spp. from flies with the presence of fecal shedders in the chicken flocks and pig herds in the sampled farms.

Three of ten houseflies yielded *Campylobacter* spp. in a study conducted in Yugoslavia in 1988 (3). The presence of contaminated flies was correlated with infected broiler flocks and the isolation of the organism from other livestock and free-living birds. A contamination rate of only 2.4% was determined in 210 flies (*Mucsa, Fannia, Lucilia*, and *Calliphora*) in a study conducted in England in 1983 (100). The survey investigated flies obtained from domestic gardens and homes remote from piggeries and poultry farms.

The transmission of *C. jejuni* by flies was clearly demonstrated in a controlled experiment conducted in Louisiana during 1984 (77). Laboratory-hatched houseflies were contaminated with *C. jejuni* by confinement in a Horsfall isolator containing infected twenty five-day-old chickens. The birds had been previously dosed with 10^8 organisms by the oral route, and were shown to be excretors of *C. jejuni* at the time of introduction of the batch of 400 flies. On the fifth day of the experiment, *C. jejuni* contamination of flies and litter was demonstrated. Transfer of 300 flies to a second unit containing SPF chickens resulted in fecal shedding of the organism by the birds after eight days, in addition to recovery of the organism from litter and drinking water. A concurrent feeding trial was initiated, in which 32 laboratory-bred houseflies were provided with a contaminated nutrient solution containing 10^8 *C. jejuni*/ml. The organism was recovered from the feet and from the abdominal contents of 10% and 95%, respectively, of the flies examined. These trials did not differentiate between the possible mechanisms of transmission by flies. Infection of the SPF chickens could have occurred following regurgitation or defecation by flies onto litter, feed, or water. Ingestion of contaminated flies by chickens obviously occurred during the period of exposure.

Free-Living Birds. Birds are frequently carriers of *C. jejuni* and represent a possible source of infection for commercial poultry. Surveys have shown that pigeons in both the United States (45) and Japan (39) are infected with *C. jejuni*. In addition, the organism has been isolated from the intestinal tract of Galliformes such as pheasants (97) and quail (50), and from Anseriformes (57). Omnivorous species and scavengers are more frequently carriers than granivores (34, 24).

Sparrows captured inside a turkey brooding unit housing an infected flock in California yielded *C. jejuni* from fecal swabs (1). The isolation of *C. jejuni* from broilers during summer in California was attributed to the management practice of opening the doors of houses, permitting entry of free-living birds (88).

In a detailed study conducted in Yugoslavia, *C. jejuni* was isolated from six of ten pigeons, two of four sparrows, and two of four swallows (3). Since free-living birds have the potential to transmit *Campylobacter* spp., direct transmission can be prevented by adequate birdproofing of convection-ventilated houses. Flocks in controlled-environment units are generally protected from intrusion of free-living birds (54).

Fomites. Commercial poultry production is associated with considerable movement of personnel among and between production units. Managers, supervisors, veterinarians, repairmen, and vaccination, insemination, and cleaning crews all enter farms, come into contact with stock, and have the potential to transmit *C. jejuni*. The standards of biosecurity vary widely in the poultry industry. Despite the 1984 outbreak of highly pathogenic avian influenza and recent episodes of mycoplasmosis in broiler grandparent flocks, there is minimal attention to the role of clothing and footwear, equipment, vehicles, and associated fomites in the dissemination of disease. *Campylobacter jejuni* was isolated from the impacted litter between the cleats of farmers' boots in one of 18 samples examined (3). Introduction of any object contaminated with moist fecal material containing viable organisms will potentially lead to infection of individual birds, followed by rapid intra-flock transmission.

Vermin. Rats have been implicated in the transmission of campylobacter in a broiler operation in Yugoslavia (3). One out of 17 trapped rats yielded *C. jejuni* from intestinal contents. An extensive infection in a multi-age duck farm in California was ascribed to environmental contamination. Rodent droppings were present in feeder troughs, and examination of the cecal contents of 15 trapped rats yielded an 87% prevalence rate of *C. jejuni* (38).

Companion Animals. Companion animal species, although carriers of *C. jejuni*, seldom come into direct contact with poultry. One exception is the use of domestic cats to control rodents. A survey conducted on 430 domestic cats from private homes and institutions yielded a 1% *C. jejuni* prevalence rate (23). A review of feline and canine campylobacteriosis (76) confirmed that immature or newly-acquired cats and dogs, especially if diarrheic (64), are the most likely to be infected with *C. jejuni* (17), and therefore present the greatest risk of infection of flocks.

Feed. Feed is unlikely to serve as a source of infection because of its low moisture content. Surveys of flocks excreting *C. jejuni* consistently fail to demonstrate the organism in feed from the supplier (12, 15, 38, 3). In one survey, *C. jejuni* was isolated from a feeder. This may have been due to contamination by introduction of litter into the feeder, by regurgitation of feed, or beak transfer of infection (41). In the controlled studies previously reviewed (51, 77), *C. jejuni* could not be isolated from feed in receptacles, despite the presence of the organism in water in the Horsfall isolators housing infected batches.

Other Mechanisms. Recent studies have demonstrated the existence of a viable but non-culturable state for *C. jejuni* in aqueous environments (67). Low ambient temperature and stagnant water with a low oxygen content favor the prolonged survival of *C. jejuni*, permitting overwintering in some aquifers. Conversion to the non-culturable but viable state represents an adaptation to survival for *Campylobacter* spp., *Escherichia coli*, and *Salmonella enteritidis* under suboptimal conditions. The epidemiological significance of the non-culturable state relates to the ability of the organism to remain viable during winter, thereby contributing to infection of flocks following restoration of more appropriate environmental conditions, or introduction into a susceptible population.

PROCESSING PLANT CONTAMINATION

The significance of campylobacter colonization of poultry relates to the potential for infection of consumers at the end of the production chain extending from growing through processing to food preparation (95). Surveys have indicated consistent but variable levels of contamination in processed carcasses and parts. A study in Canada in 1988 showed prevalence rates of 74% and 38% in turkey and chicken meat, respectively, as compared to 17% in pork and 3% in beef (40). The results of surveys conducted in poultry processing plants and on products are summarized in Table 1. Contamination is generally introduced into the plant by the feces of carrier flocks of chickens (48) or turkeys (46). There is an obvious association between the level of flock contamination as indicated by fecal carriage of *C. jejuni* and the subsequent recovery of the organism from carcasses and the environment of the processing plant (29).

Soiling of feathers in coops and on the hanging line results in lateral contamination of birds during the transport and pre-scalding phases of processing. Immersion in scalding water at 58 ° C may reduce the level of *C. jejuni* surface contamination, but at lower temperatures (52 ° C) consistent with industry practice there is little inhibitory effect (56, 99). Addition of a quaternary ammonium disinfectant to scald water at 50 ppm reduced the survival rate of *C. jejuni* at 50 ° C (32), but increasing pH of scald water to 9.0 did not reduce carcass contamination (33). The number of *C. jejuni* recovered from carcasses increased during defeathering and evisceration (56, 99, 35). Spray washing of carcasses and immersion in spin chillers lowered the level of *C. jejuni* contamination (56). Studies have shown that spin chillers serve as a significant means of lateral transmission between birds and among flocks (99). Storing eviscerated turkey carcasses in tubs of chlorinated water (50 ppm) for ten hours reduced the level of contamination from 94% to 34% (46). The public health problem of *C. jejuni* on poultry carcasses, giblets, and portions is complicated by the persistence of the organism at sub-freezing temperature (-20 ° C) for up to 26 weeks (101), and in incompletely heated foods (42 ° C) for three days. Destruction was achieved by heating to 60 ° C for 15 minutes (93).

Serologic correlation between *C. jejuni* in poultry and diarrheic consumers has been demonstrated by epidemiological surveys conducted in Australia (79) and Yugoslavia (4). Clinical campylobacteriosis has been documented in workers in a poultry processing plant in which young substitute workers during summer showed a higher attack rate than regular employees (7). Surveys on exposure to *C. jejuni* have shown a significantly higher prevalence among workers dealing with poultry than in the general farming community (36, 25), suggesting that campylobacteriosis can be regarded as an occupational disease.

Table 1. Summary of published surveys on the prevalence of *C. jejuni* on poultry.

Year	Country	Type	Poultry Product	Prevalence of *C. jejuni*	Reference
1981	USA	Turkey	Carcasses	94%	(46)
1981	USA	Broilers	Carcasses	54%	(58)
	Canada	Broilers	Carcasses	62%	
1982	USA	Broilers	Gizzards	89%	(8)
			Livers	85%	
1983	Holland	Broilers	Carcasses	50%	(56)
			Livers	75%	
1983	USA	Turkey	Skin	37%	(103)
1983	USA	Turkey	Carcasses	68%	(99)
1984	Norway	Turkey	Carcasses	56%	(70)
		Hens	Carcasses	48%	
		Broilers	Carcasses	14%	
1985	USA	Broilers	Carcasses	30-100%	(92)
1986	Israel	Broilers	Carcasses	70%	(37)
1988	USA	Broilers	Carcasses	100%	(35)
			Giblets	100%	

PREVENTION

Improved methods of biosecurity will be necessary to prevent the introduction of *C. jejuni* onto breeder or growing farms, which house commercial poultry. Under current industry conditions, relatively unrestricted movement of personnel and equipment, delivery of feed and chicks, and the recycling of litter contribute to infection and subsequent intestinal colonization. Changing of litter and complete disinfection of breeder houses between cycles is advised when persistent infection occurs. This expedient is either impractical or uneconomical for broiler growers in the United States unless concurrent diseases which depress performance, such as infectious reoviral stunting syndrome, colisepticemia, or laryngotracheitis, have occurred. The possible ameliorative effect of plastic-mesh flooring, which eliminates litter and coprophagy, has not been evaluated, but may offer the possibility of raising broiler flocks free of *Campylobacter* spp. and *Salmonella* spp.

Although preliminary studies have suggested the potential of competitive exclusion to reduce levels of *C. jejuni* in the intestinal tract (90), subsequent trials have not demonstrated any practical benefit from this procedure (80).

Processing plant contamination may be reduced by washing and disinfecting transport coops and removing feed from flocks eight hours prior to slaughter. Immersion of carcasses or portions in 0.5% acetic or lactic acid (92), or in 100 ppm chlorine, succinate, or 0.5% glutaraldehyde, reduces *C. jejuni* contamination (102). Pasteurization levels of irradiation (1.0 to 5.0 kGy) using Cobalt-60 effectively eliminate *C. jejuni* and promote the wholesomeness and shelf-life of chicken meat in the absence of organoleptic changes.

ACKNOWLEDGMENTS

Appreciation is extended to Ms. Kathleen S. Harrington for collation of the references and assistance in preparation of the manuscript.

REFERENCES

1. **Acuff, G.R., C. Vanderzant, F.A. Gardner, and F.A. Golan.** 1982. Examination of turkey eggs, poults and brooder house facilities for *Campylobacter jejuni*. J. Food Prot. **45:**1279-1281.
2. **Acuff, G.R., C. Vanderzant, M.O. Hanna, J.G. Ehlers, F.A. Golan, and F.A. Gardner.** 1986. Prevalence of *Campylobacter jejuni* in turkey carcass processing and further processing of turkey products. J. Food Prot. **49:**712-717.
3. **Annan-Prah, A., and M. Janc.** 1988. The mode of spread of *Campylobacter jejuni/coli* to broiler flocks. J. Vet. Med. B. **35:**11-18.
4. **Annan-Prah, A., and M. Janc.** 1988. Chicken-to-human infection with *Campylobacter jejuni* and *Campylobacter coli*: biotype and serotype correlation. J. Food Prot. **51:**562-564.

5. **Blaser, M. J., H.L. Hardesty, B. Powers, and W-L.L. Wang.** 1980. Survival of *Campylobacter fetus* subsp. *jejuni* in biological milieus. J. Clin. Microbiol. **11**:309-313.
6. **Blaser, M.J., F.M. LaForce, N.A. Wilson, and W.-L.L. Wang.** 1980. Reservoirs for human campylobacteriosis. J. Infect. Dis. **141**:665-669.
7. **Christenson, B., Å. Ringner, C. Blucher, H. Billaudelle, K.N. Gundtoft, G. Eriksson, and M. Böttiger.** 1983. An outbreak of campylobacter enteritis among the staff of a poultry abattoir in Sweden. Scand. J. Infect. Dis. **15**:167-172.
8. **Christopher, F.M., G.C. Smith, and C. Vanderzant.** 1982. Examination of poultry giblets, raw milk and meat for *Campylobacter fetus* subsp. *jejuni*. J. Food Prot. **45**:260-262.
9. **Clark, A.G.** 1986. The effect of toxigenic and invasive human strains of *Campylobacter jejuni* on broiler hatchability and health. Proc. 35th West. Poult. Dis. Conf. 25-27.
10. **Clark, A.G., and D.H. Bueschkens.** 1985. Laboratory infection of chicken eggs with *Campylobacter jejuni* by using temperature or pressure differentials. Appl. Environ. Microbiol. **49**:1467-1471.
11. **Clark, A.G. ,and D.H. Bueschkens.** 1988. Horizontal spread of human and poultry-derived strains of *Campylobacter jejuni* among broiler chicks held in incubators and shipping boxes. J. Food Prot. **51**:438-441.
12. **Cruickshank, J.G., S.I. Egglestone, A.H.L. Gawler, and D.G. Lanning.** 1982. *Campylobacter jejuni* and the broiler chicken process, p. 263-266. *In* D. G. Newell (ed.), Campylobacter. Epidemiology, pathogenesis and biochemistry. MTP: Medical and Technical Publishing Co., Ltd., Lancaster, England.
13. **Doyle, M.P.** 1984. Association of *Campylobacter jejuni* with laying hens and eggs. Appl. Environ. Microbiol. **47**:533-536.
14. **Doyle, M.P., and D.J. Roman.** 1982. Sensitivity of *Campylobacter jejuni* to drying. J. Food Prot. **45**:507-510.
15. **Engvall, A., Å. Bergqvist, K. Sandstedt, and M-L. Danielsson-Tham.** 1986. Colonization of broilers with campylobacter in conventional broiler-chicken flocks. Acta Vet. Scand. **27**:540-547.
16. **Figueroa, G., H. Hidalgo, M. Troncoso, S. Rosende, and V. Soto.** 1983. *Campylobacter jejuni* in broilers, hens and eggs in a developing country. p. 161-162. *In* A.D. Pearson, M.B. Skirrow, B. Rowe, J.R. Davies, and D.M. Jones (ed.), Campylobacter II. Public Health Laboratory Service, London.
17. **Fleming, M.P.** 1983. Association of *Campylobacter jejuni* with enteritis in dogs and cats. Vet. Rec. **113**:372-374.
18. **Fletcher, R.D., and W.N. Plastridge.** 1964. Difference in physiology of *Vibrio* spp. from chickens and man. Avian Dis. **8**:72-75.
19. **Fox, J.G.** 1982. Campylobacteriosis - a "new" disease in laboratory animals. Lab. Animal Sci. **32**:625-637.
20. **Genigeorgis, C.** 1986. Significance of campylobacter in poultry. Proc. 35th West. Poult. Dis. Conf. 54-59.
21. **Genigeorgis, C., M. Hassuneh, and P. Collins.** 1985. Epidemiological aspects of campylobacter infection and contamination in the chain of poultry meat production, p. 268-270. *In* A.D. Pearson, M.B. Skirrow, H. Lior, and B. Rowe (ed.), Campylobacter III. Public Health Laboratory Service, London.
22. **Genigeorgis, C., M. Hassuneh, and P. Collins.** 1986. *Campylobacter jejuni* infection on poultry farms and its effect on poultry meat contamination during slaughtering. J. Food Prot. **49**:895-903.

23. **Gifford, D.H., S.M. Shane, and R.E. Smith.** 1985. Prevalence of *Campylobacter jejuni* in Felidae in Baton Rouge, Louisiana. Int. J. Zoon. **12:**67-73.
24. **Glünder, G.** 1989. Charakterisierung von *Campylobacter* spp. aus Wildvogeln. Berl. Munch. Tierarztl. Wschr. **102:**49-52.
25. **Grados, O., N. Bravo, J.P. Butzler, and G. Ventura.** 1983. Campylobacter infection: an occupational disease risk in chicken handlers, p. 162. *In* A.D. Pearson, M.B. Skirrow, B. Rowe, J.R. Davies, and D.M. Jones (ed.), Campylobacter II. Public Health Laboratory Service, London.
26. **Grant, I.H., N.J. Richardson, and V.D. Bokkenheuser.** 1980. Broiler chickens as potential source of *Campylobacter* infections in humans. J. Clin. Microbiol. **11:**508-510.
27. **Hagan, J.R.** 1964. Diagnostic techniques in avian vibrionic hepatitis. Avian Dis. **8:**428-437.
28. **Harris, N.V., D. Thompson, D.C. Martin, and C.M. Nolan.** 1986. A survey of campylobacter and other bacterial contaminants of pre-market chicken and retail poultry and meats, King County, Washington. Am. J. Public Health **76:**401-406.
29. **Hartog, B.J., G. J.A. de Wilde, and E. de Boer.** 1983. Poultry as a source of *Campylobacter jejuni*. Arch. Lebensmittelhyg. **34:**116-122.
30. **Hofstad, M.S., E.H. McGehee, and P.C. Bennett.** 1958. Avian infectious hepatitis. Avian Dis. 2:358-364.
31. **Hoop, R., and H. Ehrsam.** 1987. Ein Beitrag zur Epidemiologie von *Campylobacter jejuni* und*Campylobacter coli* in der Huhnermast. Schweiz. Arch. Tierheilk. **129:**193-203.
32. **Hudson, W.R., and G.C. Meade.** 1987. Factors affecting the survival of *Campylobacter jejuni* in relation to immersion scalding of poultry. Vet. Rec. **121:**225-227.
33. **Humphrey, T.J., and D.G. Lanning.** 1987. Salmonella and campylobacter contamination of broiler chicken carcasses and scald tank water: the influence of water pH. J. Appl. Bacteriol. **63:**21-25.
34. **Ito, K., Y. Kubokura, K. Kaneko, Y. Totake, and M. Ogawa.** 1988. Occurrence of *Campylobacter jejuni* in free-living wild birds from Japan. J. Wildl. Dis. **24:**467-470.
35. **Izat, A.L., F.A. Gardner, J.H. Denton, and F.A. Golan.** 1988. Incidence and level of *Campylobacter jejuni* in broiler processing. Poult. Sci. **67:**1568-1572.
36. **Jones, D.M. , and D.A. Robinson.** 1981. Occupational exposure to *Campylobacter jejuni* infection. Lancet **i:**440-441.
37. **Juven, B.J., and M. Rogol.** 1986. Incidence of *Campylobacter jejuni* and *Campylobacter coli* serogroups in a chicken processing factory. J. Food Prot. **49:**290-292.
38. **Kasrazadeh, M. , and C. Genigeorgis.** 1987. Origin and prevalence of *Campylobacter jejuni* in ducks and duck meat at the farm and processing plant level. J. Food Prot. **50:**321-326.
39. **Kinjo, T., M. Morishige, N. Minamoto, and H. Fukushi.** 1983. Prevalence of *Campylobacter jejuni* in feral pigeons. Jpn. J. Vet. Sci. **45:**833-835.
40. **Lammerding, A.M., M.M. Garcia, E.D. Mann, Y. Robinson, W.J. Dorward, R.B. Truscott, and F. Tittiger.** 1988. Prevalence of *Salmonella* and thermophilic *Campylobacter* in fresh pork, beef, veal and poultry in Canada. J. Food Prot. **51:**47-52.
41. **Lindblom, G.-B., E. Sjögren, and B. Kaijser.** 1986. Natural campylobacter colonization in chickens raised under different environmental conditions. J. Hyg. **96:**385-391.

42. **Lior, H.** 1984. New, extended biotyping scheme for *Campylobacter jejuni, Campylobacter coli,* and "*Campylobacter laridis*." J. Clin. Microbiol. **20:**636-640.
43. **Lior, H. , and A. Patel.** 1987. Improved toluidine blue-DNA agar for detection of DNA hydrolysis by campylobacters. J. Clin. Microbiol. **25:**2030-2031.
44. **Luechtefeld, N.W., M.J. Blaser, L.B. Reller, and W-L.L. Wang.** 1980. Isolation of-*Campylobacter fetus* subsp. *jejuni* from migratory waterfowl. J. Clin. Microbiol. **12:**406-408.
45. **Luechtefeld, N.W., R.C. Cambre, and W-L.L. Wang.** 1981. Isolation of *Campylobacter fetus* subsp. *jejuni* from zoo animals. J. Am. Vet. Med. Assoc. **179:**1119-1122.
46. **Luechtefeld, N.W., and W-L.L. Wang.** 1981. *Campylobacter fetus* subsp. *jejuni* in a turkey processing plant. J. Clin. Microbiol. **13:**266-268.
47. **Luechtefeld, N.W., W-L.L. Wang, M.J. Blaser, and L.B. Reller.** 1981. Evaluation of transport and storage techniques for isolation of *Campylobacter fetus* subsp. *jejuni* from turkey cecal specimens. J. Clin. Microbiol. **13:**438-443.
48. **Mehle, J., M. Gubina, and B. Gliha.** 1982. Contamination of chicken meat with *Campylobacter jejuni* during the process of industrial slaughter, p. 267-269. *In* A.D. Pearson, M.B. Skirrow, B. Rowe, J.R. Davies, and D.M. Jones (ed.), Campylobacter II. Public Health Laboratory Service, London.
49. **Mentzing, L-O.** 1981. Waterborne outbreaks of campylobacter enteritis in central Sweden. Lancet **ii:**352-354.
50. **Minakshi, S. C. D. and A. Ayyagari.** 1988. Isolation of *Campylobacter jejuni* from quails: an initial report. Br. Vet. J. **144:**411-412.
51. **Montrose, M. S., S. M. Shane, and K. S. Harrington.** 1985. Role of litter in the transmission of *Campylobacter jejuni*. Avian Dis. **29:**392-399.
52. **Morrison, R. M. and T. Roberts.** 1985. Potential public health benefits of irradiating fresh chicken, pork and beef, p. 1038-1064. *In* Food irradiation: new perspectives on a controversial technology. A review of technical, public health, and economic considerations. Hearing on Federal Food Irradiation Development and Control Act of 1985, Committee on Agriculture, House of Representatives. US Government Printing Office, Serial 99-14.
53. **Mossel, D. A. A.** 1985. Media for *Campylobacter jejuni* and other campylobacters. Int. J. Food Microbiol. **2:**119-122.
54. **Neill, S. D., J. N. Campbell, and J. A. Greene.** 1984. *Campylobacter* species in broiler chickens. Avian Pathol. **13:**777-785.
55. **Neill, S. D., J. N. Campbell, and J. J. O'Brien.** 1985. Egg penetration by *Campylobacter jejuni*. Avian Pathol. **14:**313-320.
56. **Oosterom, J., S. Notermans, H. Karman, and G. B. Engels.** 1983. Origin and prevalence of *Campylobacter jejuni* in poultry processing. J. Food Prot.**46:**339-344.
57. **Pacha, R. E., G. W. Clark, E. A. Williams, and A. M. Carter.** 1988. Migratory birds of central Washington as reservoirs of *Campylobacter jejuni.* Can. J. Microbiol. **34:**80-82.
58. **Park, C. E., Z. K. Stankiewicz, J. Lovett, and J. Hunt.** Incidence of *Campylobacter jejuni* in fresh eviscerated whole market chickens. Can. J. Microbiol. **27:**841-842.
59. **Patton, C. M., S. W. Mitchell, M. E. Potter, and A. F. Kaufmann.** 1981. Comparison of selective media for primary isolation of *Campylobacter fetus* subsp. *jejuni*. J. Clin. Microbiol. **13:**326-330.
60. **Peckham, M. C.** 1958. Avian vibrionic hepatitis. Avian Dis. **2:**348-358.

61. **Penner, J. L., J. N. Hennessy, and R. V. Congi.** 1983. Serotyping of *Campylobacter jejuni* and *Campylobacter coli* on the basis of thermostable antigens. Eur. J. Clin. Microbiol. **2**:378-383.
62. **Pokamunski, S., N. Kass, E. Borochovich, B. Marantz, and M. Rogol.** 1986. Incidence of *Campylobacter* spp. in broiler flocks monitored from hatching to slaughter. Avian Pathol. **15**:83-92.
63. **Pokamunsky, S., M. Horn, N. Kass, E. Borojovich, and M. Rogol.** 1984. The incidence of campylobacter in poultry flocks in northern Israel. Ref. Vet. **41**:80-84.
64. **Prescott, J. F., and C. W. Bruin-Mosch.** 1981. Carriage of *Campylobacter jejuni* in healthy and diarrheic animals. Am. J. Vet. Res. **42**:164-165.
65. **Prescott, J. F. , and D. L. Munroe.** 1982. *Campylobacter jejuni* enteritis in man and domestic animals. J. Am. Vet. Med. Assoc. **181**:1524-1530.
66. **Prescott, J. F. , and O. S. Gellner.** 1984. Intestinal carriage of *Campylobacter jejuni* and *Salmonella* by chicken flocks at slaughter. Can. J. Comp. Med. **48**:329-331.
67. **Rollins, D. M. , and R. R. Colwell.** 1986. Viable but nonculturable stage of *Campylobacter jejuni* and its role in survival in the naturalaquatic environment. Appl. Environ. Microbiol. **52**:531-538.
68. **Rogol, M., B. Shpak, D. Rothman, and I. Sechter.** 1985. Enrichment medium for isolation of *Campylobacter jejuni-Campylobacter coli*. Appl. Environ. Microbiol. **50**: 125-126.
69. **Rosef, O., and G. Kapperud.** 1983. House flies *(Musca domestica)* as possible vectors of *Campylobacter fetus* subsp. *jejuni*. Appl. Environ. Microbiol. **45**:381-383.
70. **Rosef, O., B. Gondrosen, and G. Kapperud.** 1984. *Campylobacter jejuni* and-*Campylobacter coli* as surface contaminants of fresh and frozen poultry carcasses. Int. J. Food Microbiol. **1**:205-215.
71. **Ruiz-Palacios, G. M., E. Escamilla, and N. Torres.** 1981. Experimental *Campylobacter* diarrhea in chickens. Infect. Immun. **34**:250-255.
72. **Sacks, J. J., S. Lieb, L. M. Baldy, S. Berta, C. M. Patton, M. C. White, W. J. Bigler, and J. J. White.** 1986. Epidemic campylobacteriosis associated with a community water supply. Am. J. Pub. Health **76**:424-429.
73. **Sang, F. C., S. M. Shane, K. Yogasundram, H. V. Hagstad, and M. T. Kearney.** 1989. Enhancement of *Campylobacter jejuni* virulence by serial passage in chicks. Avian Dis. **33**:425-430.
74. **Sanyal, S. C., K. M. N. Islam, P. K. B. Neogy, M. Islam, P. Speelman, and M. I. Huq.** 1984. *Campylobacter jejuni* diarrhea model in infant chickens. Infect. Immun. **43**:931-936.
75. **Schwartz, R. H., C. Bryan, W. J. Rodriguez, C. Park, and P. McCoy.** 1983. Experience with the microbiologic diagnosis of *Campylobacter* enteritis in an office laboratory. Pediatr. Infect. Dis. **2**:298-301.
76. **Shane, S. M. , and M. S. Montrose.** 1985. The occurrence and significance of *Campylobacter jejuni* in man and animals. Vet. Res. Commun. **9**:167-198.
77. **Shane, S. M., M. S. Montrose, and K. S. Harrington.** 1985. Transmission of-*Campylobacter jejuni* by the housefly *(Musca domestica)*. Avian Dis. **29**:384-391.
78. **Shane, S. M., D. H. Gifford, and K. Yogasundram.** 1986. *Campylobacter jejuni* contamination of eggs. Vet. Res. Commun. **10**:487-492.
79. **Shanker, S., J. A. Rosenfield, G. R. Davey, and T. C. Sorrell.** 1982. *Campylobacter jejuni*: incidence in processed broilers and biotype distribution in human and broiler isolates. Appl. Environ. Microbiol. **43**:1219-1220.

80. **Shanker, S., A. Lee, and T. C. Sorrell.** 1986. *Campylobacter jejuni* in broilers: the role of vertical transmission. J. Hyg. **96:**153-159.
81. **Shanker, S., A. Lee, and T. C. Sorrell.** 1988. Experimental colonization of broiler chicks with *Campylobacter jejuni.* Epidemiol. Infect. **100:**27-34.
82. **Simpson, J. W. , and A. G. Burnie.** 1983. Campylobacter excretion in canine feces. Vet. Rec. **112:**46.
83. **Skirrow, M. B. , and J. Benjamin.** 1980. '1001' Campylobacters: cultural characteristics of intestinal campylobacters from man and animals. J. Hyg. **85:**427-442.
84. **Skirrow, M. B. , and J. Benjamin.** 1980. Differentiation of enteropathogenic campylobacter. J. Clin. Pathol. **33:**1122.
85. **Smibert, R. M.** 1978. The genus *Campylobacter.* Ann. Rev. Microbiol. **32:**673-709.
86. **Smibert, R. M.** 1984. Genus *Campylobacter* Sebald and Veron 1963, 907[AL], p. 111-118. *In* N. R. Krieg and J. G. Holt (ed.), Bergey's manual of systematic bacteriology, vol. 1. The Williams & Wilkins Co., Baltimore.
87. **Smith, M. V., II , and P. J. Muldoon.** 1974. *Campylobacter fetus* subsp. *jejuni (Vibrio fetus)* from commercially processed poultry. Appl. Microbiol. **27:**995-996.
88. **Smitherman, R. E., C. A. Genigeorgis, and T. B. Farver.** 1984. Preliminary observations on the occurrence of *Campylobacter jejuni* at four California chicken ranches. J. Food Prot. **47:**293-298.
89. **Soerjadi, A. S., G. H. Snoeyenbos, and O. M. Weinack.** 1982. Intestinal colonization and competitive exclusion of*Campylobacter fetus* subsp. *jejuni* in young chicks. Avian Dis. **26:**520-524.
90. **Soerjadi-Liem, A. S., G. H. Snoeyenbos, and O. M. Weinack.** 1984. Comparative studies on competitive exclusion of three isolates of *Campylobacter fetus* subsp. *jejuni* in chickens by native gut microflora. Avian Dis. **28:**139-146.
91. **Stern, N. J.** 1981. Recovery rate of *Campylobacter fetus* subsp. *jejuni* on eviscerated pork, lamb and beef carcasses. J. Food Sci. **46:**1291, 1293.
92. **Stern N. J., P. J. Rothenberg, and J. M. Stone.** 1985. Enumeration and reduction of *Campylobacter jejuni* in poultry and red meats. J. Food Prot. **48:**606-610.
93. **Svedhem, Å., B. Kaijser, and E. Sjogren.** 1981. The occurrence of *Campylobacter jejuni* in fresh food and survival under different conditions. J. Hyg. **87:**421-425.
94. **Tauxe, R. V., D. A. Pegues, and N. Hargrett-Bean.** 1987. *Campylobacter* infections: the emerging national pattern. Am. J. Public Health **77:**1219-1221.
95. **Tauxe, R. V., N. Hargrett-Bean, C. M. Patton, and I. K. Wachsmuth.** 1988. *Campylobacter* isolates in the United States, 1982-1986. Morbid. Mortal. Weekly Rep. **37 (SS-2):**1-13.
96. **Truscott, R. B. and P. H. G. Stockdale.** 1966. Correlation of the identity of bile and cecal vibrios from the same field cases of avian vibrionic hepatitis. Avian Dis. **10:**67-73.
97. **Volkheimer, A., and H.-H. Wuthe.** 1986. *Campylobacter jejuni/coli* bei Rebhuhnern (*Perdix perdix L.*) und Fasanen (*Phasianus colchicus L.*). Berl. Munch. Tierarztl. Wschr. **99:**374.
98. **Welkos, S. L.** 1984. Experimental gastroenteritis in newly-hatched chicks infected with *Campylobacter jejuni*. J. Med. Microbiol. **18:**233-248.
99. **Wempe, J. M., C. A. Genigeorgis, T. B. Farver, and H. I. Yusufu.** 1983. Prevalence of *Campylobacter jejuni* in two California chicken processing plants. Appl. Environ. Microbiol. **45:**355-359.

100. **Wright, E. P.** 1983. The isolation of *Campylobacter jejuni* from flies. J. Hyg. **91:**223-226.
101. **Yogasundram, K. , and S. M. Shane.** 1986. The viability of *Campylobacter jejuni* on refrigerated chicken drumsticks. Vet. Res. Commun. **10:**479-486.
102. **Yogasundram, K., S. M. Shane, R. M. Grodner, E. N. Lambremont, and R. E. Smith.** 1987. Decontamination of *Campylobacter jejuni* on chicken drumsticks using chemicals and radiation. Vet. Res. Commun. **11:**31-40.
103. **Yusufu, H. I., C. Genigeorgis, T. B. Farver, and J. M. Wempe.** 1983. Prevalence of *Campylobacter jejuni*, at different sampling sites in two California turkey processing plants. J. Food Prot. **46:**868-872.

DISCUSSION

N. STERN: I think that there is a possibility of control of the organism in the live bird vs. treating with chemicals and irradiation at the end process. Frank Jones had some interesting data showing the marked increase in the excretion of *Salmonella* between the time the birds left the farm and when the birds went to processing. Are there any data that you might know of which indicates that stress of the birds also increases the excretion levels of *Campylobacter*?

S. SHANE: No, and I am very fascinated by Frank's observation which I think is fairly unique among those of us that work with enteric organisms; it really is at the industry face of practice and the intestinal colonization. I would think that they probably would be because I don't think that what Frank has demonstrated is truely a multiplication phenomena. I think it is pysiological with the period of eight to ten hour feed withdrawal allowing passage of material down the intestine. We have a lot of stimulation of the bird, and I think we may have excretion of cloacal material or cecal material into the environment of the coop and onto the feathers. Looking at birds, we seem to have more brown material of the darker variety around the vent. What we are looking at is an expulsion of the material which is there.

R. TAUXE: When we have an outbreak of *Campylobacter* infection in humans that is waterborne, our most immediate control measure is to hyperchlorinate the town's water system so the water coming out of people's taps actually has active residual chlorine and may even smell of chlorine. People don't like the smell, but they appreciate what is being done. To what extent is it possible to chlorinate, or is hyperchlorination of the water supply something chickens could object to?

S. SHANE: I think that the human public health people have a much better situation in that they have relatively clean water supplies. Where *Campylobacter* has occurred in a water supply in Florida, there was an open resevoir and birds were deficating in the tank, and there was obviously a breakdown in that chlorination system. You talk about hpyerchlorination, probably residual chlorine of one to two parts per million should do the trick in domestic drinking water. In the chicken situation, it may be different. We may have a lot more organic material in that water, and we also have a tremendous amount of organic material in the drinking water. This is denoted by the tremendous amount of *E. coli* that we have and particular material in the drinkers. It is going to be very necessary to look at the difference between nipple drinkers and conventional drinkers.

POTENTIAL FOR REDUCTION IN COLONIZATION OF POULTRY BY *CAMPYLOBACTER* FROM ENVIRONMENTAL SOURCES

D. M. Rollins

Infectious Diseases Department
Naval Medical Research Institute
Bethesda, Maryland 20814

ABSTRACT

The isolation rate of *Campylobacter jejuni* from cloacal vent swabs of 49-day-old broiler chickens at a poultry farm in southern England was 1,194/1,500 (80%) in a 0.25%-0.5% sample taken weekly between May and September 1986. *Campylobacter* was cultured from birds housed in 18 of the 22 poultry sheds tested and from several species of small mammals. No *Campylobacter* was isolated from the air, water, feed, litter, equipment surfaces, or other environmental samples. However, direct microscopy revealed spiral bacteria in the drinking water, and an indirect fluorescent antibody approach demonstrated *Campylobacter* organisms in the water supply system and in feed and litter that had been exposed to the birds. An investigation was undertaken to determine the effectiveness of reducing colonization by *Campylobacter*. Efforts were directed at improving hygienics, including: increased chlorination of the water supply, descaling or replacement of reservoirs, supply lines, and drinkers, high pressure/hot water washing, treatment with quaternary ammonium compound, and the withdrawal of the prophylactic antibiotic, Furazolidone, from the feed. The isolation rate of *C. jejuni* from broiler chicken was reduced from 80% to less than 2% (13/800), a 50-fold reduction ($p < 0.001$), during the intervention period. An analogous reduction was observed for the isolation of the endemic strain in the broilers and in human case reports in the region. After the interventions were withdrawn, the broilers gradually became recolonized, but not immediately with the same endemic serotype.

INTRODUCTION

Campylobacter jejuni and *Campylobacter coli* cause human enteric illness at a rate previously demonstrated for only *Salmonella* spp. (14, 18). As with salmonellosis, the predominant mode of transmission for campylobacteriosis is a fecal/oral route via contaminated food or water. *Campylobacter* spp. are commonly isolated from commercial poultry and meat products, unchlorinated water, unpasteurized milk, and other foodstuffs implicated in foodborne outbreaks (3, 8, 26, 28, 29). In addition to chicken, swine, sheep, and cattle, *Campylobacter* spp. are associated with a diversity of wild and domestic animals, both as commensals and pathogens (1, 19, 22). The economic significance of *Campylobacter* spp. among poultry and livestock has been described and its relevance in today's mass production animal facilities continues to be critical, both from a financial standpoint as well as from the aspect of transmission of disease to humans (7, 9, 15).

Chickens are, potentially, a major vehicle for the transmission of campylobacteriosis to humans. Improper handling during food preparation and insufficient cooking contribute to the ingestion of pathogenic *Campylobacter*. Studies indicate that the most frequently isolated serotypes from retail poultry are identical to those from humans (24, 25). Most sporadic campylobacteriosis remains unreported to medical authorities and is most commonly recognized in point-outbreaks, where large numbers of individuals consume a common meal (3). However, epidemiological evidence suggests that chickens may be responsible for 48% of the estimated 2,000 campylobacteriosis cases per 100,000 persons per year in the U. S. (15, 25).

Carrier rates for chickens sampled at slaughter and market range from 22%-87% (2, 15, 16, 23). Although, newborn chickens are known to be *Campylobacter*-free and eggs have been demonstrated impenetrable to the bacteria, the route(s) by which these animals are exposed and become colonized during the grow-out stage of production are not understood (6). If an effective intervention regimen can be formulated, the result would be a cleaner product at market and potentially, a reduction in campylobacteriosis among consumers. To accomplish this goal, modes of transmission and mechanisms of survival of *Campylobacter* in the environment need to be determined.

THE STUDY

Epidemiology of Poultry Colonization. An outbreak among the citizens of Bournemouth, England, began in November, 1984, and was retrospectively deduced and studied (17). Ungutted chicken was established as a possible source of infection after a study of *Campylobacter enteritis* at a local catering school. Investigations implicated the chickens, which were delivered fresh, but unrefrigerated, as one of three major risk factors. Resistance typing, serotyping, biotyping, and electrophoretic techniques identified an "endemic strain" (Lior serotype 1, Penner serotype 4; LIP4) among tested isolates from human cases and from food outlets in the region.

The evidence suggested a common source of infection that was presumptively traced to a single chicken wholesaler (Wholesaler A), receiving its poultry only from Farm X (Table 1). The latter supplied approximately 30% of all chickens in the region and 80% of the "New York-dressed," i.e., noneviscerated, birds. In addition to poultry grow-out facility, Farm X also functioned as breeder, hatchery, slaughterhouse, processor, packer, and supplier.

Table 1. Isolation rates of *C. jejuni* at Wholesaler A and other poultry outlets.

Chicken Supplier	*Camplylobacter* Incidence	Endemic Strain L1P4/Total Typed
Wholesaler A	216/232 (93%)	123/123 (100%)
All Other Outlets	102/258 (40%)	7/72 (10%)

C. jejuni was cultured from 1,194 of 1,500 (80%) cloacal vent swabs obtained from 49-day-old broiler chicken in a 15 week period at Farm X. The endemic LIP4 serotype of *C. jejuni* was the most common type isolated from chickens at Wholesaler A and Farm X. To further establish the relationship with sporadic cases in the region, Farm X and the surrounding area were investigated in an effort to determine the source(s) of contamination of the chickens with *Campylobacter*.

Nearly 900 eggs were examined in the hatchery and no evidence of *Campylobacter* contamination by vertical transmission could be identified. It was quickly determined that newborn and infant birds did not shed *Campylobacter*. Conclusions concerning the first contact with the bacterium were made more difficult because the newborn chickens were conventionally maintained on an antibiotic and special feed regimen to promote growth and to limit infection. In fact, the birds did not typically become colonized or "shed" campylobacters prior to withdrawal of the antibiotic, 28 days after hatching.

At slaughter, usually *ca.* 49 days, either 100% of the chickens were culture-positive or all were culture-negative within individual poultry sheds. This was not unanticipated, given the crowded conditions and common food and watering sources. Once a number of birds became colonized, transmission among all birds within a shed was rapid. At this particular farm, nearly 500,000 birds are housed in 19 sheds.

Extensive attempts at isolating *Campylobacter* from a diversity of sites proved fruitless. Negative cultures of well water, poultry drinking water, and fresh feed and litter were obtained by three separate laboratories. *Campylobacter* was cultured only from the nearby River Frome, fresh fecal droppings, birds at slaughter, and the abattoir immersion waters. Negative samples included the well water, poultry water supply, feed, litter, 115 of 118 trapped rodents, filtered shed air, and swabs of shed floors, walls, ceilings, and fans. *Pseudomonas* spp., *E. coli*, and *Yersinia enterocolitica* were cultured from the chicken watering troughs (drinkers) in the poultry sheds at Farm X.

Table 2. Summary of flourescent antibody (IFA) and culture results on samples processed during the chicken-borne transmission study.

Source	Sampling Site	IFA	Culture
Drinking Water	Borehole	P	-
	Resevoir	P	-
Hatchery	Eggs	ND	-
	Newborn Chicks	ND	-
Poultry Farm	Adult Broilers	ND	+
	Fresh Fecal Droppings	ND	+
	Rising Main	P	-
	Header Tank	P	-
	End of Drinking Line	P	-
	Drinker	P	-
	Litter (Fresh)	NP	-
	Litter (Used)	P	-
	Feed (Fresh)	NP	-
	Feed (Exposed)	P	-
	Shed Surfaces	ND	-
	Shed Fan, Air	ND	-
River Frome Water	Frampton	P	+
	Grimstone	P	+
	Dorchester	P	+
	Bockhampton	P	+
Abattoir	Immersion Waters	P	+
	Chicken Carcasses	ND	+
	49-Day Broilers	ND	+

P, *Campylobacter* present by IFA in samples prior to intervention; NP, *Campylobacter* not present; ND, not done.

In the absence of conclusive culture data, an indirect fluorescent antibody (IFA) method not requiring isolation and growth of the organism was utilized (4, 10). Anti-campylobacter hyperimmune sera was raised in rabbits via multiple-site injections of both formalinized and heat-killed, whole cell, antigen preparations. Titers of all sera were quantified to > 10,000 by ELISA. Briefly, filtered samples were moist-heat-fixed on slides at 56° C for 30 minutes. Where filtration was not possible, a suspension of sample was alcohol-fixed prior to IFA staining. These samples were successively incubated in a moist chamber at 37° C for 30 minutes, using FA rhodamine (Sigma Chemical Co., St. Louis, MS), an appropriate dilution of the pooled anti-campylobacter hyperimmune serum, and fluorescein-labeled goat anti-rabbit immunoglobulin G (DifCo Laboratories). A minimum of 50 microscopic fields per sample were examined for cellular fluorescence.

The presence of *Campylobacter* was demonstrated by the IFA method in the drinking water supply, the chicken watering system in certain sheds, used chicken litter, exposed chicken feed, the River Frome at four sites, abattoir immersion water and other sites at the slaughterhouse. An abbreviated summary of IFA and culture results are shown in Table 2.

Ecology at Farm X. The source of drinking water for the poultry at Farm X is a 29-meter deep "bore hole", i.e., small-bore well, that was only partially cased, possibly allowing leaching of groundwater into the well. Intermittent chlorination was accomplished by manual delivery of one cup of "Chloros," a sodium hypochlorite solution, into the top of the bore hole each week. Monitoring of residual chlorine levels in the Farm X supply revealed that levels of up to 3 ppm chlorine could be detected shortly after chlorination, and slowly diminished to zero ppm within 48 hours. The River Frome, known to be contaminated with culturable *C. jejuni* is situated nearby, although geological data supports the conclusion that the water source for the poultry farm, is not interconnected with the river water. Interestingly, the Farm X abattoir was located upriver from the grow-out facility and disposed of untreated slaughter waste directly into the river, thereby continuously "seeding" the aquifer with organic material and viable campylobacters of animal origin. This abattoir is no longer in service.

Drinking water pumped from the bore hole is stored in a 12,500 gallon field reservoir, and gravity fed to the poultry farm where it is distributed, via a series of "rising mains," to small, overhead reservoirs (header tanks) located inside each chicken shed. Drinking water flows, on demand, from the header tanks to plastic watering troughs (drinkers) via a gravity feed system.

Typically, the sheds were cleaned out and fresh litter distributed, after each generation of birds was slaughtered and prior to the introduction of another flock into the shed. The header tanks, tubing and piping, connectors, and watering troughs (drinkers) were characterized by a distinctive, thick biofilm and mineral accumulation that was not effectively cleaned before the introduction of a new flock.

The IFA results conclusively indicated that Campylobacter cells were present in many of the samples that were not culturable by standard methods. Crude estimates of cell numbers in positive samples ranged from a few cells to 10^6 cells per milliliter. These cells often appeared in clumps on the filtered preparations and may have been "microcolonies" dislodged from the biofilm encrusting the water supply system.

Intervention Attempts. Improvement of hygienics at Farm X was the focus of an effort to intervene in the passage of *Campylobacter* from environmental sources to poultry. Intervention included vigorous maintenance of effective levels of free chlorine in the water supply, and the implementation of an extensive sanitation effort at the farm.

Before the introduction of new flocks, used feed and litter was removed. Header tanks, supply lines, and drinkers were descaled or replaced, high pressure/hot water washed, and soaked with quaternary ammonium compounds. Results of eliminating the prophylactic antibiotic, upon which newborn chicks are reared, did not conclusively denote its role in colonization susceptibility.

The collective intervention regimen was highly successful (Table 3). A 50-fold reduction in the colonization rate of the chickens was observed during the intervention period. Upon termination of these measures, the incidence of *Campylobacter* rose steadily to pre-intervention levels as new strains of environmental *Campylobacter* reestablished the ecological niche provided in the poultry drinking water system.

Table 3. Effect of intervention on poultry colonization by *Campylobacter* in 49-day-old broiler chicken at Farm X.

Trial Period	Duration	*Campylobacter* Incidence
Pre-Intervention	15 weeks	1194/1500 (80%)
During Intervention	8 weeks	13/800 (<2%)
Post-Intervention	Weeks 2-5	95/350 (27%)
Post-Intervention	Weeks 6-10	358/450 (80%)

Interestingly, the endemic LlP4 strain did not immediately reestablish as the dominant serotype, as it was prior to the intervention (Table 4). With the ensuing struggle to establish dominance in the niche, a diversity of strains were isolated from the birds. The reduced colonization of chicken was reflected as a decrease in the total number of case reports of campylobacteriosis in the region. An analogous reduction in the proportion of LlP4 was also observed.

Table 4. Effect of intervention on poultry colonization and poultry-borne infection by endemic L1P4 strain of *C. jejuni*.

Trial Period	Chicken From Farm L1P4/Total Typed	Regional Cases L1P4/Total Typed
Pre-Intervention	126/139 (91%)	108/352 (31%)
Post-Intervention	8/110 (7.3%)	7/108 (5.9%)

DISCUSSION

An effort to establish the epidemiology of a regional outbreak of campylobacteriosis identified a number of preventable risk factors. Chickens proved to be the means by which humans became infected with *Campylobacter jejuni*, but the source for poultry contamination was sought. Proper implementation of a few simple measures may have limited chicken-borne infection in the consumer population. Refrigeration of poultry immediately after processing and continuing through consumption is a generally recognized health measure to minimize multiplication of contaminating bacteria. At the consumer level, proper storage, handling, and preparation of raw poultry is imperative. At the growout facility maintenance of sufficient levels of chlorination in the farm drinking water and thorough sanitation appeared to play a key role in the successful intervention of colonization of the broilers. Undoubtedly, the simplistic design of the local drinking water supply system contributed to the ecology of poultry colonization. The well was incompletely cased and contamination from outside sources was possible at any of several sites. Proper disposal of slaughter waste and improved hygienics at the abattoir were addressed at Farm X by permanently closing this facility and birds are now transported to an independent slaughterhouse.

Streamwater and inadequately chlorinated well and municipal water may contain potentially pathogenic *Campylobacter*, but conventional culture methods have proven inadequate for confirmation of suspected waterborne sources of contamination (27, 29). In this study, potential sources for chicken colonization were evaluated by an indirect fluorescent antibody technique after extensive culture efforts failed. *Campylobacter* cells were detected by the IFA technique in many samples from which the organism could not be cultured, including the poultry drinking water supply and poultry-exposed litter, feed, and equipment surfaces. The organisms appear to have established an ecological niche, persisting in an adherent biofilm on moist equipment surfaces and in the water supply lines (13). Although three rodents were culture-positive during the study, it is highly unlikely that such a low incidence could account for the overwhelming colonization rates in the broiler flocks.

The inability to culture *Campylobacter* from the environmental samples may be a consequence of nutritional and cultural conditions not being met by currently available media and methods. In fact, the effect of "spiking" the water with chlorine at semi-regular intervals may have resulted in a chlorine shock to the microorganisms, thereby favoring the nonculturable stage (11). *Campylobacter* cells may enter such a stage under conditions less than optimum for growth (12, 20, 21). Though nonculturable, *Campylobacter* cells apparently retained viability and potential pathogenicity, becoming culturable after ingestion by the chickens.

Enumeration of the minute, fluorescing cells often occurring in clumps was difficult and purposefully conservative, restricted to recognizable *Campylobacter* vibrioids. The limitations of the fluorescent antibody method have been reviewed elsewhere (5). The success attained here with the IFA technique was attributable to the high titers of antiserum obtained with a rigid hyperimmunization schedule. Since pooled polyvalent sera was used, a broad range of campylobacters could be identified. Monospecific adsorbed serum is a valuable tool for tracing the source of known serotypes associated with infectious outbreaks. However, the diversity of serotypes of the genus *Campylobacter* and the expense incurred to produce such reagents may make such a monospecific method cost-prohibitive for most laboratories. For our purposes, a pool of polyclonal antisera proved adequate for detecting *Campylobacter* in environmental samples.

CONCLUSIONS

Intervention efforts suggest that a reduction can be achieved in the total numbers of campylobacters contaminating poultry at market by interfering in the initial contact between the birds and environmental strains of *Campylobacter*. These microorganisms are apparently able to establish a niche in the drinking water and/or on the equipment surfaces. Consistent chlorination and good hygienics at the grow-out facility may reduce the numbers of organisms ingested by poultry. Cross-contamination of birds will continue to occur during slaughter and processing, but with fewer flocks colonized, a healthier consumer product is possible. Where culture methods may be ineffective, the IFA technique offers a powerful tool for detecting the organism in water, feed, and litter samples and permits gathering data for epidemiological analyses.

ACKNOWLEDGEMENTS

This study was conducted as a collaborative project with Dr. Andrew Pearson and colleagues of the Public Health Laboratory Service, U.K., and Dr. Rita Colwell of the University of Maryland. A thorough description of the work is in preparation for publication elsewhere.

REFERENCES

1. **Blaser, M.J., F.M. LaForce, N.A. Wilson, and W.-L.L. Wang.** 1980. Reservoirs for human campylobacteriosis. J. Infect. Dis. **141:**665-69.
2. **Bolton, F.J., H.C. Dawkins, and L. Robertson.** 1982. *Campylobacter jejuni/coli* in abattoirs and butcher shops. J. Inf. **4:**243-245.
3. **Brouwer, R., M.J.A. Mertens, T.H. Siem, and J. Katchaki.** 1979. An explosive outbreak of *Campylobacter enteritis* in soldiers. Antonie van Leeuwenhoek. **45:**517-19.
4. **Daley, R.J., and J.E. Hobbie.** 1975. Direct counts of aquatic bacteria by a modified epifluorescence technique. Limnol. Oceanog. **20:**875-82.
5. **Daniellson, D.** 1965. A membrane filter method for the demonstration of bacteria by the fluorescent antibody technique. ACTA Pathol. Microbiol. Scand. **63:**597-603.
6. **Doyle, M.P.** 1986. Association of *Campylobacter jejuni* with laying hens and eggs. Appl. Environ. Microbiol. **47:**533-36.
7. **Doyle, M.P.** 1985. Food-borne pathogens of recent concern. Annu. Rev. Nutr. **5:**25-41.
8. **Finch, M.J., and Blake, P.A.** 1985. Foodborne outbreaks of campylobacteriosis: the United States experience, 1980-1982. Amer. J. Epidemiol. **122:**262-268.

9. **Grant, I.H., Richardson, N.J., and Bokkenheuser, V.D.** 1980. Broiler chickens as a potential source of *Campylobacter* infections in humans. J. Clin. Microbiol. **11**:508-510.
10. **Hobbie, J.E., R.J. Daley, and S. Jasper.** 1977. Use of Nuclepore filters for counting bacteria by fluorescence microscopy. Appl. Environ. Microbiol. **33**:1225-1228.
11. **Klein, D.A., and S. Wu.** 1974. Stress: a factor to be considered in heterotrophic microorganism enumeration from aquatic environments. Appl. Microbiol. **27**:429-31.
12. **Grimes, D.J., R.W. Atwell, P.R. Brayton, L.M. Palmer, D.M. Rollins, D B. Roszak, F.L. Singleton, M.L. Tamplin, and R.R. Colwell.** 1986. The fate of enteric pathogenic bacteria in estuarine and marine environments. Microbiol. Sci. **3**:324-29.
13. **Marshall, K.C.** 1986. Bacterial adhesion in oligotrophic habitats. Microbiol. Sci. **2**:321-26.
14. **Newell, D.G., (ed.).** 1982. Section I. Geographical epidemiology, p. 1-32. *In Campylobacter*: epidemiology, pathogenesis, biochemistry. MTP Press Limited, Lancaster, England.
15. **NRC (National Research Council).** 1987. *In* Poultry inspection: The basis for a risk assessment approach. Report of the Committee on Public Health Risk Assessment of Poultry Inspection Programs, Food and Nutrition Board. National Academy Press, Washington, D.C.
16. **Park, C.E., Z.K. Stankiewicz, J. Lovett, and J. Hunt.** 1981. Incidence of *Campylobacter jejuni* in fresh eviscerated whole market chickens. Can. J. Microbiol. **27**:841-42.
17. **Pearson, A.D., W.L. Hooper, H. Lior, M. Greenwood, P.A.C. Donaldson, and P. Hawtin.** 1985. Why investigate sporadic cases? The significance of fresh and New York dressed chicken as a source of Campylobacter infection, p. 290-91. *In* A.D. Pearson, M.B. Skirrow, H. Lior, and B. Rowe (ed.), Campylobacter III. Public Health Laboratory Service, London.
18. **Ibid.**, Session VII. Epidemiology, developing countries and prevention, p. 239-91.
19. **Prescott, J.F., and C.W. Bruin-Mosch.** 1981. Carriage of *Campylobacter jejuni* in healthy and diarrheic animals. Amer. J. Vet. Res. **42**:164-165.
20. **Rollins, D.M.** 1987. Characterization of growth, decline, and the viable but nonculturable stage of *Campylobacter jejuni*. University of Maryland Ph.D. dissertation. University Microfilms International, Ann Arbor, Michigan. United States Copyright issued 1988.
21. **Rollins, D.M., and R.R. Colwell.** 1986. Viable but nonculturable stage of *Campylobacter jejuni* and its role in survival in the natural aquatic environment. Appl. Environ. Microbiol. 52:531-38.
22. **Rosef, O., B. Gondrosen, G. Kapperud, and B. Underdal.** 1983. Isolation and characterization of *Campylobacter jejuni* and *Campylobacter coli* from domestic and wild animal mammals in Norway. Appl. Environ. Microbiol. **46**:855-59.
23. **Sarkar, R.K.S., S. Chowdhury, G.B. Nair, and S.C. Pal.** 1984. Prevalence of *Campylobacter jejuni* in an abattoir. Ind. J. Med. Res. **80**:417-20.
24. **Shankers, S., Rosenfeld, J.A., Davey, G.R., and Sorrell, T.C.** 1982. *Campylobacter jejuni*: incidence in processed broilers and biotype distribution in human and broiler isolates. Appl. Environ. Microbiol. **43**:1219-1220.

25. **SKCDH (Seattle-King County Department of Health).** 1984. *In* Surveillance of the flow of *Salmonella* and *Campylobacter* in a community. Prepared for the Bureau of Veterinary Medicine, U.S. Food and Drug Administration. Seattle, Washington: Communicable Disease Control Section, Seattle-King County Department of Public Health.
26. **Stern, N.J.** 1981. Recovery rate of *Campylobacter fetus* spp. *jejuni* on eviscerated pork, lamb, and beef carcasses. J. Food Sci. **46:** 1291-1292.
27. **Stevenson, L.H.** 1978. A case for bacterial dormancy in aquatic systems. Microb. Ecol. **4:**127-33.
28. **Taylor, P.R., W.M. Weinstein, and J.H. Bryner.** 1979. *Campylobacter fetus* infection in human subjects: association with raw milk. Am. J. Med. **66:**779-83.
29. **Vogt, R.L., H.E. Sours, T. Barrett, R.A. Feldman, R.J. Dickinson, and L. Witherell.** 1982. *Campylobacter enteritis* associated with contaminated water. Ann. Int. Med. **96:**292-96.

DISCUSSION

J. BAILEY: Have you recognized the possibility in that study in England, that although you have definitely demonstrated you have the non-culturable forms that there are other potential sources of *Campylobacter*? We have had some discussions this week that even if only one bird in a house picks up *Campylobacter* from a fly or anything else, that all of the rest of the birds in that house would be carrying *Campylobacter* within 48 to 72 hours.

D. ROLLINS: We recognize that scenario as a possibility, but we did not identify any other sources of *Campylobacter* in this study.

SESSION II: COMPETITIVE EXCLUSION

Convener: *Esko Nurmi*

National Veterinary Institute
Helsinki, Finland

The last two to three decades have witnessed an ever sharpening attention towards the elimination of *Salmonella* from poultry. Of the measures taken, none has been sufficiently effective. On the contrary, the modern methods of production have increased the numbers of chronic *Salmonella* carriers in poultry. In the decade of the eighties, poultry has also been the most important source of human *Campylobacter* infections.

The traditional preventive measures are either cost-prohibitive or practically impossible to take. During the last two decades, methods based on new ideas and concepts have been sought. One of the major and promising methods is that of competitive exclusion, which arises from a Finnish discovery published in *Nature*, (Nurmi and Rantala, 1973). They demonstrated that oral introduction of intestinal bacterial flora could protect chickens from salmonellae colonization. Several research groups set out to identify the organisms that gave the protection. Pure cultures of many bacteria have given reasonably good results. These results have, however, never been comparable to those with whole flora or manipulated undefined flora. Unraveling the mechanisms by which the enteropathogens are competitively excluded would be of paramount importance.

An undefined preparation has been used routinely in Finland and in Sweden for many years to treat tens of millions of broilers. The results have been promising and the prevalence of *Salmonella* infections has significantly decreased. The safety of undefined preparations has been extensively discussed. Experiences show that this kind of product can be rendered free of pathogenic microorganisms.

The success of protecting chicks against salmonellae by competitive exclusion has encouraged several working groups to apply the same method to *Campylobacter*. Today it appears feasible that a comparable preparation against *Campylobacter* can be developed. Evidently, the efficacy is due, at least in part, to different strains of bacteria than those that exclude *Salmonella*.

Competitive exclusion is a complicated phenomenon. There are several factors that bear on the application which need to be clarified. It is important to realize the complexity of the alimentary ecosystem, the mechanisms of colonization, the different ways of administration, and product safety. The competitive exclusion session covers these topics in four excellent contributions.

POPULATION DYNAMICS OF THE INTESTINAL TRACT

R. D. Rolfe

Department of Microbiology
School of Medicine
Texas Tech University Health Sciences Center
Lubbock, Texas 79430

ABSTRACT

The animal host and its intestinal microbial flora constitute an enormously complex ecosystem in which there is a significant impact of the host on the microbial flora and components of the microbial flora on one another. Numerous factors influence the interactions which occur between the microbial components of the normal flora as well as the interactions between the normal flora and the host. The qualitative and quantitative stability of the bacterial flora in the intestinal tract is remarkable considering the hundreds of bacterial species which are present and the different types of interactions which are occurring. The prevalent role played by anaerobic bacteria in this dynamic ecosystem is evident by the fact that over 99% of the bacteria isolated from a human fecal specimen will not grow in the presence of atmospheric oxygen. Perhaps the most important function of the indigenous intestinal microflora to the host is its ability to inhibit the colonization of invading microorganisms (pathogenic or other) in the intestinal tract. Terms such as bacterial antagonism, competitive exclusion, bacterial interference, and colonization resistance have been used to describe this function. The objective of this manuscript is to give an overview of the complex factors which are responsible for maintaining homeostasis of the intestinal microbial flora and thus prevent colonization by pathogenic organisms.

Colonization Control of Human Bacterial Enteropathogens in Poultry

INTRODUCTION

The animal host and its intestinal microbial flora constitute an enormously complex ecosystem in which there is a significant impact of the host on the microbial flora and components of the microbial flora on one another. Numerous factors influence the interactions which occur between the microbial components of the normal flora as well as the interactions between the normal flora and the host. The qualitative and quantitative stability of the bacterial flora in the intestinal tract is remarkable considering the number of bacterial species which are present and the different types of interactions which are occurring. For example, it has been estimated that a single human fecal specimen contains 400 to 500 different species of culturable bacteria (66). The prevalent role played by anaerobic bacteria in this dynamic ecosystem is evident by the fact that over 99% of the bacteria isolated from a human fecal specimen will not grow in the presence of atmospheric oxygen (66).

Perhaps the most important function of the indigenous intestinal microflora to the host is its ability to inhibit the colonization of invading microorganisms (pathogenic or other) in the intestinal tract. Terms such as bacterial antagonism (36), competitive exclusion (60), bacterial interference (3) and colonization resistance (94) have been used to describe this function. The objective of this manuscript is to give an overview of the complex factors which are responsible for maintaining homeostasis of the intestinal microbial flora and thus prevent colonization by pathogenic organisms. These factors are only now being defined.

DEVELOPMENT OF COLONIZATION RESISTANCE

As long as the amniotic membrane remains intact, the normal fetus is sterile until shortly before birth (78). As a result of passing through the vagina, the neonate is seeded with a wide variety of microorganisms originating from the vagina and external genitalia of the mother and other environmental sources to which it is exposed (77). Initial colonization is fortuitous, depending of the first suitable organisms to arrive at a particular site. Despite the wide variety of bacteria constantly infiltrating the intestinal tract of infants, the sequence of colonization is predictable and dependent on the animal species (50, 75, 76, 80), the type of nourishment received (breast milk or formula) (61), and its environment. Many of the microbes are not able to colonize habitats in the neonatal intestinal tract and disappear from it soon after birth. Organisms best suited for survival in the intestinal environment become established by a process of natural selection which is known as ecologic succession (78, 101). In general, the newborn animal is first colonized with non-fastidious organisms such as enteric gram-negative bacilli and streptococci (57, 73). These organisms are followed by predominantly anaerobic species, which slightly suppress the population sizes of the initial colonizers. Thus, one group of organisms after another become dominate only later to be suppressed by organisms which are in turn suppressed. This process continues until a stable, climax flora develops. This generally occurs at the time the animals are weaned from a predominate milk diet (57, 61). Although considerable variation in the composition of the climax flora is found among individuals, studies in single individuals have shown that it is quite stable over prolonged periods of time as long as the environmental, nutritional, and physiologic conditions of the individual remain stable (21, 46, 51, 78, 90, 94, 103). Furthermore, only the most extreme stress situations, such as antibiotic administration and dramatic diet changes (e.g., starvation), have a major effect on the stability of the climax microflora (46, 90).

MODELS USED TO EXAMINE COLONIZATION RESISTANCE

Several different experimental models have been used by investigators to show that the normal bacterial flora interferes with the establishment of enteric pathogens and provides natural protection against infection. Table 1 lists some of the models which have been used to study this phenomenon.

A. *IN VIVO* MODELS

1. Infant Animals

Table 1. Models used to study colonization resistance.

IN VIVO Models	*IN VITRO* Models
Gnotobiotic Animals	Broth Cultures
Antibiotic-Treated Animals	Agar Cultures
Infant Animals	Continuous Flow Cultures

It has been speculated that the intestinal tract of infant animals, which is frequently susceptible to colonization with enteric pathogens, lacks essential microbial components present in the adult flora that prevents colonization by pathogens (22). For example, differences between bowel flora of infants and that of adults appear to be sufficient to influence colonization by clostridia. Two examples are described below.

Infant botulism is a disease in which *Clostridium botulinum* multiplies in the intestinal tract of infants and produces its potent neurotoxin (55). A similar type of intestinal infection with *C. botulinum* is extremely rare in adults. There is experimental evidence in animals which suggests that variations between bowel flora of infants and adults may account for differences in their susceptibility to intestinal botulism. Using infant mice as a model, Sugiyama and Mills (86) experimentally reproduced the limited age susceptibility to *C. botulinum* intestinal overgrowth. These investigators showed that *C. botulinum* spores, when injected intracecally, would germinate in the intestine of mice only between the ages of seven and 13 days. Younger and older mice were resistant to the challenge. The resistance of older mice correlated with the time they began to sample solid food. It is during the time of susceptibility to intestinal colonization with

C. botulinum that the intestinal flora of mice is undergoing dramatic quantitative and qualitative changes associated with ecologic succession (81).

Clostridium difficile is an important etiologic agent of antimicrobial agent-associated diarrheal disease in adults (42). Asymptomatic adults seldom have toxigenic *C. difficile* in their intestinal tract (71). On the other hand, up to 90% of infants less than one year of age are asymptomatically colonized with toxigenic *C. difficile* (71). *In vivo* experiments in hamsters and *in vitro* experiments suggest that the developmental changes which occur in the intestinal flora of infants accounts for their susceptibility to *C. difficile* intestinal colonization (70, 72, 73).

2. Germfree Animals

The importance of the indigenous microbial flora in protecting against colonization by exogenous bacteria has also been demonstrated in gnotobiotic animals. Many bacteria (including species not indigenous to the intestinal flora) are able to colonize germfree animals, whereas the colonization of conventional animals with the same microorganisms is difficult (83). For example, Moberg and Sugiyama (65) showed that the intestines of germ-free mice are colonized with *C. botulinum* when as few as ten spores of this microorganism are orally administered. On the other hand, adult mice with a conventional microflora are resistant to *C. botulinum* intestinal colonization even when 10^5 spores are inoculated orally. When adult germfree mice are housed with conventional animals, they become resistant to challenge with 10^5 spores in about three days, presumably due to the establishment of a protective intestinal flora (65).

3. Animals Administered Antimicrobial Agents

Treatment of conventional animals and humans with antibiotics often causes an increase in susceptibility to intestinal colonization with pathogens. This is presumably due to the elimination of key flora components responsible for its homeostasis (1, 18, 29, 44, 93, 94). For example, investigators have demonstrated that oral administration of antibiotics to mice and guinea pigs renders the animals susceptible to infection with antibiotic-resistant strains of *Shigella flexneri* and *Vibrio cholerae* (35, 36). Introduction of antibiotic-resistant *Escherichia coli* into the gastrointestinal tract of these animals results in multiplication of *E. coli* and subsequent elimination of the pathogens. In humans, treatment with antibiotics frequently gives rise to secondary intestinal infections (e.g., pseudomembranous colitis due to *C. difficile*) (5).

B. *IN VITRO* MODELS

Perhaps less relevant than the above experiments with animals, but offering supportive evidence for a protective effect of commensal microbes, are the results of mixed cultures *in vitro* involving pathogens and members of the intestinal flora (52, 58, 72, 83). These experiments have established the capacity of resident microorganisms of the gastrointestinal tract to inhibit the growth of such important pathogens as *V. cholerae, Salmonella, C. difficile* and *Shigella*. However; *in vitro* culture model systems cannot be relied upon to reflect mechanisms of *in vivo* interactions unless elaborate precautions are taken. Numerous investigators have reported the lack of correlation between microbial interactions observed *in vitro* and the interactions of the same bacteria *in vivo* (39, 51, 78).

MECHANISMS OF COLONIZATION RESISTANCE

The population levels and types of microbes in many climax communities of the gastrointestinal tract, and the succession of these communities, are regulated by multifactorial processes (78). Colonization resistance includes all factors that hamper the colonization of the intestinal tract by exogenous microorganisms. Some of the regulatory forces are exerted by the animal host, its diet, and environment. Some are exerted by the microbes themselves. Mechanisms regulating the indigenous microflora are often redundant such that two or more of these inhibitory mechanisms usually function synergistically in controlling the growth of microorganisms.

A. INFLUENCE OF HOST ON MICROBIAL FLORA

The physiology of the animal host influences the nature of the normal intestinal microbial flora by a variety of mechanisms (Table 2). Some of these host-controlled regulatory mechanisms are described in greater detail below.

Table 2. Host regulatory forces on the normal intestinal flora.

Bile Acids	Antibody
Cellular Immunity	Antimicrobial Agents
Receptor Analogues	Oxidation-Reduction Potential
Peristalsis	Epithelial Cell Turnover
pH	Electrolyte Composition
Hormones	Diet
Receptor Specificity	

1. Immunological Factors

There is considerable controversy as to whether a host immunological response by itself influences the composition of the indigenous flora. Most components of the gastrointestinal microbial flora are poor inducers of immunoglobulins (8, 34). This is undoubtedly due to the close antigenic similarities between intestinal microorganisms and tissues of the gastrointestinal tract. Obviously, being recognized as "self" would give indigenous microbes enormous advantage in colonizing their habitats. It is interesting to note, however, that patients with hypogammaglobulinemia often have increased numbers of anaerobic bacteria in their small intestines as compared to healthy controls (17).

Even though the host immunologic response by itself does not appear to be an important regulator of the microbial flora, there is evidence that synergistic interactions between the immunologic response of the host and the direct antagonistic effects of indigenous microorganisms may be important in inhibiting certain pathogens. Shedlofsky and Freter (83) have shown that the local intestinal antibacterial effect against *V. cholerae* in immunized mice is negligible when this is the only microorganism present. On the other hand, the numbers of *V. cholerae* are appreciably reduced in the intestinal tract of mice in the presence of both local immunity and other bacteria capable of suppressing *V. cholerae* growth *in vitro*. It was concluded that local immunity functions optimally if the numbers of the exogenous microorganisms are kept low by mechanisms such as peristalsis or bacterial antagonistic activity.

2. Peristalsis

The continuous unidirectional flow of material through the lumen of the upper and middle regions of the small intestine is a strong influence preventing microbial communities from developing unless they can attach to underlying epithelial structures (78).

3. Antimicrobial Agents

Antimicrobial agents are capable of causing rapid and radical changes in the normal flora. These agents promote colonization by resistant exogenous microorganisms through their inhibition of the growth of sensitive microbial competitors (51). Even when present in subinhibitory concentrations antimicrobial agents may impair the adherence of microorganisms to epithelial cells and therefore their ability to colonize the host (30, 88, 96). It may take several weeks for the intestinal flora to return to normal following discontinuation of an antimicrobial agent (51).

4. pH

Hydrogen ion concentration is a major factor dictating what types of microbes can colonize habitats in the stomach and upper small intestine (4).

5. Diet

During the last several years investigators have examined diet as a factor regulating the composition of the intestinal flora (32). These studies have been primarily concerned with the interrelationships between diet, intestinal flora, and cancer of the bowel (99). Epidemiological data have indicated a correlation between intake of dietary fat and colon cancer (99). Diet-induced changes in the composition or metabolic activity of the intestinal flora may favor an indigenous flora which is capable of transforming intestinal contents into precarcinogens, carcinogens, and co-carcinogens. Studies with experimental animals have shown that diet can influence the overall composition of the intestinal flora (67). On the other hand, there is very little evidence which shows that diet influences the composition of the intestinal flora in human adults (32).

B. BACTERIAL ANTAGONISM

Bacterial antagonism is the inhibition of growth or reduction in number of one bacterial species by one or more other bacterial species. *In vivo* and *in vitro* studies, using the kinds of methods listed in Table 1, have defined several mechanisms by which one bacterium may inhibit the growth of another. These mechanisms are listed in Table 3 and are described in greater detail below.

Table 3. Mechanisms of bacterial antagonism.

INDIRECT ANTAGONISM	DIRECT ANTAGONISM
Modification of Bile Salts	Depletion of or Competition for Essential Substrates
Induction of Immunologic Processes	Competition of Bacterial Receptor Sites
Stimulation of Peristalsis	Creation of a Restrictive Physiologic Environment
	Elaboration of an Antibiotic-Like Substance

1. Indirect Antagonism

Indirect antagonism is a result of the normal microbial flora altering the physiologic response of the host which in turn affects the interactions between the host and the microorganism. Examples of indirect interactions include chemical modification of bile salts, the induction of immunological responses, and the stimulation of peristalsis.

Bile Acids. The same bile acids which are essential for digestion and absorption of dietary fats in the intestinal tract, as well as the metabolism of cholesterol, may play an important role in regulating the composition of the normal intestinal microflora (9, 78). The primary bile acids, cholic acid and chenodeoxycholic acid, are synthesized by the liver and are conjugated to either taurine or glycine. Human bile also contains conjugates of a secondary bile acid, deoxycholic acid, which is formed by the dehydroxylation of cholic acid. In the large intestine, the bile acid conjugates are hydrolyzed to release free acids by a variety of bacteria, particularly anaerobes (27). Only free bile acids are present in the feces of healthy adults. The *in vitro* growth of many different intestinal facultative and anaerobic bacteria is inhibited by low concentrations of unconjugated bile acids (9, 33). However, the majority of *in vivo* investigations have failed to demonstrate significant antibacterial activity of unconjugated bile acids (33).

Antibodies. The normal microbial flora plays a major role in stimulating immunological host defense mechanisms. For example, secretory IgA is the only antibody in the lumen of germfree mice immunized either parenterally or locally (41). On the other hand, similarly immunized mice with a conventional microflora have appreciable titers of IgG, IgM as well as secretory IgA in their intestine. The increased titer of immunoglobulins in the intestines of these conventional animals may in turn influence the ability of enteric pathogens to colonize the intestinal tract.

2. Direct Antagonism

Direct bacterial antagonism can occur through four different mechanisms (Table 3). Each of these mechanisms has been shown to occur *in vitro*, and there is some evidence for their occurrence *in vivo*. Although the importance of these mechanisms in regulating the composition of the normal flora is unclear, there is strong evidence that strictly anaerobic bacteria play an essential role in each of these mechanisms (32).

Depletion of or Competion for Essential Substrates. The importance of competition between microorganisms for growth limiting nutrients as a mechanism controlling microbial populations is difficult to assess in relation to the other inhibitory factors which may be functioning in the environment. Freter (38) examined the inhibition of *S. flexneri* in static and continuous flow cultures. Inhibition under these conditions was reversed either by addition of glucose to the culture medium or aeration. He concluded that the inhibitory activity of coliforms against *S. flexneri* was due to competition for carbon and energy sources in a highly reduced environment. From these and other data, investigators have postulated that a "protective" normal bacterial flora consists of a diverse group of indigenous microorganisms capable of using all the potential carbon sources in the environment (38). An exogenous organism attempting to enter such an environment would not have an available energy source and would be unable to colonize the particular environment. However, if some bacterial strains are removed from the indigenous microflora, such as through the use of antimicrobial agents, the limiting nutrients that normally supported these strains will then increase in concentration and will be able to support the growth of other bacteria, including enteric pathogens.

Competition for Bacterial Receptor Sites. Mucosal attachment is a prerequisite for successful colonization of the intestinal tract by both the indigenous microflora and pathogens (37, 43). Zilberg *et al.* (102) demonstrated a correlation between affinity of adherence to mouse intestinal segments *in vitro* and infectivity *in vivo* for enterotoxigenic *E. coli*. Bacterial competition for attachment sites in the intestinal tract is a mechanism by which the normal flora can prevent colonization and therefore subsequent pathogenicity of invading microorganisms (79). For example, Davidson *et al.* (24) showed that mice and pigs orally inoculated with a non-enterotoxin producing strain of *E. coli* possessing the K88 antigen were protected from subsequent oral challenge with an enterotoxin producing strain of *E. coli* possessing the K88 antigen. No protection was observed when a non-enterotoxin-producing, K88 negative strain of *E. coli* was first introduced. This protection was attributed to the blocking of the K88 antigen receptor site on the intestinal epithelium by the non-enterotoxin-producing strain of *E. coli*.

Creation of a Restrictive Physiologic Environment. Another means of direct bacterial antagonism is the creation of a physiologic environment by one microorganism which is inhibitory to another. By-products of bacterial metabolism which can contribute to the creation of a restrictive physiologic environment include hydrogen ion concentration, oxidation-reduction potential, hydrogen sulfide, volatile fatty acids, and alcohols.

Hydrogen Ion Concentration: Low pH is thought to be the major mechanism by which lactic acid producing bacteria (primarily *Lactobacillus, Bifidobacterium*, and *Streptococcus*) inhibit the *in vivo* and *in vitro* growth of various facultative and anaerobic bacteria (74, 89).

Oxidation-Reduction Potential: Oxygen utilization by the microbial components of the intestinal normal flora produces anaerobic micro-environments. Both commensal and pathogenic anaerobes can flourish under these conditions, but the growth of pathogens requiring oxygen would be restricted (56, 57).

Hydrogen Sulfide: Freter *et al.* (38) have demonstrated that hydrogen sulfide restricts the range of substrates which a given bacterial species can efficiently utilize under anaerobic conditions and this may be a factor by which the normal flora inhibits the growth of exogenous microorganisms.

Volatile Fatty Acids: Volatile and non-volatile short chain fatty acids are present throughout the intestinal tract as byproducts of the metabolism of anaerobic bacteria and have been shown by several investigators to be inhibitory to a wide variety of bacteria (12, 19, 51, 57, 58, 73, 78). For example, in the presence of adequate concentrations of nutrients, the inhibition of *S. flexneri* by coliforms in broth cultures is due to volatile fatty acid production (primarily formic and acetic acids) and a concomitant decrease in the oxidation-reduction potential of the medium (63). Either factor alone has some antagonistic effect, but the two factors function synergistically to produce a bactericidal effect against *Shigella*. Bohnhoff *et al.* (12, 64) showed that a similar mechanism exists for the inhibition of *Salmonella enteritidis* in the intestinal tract of mice. The quantity of volatile fatty acids present in the normal mouse cecum is sufficient to prevent the multiplication of *S. enteritidis in vitro* and the inhibitory activity was greatest at low pH and Eh levels. The pH of the environment was extremely important in all these experiments since at pH levels above 7.0, the volatile fatty acids are primarily in the dissociated state and unable to inhibit the growth of enteric pathogens (51). On the other hand, as the pH is lowered, the proportion of undissociated acid molecules increase and the acids are then able to enter the bacterial cell and inhibit bacterial metabolism.

Elaboration of Antibiotic-Like Substances. The fourth general mechanism of direct bacterial antagonism is the production of an antibiotic-like substance by one microorganism which inhibits the multiplication of another. The chemical nature and mode of action of these inhibitory substances are quite diverse and include ammonia, hydrogen peroxide, hemolysins, lysostaphin, bacterial enzymes, bacteriophage tails, defective bacteriophage, and bacteriocins.

The most extensively studied of the antibiotic-like compounds are the bacteriocins (53). Practically all genera of bacteria have been shown to produce bacteriocins or bacteriocin-like compounds. A bacteriocin is defined as a diffusible substance produced by a microorganism which possesses an essential biologically active protein moiety and has a bactericidal mode of action against other bacterial strains but not against the producing microorganism.

The significance of bacteriocins as regulators of bacterial populations is unclear (14, 28). Early investigators assigned a major role to bacteriocins or bacteriocin-like substances in the regulation of microbial populations (74). These investigators suggested that the stability of the intestinal flora and resistance to colonization by exogenous bacteria was due to elaboration of bacteriocins by the resident bacteria. Even though production of bacteriocins by enteric bacteria has been frequently demonstrated *in vitro*, evidence accumulated in recent years have failed to provide consistent evidence for an important ecological role of bacteriocins in the gastrointestinal tract (40, 54).

APPLICATION OF COLONIZATION RESISTANCE

Clinicians have long been intrigued by the prospect of manipulating the normal flora to prevent or cure a disorder. A variety of therapeutic approaches of this type have been reported in the literature. A few examples are given below.

A. *LACTOBACILLUS*

Lactobacillus inhibits the *in vitro* growth of several enteric pathogens including *Salmonella typhimurium, Staphylococcus aureus*, enteropathogenic *E. coli*, and *Clostridium perfringens*, and has been used in both man and animals to treat a broad range of intestinal disorders (e.g., chronic constipation, functional diarrhea, mucous colitis, sigmoid diverticulitis, antibiotic-induced colitis, and irritable colon,(45, 47). However, numerous controlled clinical studies during the last few years have failed to demonstrate any beneficial effect of *Lactobacillus* administration (74, 89).

B. *STREPTOCOCCUS FAECIUM*

Another bacterial preparation which has been used to prevent or treat intestinal disorders in man and animals is the SF-68 strain of *S. faecium* (6, 15, 59). This microorganism is characterized by a short replication time and inhibitory activity against several enteropathogens (13, 20).

C. *SACCHAROMYCES BOULARDII*

S. boulardii is a thermophilic, nontoxic yeast presently used in many countries to prevent diarrhea and other gastrointestinal disturbances. This yeast remains viable in the gastrointestinal tract and inhibits the growth of a number of bacterial pathogens *in vivo* and *in vitro* (2, 62). Double-blind-controlled clinical trials have demonstrated its efficacy against antibiotic-associated diarrhea in humans (87, 91). Also, animal studies have shown that administration of *S. boulardii* protects hamsters and gnotobiotic mice against *C. difficile*-induced intestinal disease (23, 31, 92).

D. CECAL AND FECAL SUSPENSIONS

Wilson *et al.* (100) showed that administration of cecal homogenates to vancomycin treated hamsters significantly protects them from lethal cecitis due to toxigenic *C. difficile*. It has also been shown that fecal suspension enemas are clinically effective in the normalization of bowel function in patients suffering from recurring *C. difficile* enterocolitis (16, 82).

The value of inoculation with fecal suspensions or bacterial flora components has also been demonstrated in the poultry industry. Peroral inoculation of newly hatched chickens with fecal or intestinal homogenates from healthy adult birds has been demonstrated by several research groups to increase resistance to infection by *Salmonella* species (68). The same protective effect can be gained by inoculation of newly hatched birds with non-defined, mixed cultures of obligate anaerobes and facultative anaerobes derived from adult fowls (85).

E. SELECTIVE DECONTAMINATION

In patients and animals with impaired immune competence, the digestive tract is often the site of entry for potential pathogens. Although there is a wide variety of systemic antimicrobials effective in treating infections in these patients, it is generally held that prevention is a much better approach than treatment. Decontamination of the digestive tract by treatment with oral antibiotics is being used to prevent such infections (25, 49).

Total decontamination of the intestinal tract can be achieved by oral administration of a combination of non-absorbable, antimicrobial agents active against both aerobic and anaerobic bacteria and yeasts. However, because this kind of treatment severely decreases colonization resistance, the patients must be nursed in protective isolation and the food must be sterile. Furthermore, a state of total decontamination is difficult to achieve, expensive, and very stressful to the patient (10, 69). Failure to prevent bacterial contamination in these patients will inevitably result in bacterial colonization by resistant microorganisms and frequently fatal infections.

Selective decontamination or selective antimicrobial modulation of immunosuppressed patients is designed to eliminate intestinal facultative gram-negative bacilli, known to be potential pathogens, while maintaining the anaerobic microflora, thought to play a major role in colonization resistance (7, 11, 49, 95, 98). Furthermore, the obligate anaerobic flora rarely causes infection in immunosuppressed patients. Selective decontamination of the intestinal flora has been repeatedly associated with decreased incidence of complicating infections in immunosuppressed patients (26, 48, 49, 84, 97).

CONCLUSIONS

There is considerable evidence that a variety of mechanisms operate to exclude pathogens from the intestinal tract. The mechanisms involved in colonization resistance against infection by opportunistic and pathogenic organisms are complex, as are the interaction between the hundreds of bacterial species present in the intestine that are responsible for these protective mechanisms. It is unlikely than a single species is responsible for the inhibitory effect of the normal flora on potential pathogens. Instead, synergistic relations between members of the normal flora appear to be important in suppressing the intestinal colonization by potential pathogens. Further work will be necessary to reveal the relative contributions of the various factors in maintaining the integrity of the intestinal flora. However, possibilities exist for the controlled manipulation of the normal microflora so that health promoting activities of the microbes are emphasized.

REFERENCES

1. **Abrams, G.D., and J.E. Bishop.** 1966. Effect of the normal microbial flora on the resistance of the small intestine to infection. J. Bacteriol. **92:**1604-1608.
2. **Albert, 0., J. Massot, and M.C. Courtois.** 1977. Etude cinetique quantitative de la repartition d'une levure vivante du genre *Saccharomyces* a differents niveaux du tractus digestif. Vie. Med. **18:**1604-1606.
3. **Aly, R., and H.R. Shinefield.** 1982. Bacterial Interference, CRC Press, Boca Raton.
4. **Artwohl, J.E., and D.C. Savage.** 1979. Determinants in microbial colonization of the murine gastrointestinal tract: pH, temperature, and energy-yielding metabolism of *Torulopsis pintolopesii*. Appl. Environ. Microbiol. **37:**697-703.
5. **Bartlett, J.G., T.W. Chang, M. Gurwith, S.L. Gorbach, and A.B. Onderdonk.** 1978. Antibiotic associated pseudomembranous colitis due to toxin producing Clostridia. N. Eng. J. Med. 298:531-534.
6. **Bellomo, G., G. Mangiagli, L. Nicastro, and G. Frigerio.** 1980. A controlled double-blind study of SF 68 strain as a new biological preparation for the treatment of diarrhea in pediatrics. Curr. Ther. Res. **28:**927-936.
7. **Berg, R.D.** 1981. Promotion of the translocation of enteric bacteria from the gastrointestinal tracts of mice by oral treatment with penicillin, clinicalmycin, or metronidazole. Infect. Immun. *33:*854-861.
8. **Berg, R.D., and D.C. Savage.** 1975. Immune response of specific pathogen-free and gnotobiotic mice to antigens of indigenous and non-indigenous microorganisms. Infect. Immun. **11:**320-329.
9. **Binder, H.J., B. Filburn, and M. Flock.** 1975. Bile and inhibition of intestinal microorganisms. Am. J. Clin. Nutr. **28:**119-125.
10. **Bodey, G.P.** 1984. Current status of prophylaxis of infection with protected environments. Am. J. Med. **76:**678-684.
11. **Bodey, G.P., V. Rodriquez, H. Chang, and G. Narboni.** 1978. Fever and infection in leukemic patients.Cancer **41:**1610-1621.
12. **Bohnhoff, M., C.P. Miller, and W.R. Martin.** 1964. Resistance of the mouse's intestinal tract to experimental salmonella infection. II.Factors responsible for its loss following streptomycin treatment. J. Exp. Med. **120:**817-828.
13. **Bongetta, R., T. Quirino, G. Ortisi, G. Privitera, D. Foschi, G. Cavagna, M. Moroni, and V. Rovati.** 1981. The colonization of *Streptococcus faecium* in human intestinal tract after oral administration. Boll. Ist Sieroter. Milanese. **60:**381-385.
14. **Booth, S.J., J.L. Johnson, and T.D. Wilkins.** 1977. Bacteriocin production by strains of *Bacteroides* isolated from human faeces and the role of these strains in the bacterial ecology of the colon. Antimicrob.Agents. Chemother. **11:**718-724.
15. **Borgia, M., N. Sepe, V. Brancato, and R. Borgia.** 1982. A controlled clinical study on *Streptococcus faecium* preparation for the prevention of side reactions during long-term antibiotic treatments. Curr. Ther. Res. **31:**265-271.
16. **Bowden, T.A., A.R. Mansberger, and L.E. Lykins.** 1978. *Pseudomembranous enterocolitis*: mechanism of restoring flora homeostasis. Am. Surg. **74:**178-183.
17. **Brown, W.R., D.C. Savage, R.S. Dubois, A. Mallory, and F. Kern, Jr.** 1972. Intestinal microflora of immunoglobulin-deficient and normal human subjects. Gastroenterology **62:**1143- 1152.

18. **Burr, D.H., and H. Sugiyama.** 1982. Susceptibility to enteric botulinum colonization of antibiotic-treated adult mice. Infect. Immun. **36**:103-106.
19. **Byrne, B.M., and J. Dankert.** 1979. Volatile fatty acids and aerobic flora in the gastrointestinal tract of mice under vanous conditions. Infect. Immun. **23**:559-563.
20. **Carbone, M., L. Bonina, and M.T. Fera.** 1980. Microbiological properties of *Streptococcus faecium* SF 68 strain and its relationship with other microorganisms. Boll. Ist. Sieroter. Milanese **6**:591-598.
21. **Caugant, D.A., B.A. Levin, and R.K. Selander.** 1981. Genetic diversity and temporal variation in the *E. coli* population of a human host. Genetics **98**:467-490.
22. **Cooperstock, M.S., and A.J. Zedd.** 1983. Intestinal flora of infants, p. 79-99. *In* D.J. Hentges (ed.), Hurnan Intestinal Microflora in Health and Disease. Academic Press, New York.
23. **Corthier, G., F. Dubos, and R. Ducluzeau.** 1986. Prevention of *Clostridium difficile* induced mortality in gnotobiotic mice by *Saccharomyces boulardii*. Can. J. Microbiol. **32**:894896.
24. **Davidson, J., and D.C. Hirsh.** 1975. Use of the K88 antigen for *in vivo* bacterial competition with porcine strains of enteropathogenic *Escherichia coli*. Infect. Immun. **12**:134-136.
25. **de Vries-Hospers, H.G., D.T. Sleijfer, N.H. Mulder, D. van der Waaij, H.O. Nieweg, and H.F.K. van Saene.** 1981. Bacteriological aspects of selective decontamination of the digestive tract as a method of infection prevention in granulocytopenic patients. Antimicrob. Agents Chemother. **19**:813-820.
26. **Dekker, A.W., M. Rozenberg-Arska, J.J. Sixma, and J. Verhoef.** 1981. Prevention of infection by trimethoprim-sulfamethoxazole plus amphotericin B in patients with acute non-lymphocytic leukemia. Ann. Intem. Med. **95**:555-559.
27. **Drasar, B.S., and M.J. Hill.** 1974. Human Intestinal Flora, Academic Press, New York.
28. **Ducluzeau, R., F. Dubos, R. Raibaud, and G.D. Abrams.** 1976. Inhibition of *Clostridium perfringens* by an antibiotic substance produced by *Bacillus licheniformis* in the digestive tract of gnotobiotic mice: Effect on other bacteria from the digestive tract. Antimicrob. Agents. Chemother. **9**:20-25.
29. **Ebright, J.R., R. Fekety, J. Silva, and K.H. Wilson.** 1981. Evaluation of eight cephalosporins in hamster colitis model. Antimicrob. Agents. Chemother. **19**:980-986.
30. **Eisenstein, B.I., I. Ofek, and E.H. Beachey.** 1979. Interference with the mannose binding and epithelial cell adherence of *Escherichia coli* by sublethal concentrations of streptomycin. J. Clin. Invest. **63**:1219-1228.
31. **Elmer, G.W., and L.V. McFarland.** 1987. Suppression by *Saccharomyces boulardii* of toxigenic *Clostridium difficile* overgrowth after vancomycin treatment in hamsters. Antimicrob. Agents. Chemother. **31**:129-131.
32. **Finegold, S.M., V.L. Sutter, and G.B. Mathisen.** 1983. Normal indigenous intestinal flora, p. 3-31. *In* D.J. Hentges (ed.), Human Intestinal Microflora in Health and Disease. Academic Press, New York.
33. **Floch, M.H., J.J. Binder, B. Filburn, and W. Gershengoren.** 1972. The effect of bile acids on intestinal microflora. Am. J. Clin. Nutr. **25**:1418-1426.
34. **Foo, M.C., and A. Lee.** 1974. Antigenic cross-reaction between mouse intestine and a member of the autochthonous microflora. Infect. Immun. **9**:1066-1069.
35. **Freter, R.** 1955. The fatal enteric cholera infection in the guinea pig, achieved by inhibition of normal enteric flora. J. Infect. Dis. **97**:57-65.

36. **Freter, R.** 1956. Experimental enteric *Shigella* and *Vibrio* infections in mice and guinea pigs. J. Exp. Med. **104:**411-418.
37. **Freter, R.** 1980. Prospects for preventing association of harmful bacteria with mucosal surfaces, p. 441-458. *In* Bacterial Adherence. Chapman Hall, London.
38. **Freter, R., H. Brickner, M. Botney, D. Cleven, and A. Aranki.** 1983. Mechanisms that control bacterial populations in continuous-flow culture models of mouse large intestinal flora. Infect. Immun. **39:**676-685.
39. **Freter, R., E. Stauffer, D. Cleven, L.V. Holdeman, and W.E.C. Moore.** 1983. Continuous-flow cultures as in vitro models of the ecology of large intestinal flora. Infect. Immun. **39:**666-675.
40. **Friedman, D.R., and S.P. Halbert.** 1960. Mixed bacterial infections in relation to antibiotic activities. J. Immunol. **84:**11-19.
41. **Fubara, E.S., and R. Freter.** 1972. Availability of locally synthesized and systemic antibodies in the intestine. Infect. Immun. **6:**965-981.
42. **George, W.L., R.D. Rolfe, G.K.M. Harding, R. Klein, C.W. Putnam, and S.M. Finegold.** 1982. *Clostridium difficile* and cytotoxin in feces of patients with antimicrobial agent-associated pseudomembranous colitis. Infect. **10:**205-207.
43. **Gibbons, R.J.** 1977. Adherence of bacteria on host tissue, p. 395-406. *In* D. Schlessinger (ed.), Microbiology. American Society for Microbiology, Washington, D.C..
44. **Gorbach, S.L., M. Barza, M. Giuliano, and N.V. Jacobus.** 1988. Colonization resistance of the human intestinal microflora: Testing the hypothesis in normal volunteers. Eur. J. Clin. Microbiol. Infect. Dis. **7:**98-102.
45. **Gorbach, S.L., T.W. Chang, and B. Goldwin.** 1987. Successful treatment of relapsing *Clostridium difficile* colitis with *Lactobacillus* GG. Lancet. **2:**1519.
46. **Gorbach, S.L., L. Nahas, and L. Weinstein.** 1967. Studies of intestinal microflora. I. Effects of diet, age, and periodic sampling on numbers of fecal microorganisms in man. Gastroenterology. **53:**845-855.
47. **Gotz, V., J.A. Romankiewicz, J. Moss, and H.W. Murray.** 1979. Prophylaxis against ampicillin associated diarrhea with a *Lactobacillus* preparation. Am. J. Hosp. Pharm. **36:**754-757.
48. **Guiot, H.F.L., A.V. Helmig-Schurter, J.W.M. van der Meer, and R. van Furth.** 1986. Selective antimicrobial modulation of the intestinal microbial flora for infection prevention in patients with hematologic malignancies. Evaluation of clinical efficacy and the value of surveillance cultures. Scand. J. Infect. Dis. **18:**153-160.
49. **Guiot, H.F.L., J.W.M. van der Meer, and R. van Furth.** 1981. Selective antimicrobial modulation of the human microbial flora: infection prevention in patients with decreased host defense mechanisms by selective elimination of potentially pathogenic bacteria. J. Infect. Dis. **143:**644-654.
50. **Harris, M.A., C.A. Reddy, and G.R. Carter.** 1976. Anaerobic bacteria from the large intestine of mice. Appl. Environ. Microbiol. **31:**907-912.
51. **Hentges, D.J.** 1983. Role of the intestinal microflora in host defense against infection, p. 311-331. *In* D.J. Hentges (ed.), Human Intestinal Microflora in Health and Disease. Academic Press, New York.
52. **Hentges, D.J., and B.R. Maier.** 1970. Inhibition of *Shigella flexneri* by the normal intestinal flora. III. Interactions with *Bacteroides fragilis* strains *in vitro*. Infect. Immun. **2:**364-376.

53. **Iglewski, W.J., and N.B. Gerhardt.** 1978. Identification of an antibiotic- producing bacterium from the human intestinal tract and characterization of its antimicrobial product. Antimicrob. Agents. Chemother. **13:**81-89.
54. **Ikari, N.S., D.M. Kenton, and V.M. Young.** 1969. Interaction in the germfree mouse intestine of colicinogenic and colicin-sensitive microorganisms. Proc. Soc. Exp. Biol. Med. **130:**1280-1284.
55. **Johnson, R.O., S.A. Clay, and S.S. Arnon.** 1979. Diagnosis and management of infant botulism. Am. J. Dis. Child. **133:**586-593.
56. **Koopman, J.P., F.G.J. Janssen, and J.A.M. van Druten.** 1975. Oxidation-reduction potentials in the cecal contents of rats and mice. Proc. Soc. Exp. Biol. Med. **149:**995-999.
57. **Lee, A., and E. Gemmell.** 1972. Changes in the mouse intestinal rnicroflora during weaning: role of volatile fatty acids. Infect. Immun. **5:**17.
58. **Levison, M.E.** 1973. Effect of colon flora and short-chain fatty acids on growth in vitro of *Pseudomonas aeruginosa* and *Enterobactericeae*. Infect. Immun. **8:**30-35.
59. **Lewenstein, A., G. Frigerio, and B. Moroni.** 1979. Biological properties of SF 68, a new approach for the treatment of diarrhea diseases. Curr. Ther. Res. **26:**967-981.
60. **Lloyd, A.B., R.B. Cumming, and R.D. Kent.** 1977. Prevention of *Salmonella typhimurium* infection in poultry by pretreatment of chickens and poults with intestinal extracts. Aust. Vet. J. **53:**82-87.
61. **Long, S.S., and R.M. Swenson.** 1977. Development of anaerobic fecal flora in healthy newborn infants. J. Pediatr. **91:**298-301.
62. **Massot, J., M. Desconclois, and F. Patte.** 1977. Effet protecteur d'un *Saccharomyces* lors d'une infection bacterienne experimentale chez la souris. Boll. Soc. Mycol. Med. **6:**45-48.
63. **Meynell, G.G.** 1963. Antibacterial mechanisms of the mouse gut. II. The role of Eh and volatile fatty acids in the normal gut. Brit. J. Expt. Pathol. **44:**209-219.
64. **Miller, C.P., and M. Bohnhoff.** 1963. Changes in the mouse's enteric flora associated with enhanced susceptibility to *Salmonella* infection during streptomycin treatment. J. Infect. Dis. **113:**59-66.
65. **Moberg, L.J., and H. Sugiyama.** 1978. Microbial ecological basis of infant botulism as studied with germfree mice. Infect. Immun. **25:**653-657.
66. **Moore, W.E.C., and L.V. Holdeman.** 1974. Human fecal flora: The normal flora of 20 Japanese-Hawaiians. Appl. Microbiol. **27:**961-979.
67. **Onderdonk, A.B., W.M. Weinstein, N.M. Sullivan, J.G. Bartlett, and S.L. Gorbach.** 1974. Experimental intraabdominal abscesses in rats: Quantitative bacteriology of infected animals. Infect. Immun. **10:**1256-1259.
68. **Rantala, M., and E. Nurmi.** 1973. Prevention of the growth of *Salmonella infantis* in chicks by the flora of the alimentary tract of chickens. Br. Poult. Sci. **14:**627-630.
69. **Ribas-Mundo, M., A. Franeda, and C. Rozman.** 1981. Evaluation of a protective environment in the management of granulocytopenic patients: a comparative study. Cancer **48:**419-424.
70. **Rolfe, R.D.** 1984. Role of volatile fatty acids in colonization resistance to *Clostridium dificile*. Infect. Immun. 45:185-191.
71. **Rolfe, R.D.** 1988. Asymptomatic intestinal colonization by *Clostridium difficile*, pp. 201-225. *In* R.D. Rolfe and S.M. Finegold (ed.), *Clostridium difficile*: Its Role in Intestinal Disease. Academic Press, New York.

72. **Rolfe, R.D., S. Helebian and S.M. Finegold.** 1981. Bacterial interference between *Clostridium difficile* and normal fecal flora. J. Infect. Dis. **143:**470-475.
73. **Rolfe, R.D. and J.P. Iaconis.** 1983. Intestinal colonization of infant hamsters with *Clostridium difficile*. Infect. Immun. **42:**480-486.
74. **Rusch, V.** 1980. Medicine and the microbial world. Microecol. Ther. **10:**163-172.
75. **Salanitro, J.P., I.G. Blake, and P.A. Muirhead.** 1977. Isolation and identification of fecal bacteria from adult swine. Appl. Environ. Microbiol. **33:**79-84.
76. **Salanitro, I.P., I.G. Blake, P.A. Muirhead, M. Maglio, and J.R. Goodman.** 1978. Bacteria isolated from the duodenum, ileum, and cecum of young chicks. Appl. Environ. Microbiol. **35:**782-790.
77. **Sarkany, I., and C.C. Gaylarde.** 1968. Bacterial colonisation of the skin of the newborn. J. Pathol. Bacteriol. **95:**115-122.
78. **Savage, D.C.** 1977. Microbial ecology of the gastrointestinal tract. Ann. Rev. Microbiol. **31:**107-133.
79. **Savage, D.C.** 1980. Adherence of normal flora to mucosal surfaces, p. 31-60. *In* E.H. Beachey (ed.), Bacterial Adherence. Chapman and Hall, London.
80. **Savage, D.C.** 1983. Morphological diversity among members of the gastrointestinal microflora. Intern. Rev. Cytol. **82:**305-334.
81. **Savage, D.C., R. Dubos, and R.W. Schaedler.** 1968. The gastrointestinal epithelium and its autochthonous bacterial flora. J. Exp. Med. **127:**67-76.
82. **Schwan, A., S. Sjolin, U. Trottestan, and B. Aronsson.** 1983. Relapsing *Clostridum difficile* enterocolitis cured by rectal infusion of homologous faeces. Lancet. **2:**845.
83. **Shedlofsky, S., and R. Freter.** 1974. Synergism between ecologic and immunologic control mechanisms of intestinal flora. J.Infect Dis. **129:**296-303.
84. **Sleijfer, D.T., N.H. Mulder, H.G. de Vries-Hospers, V. Fidler, H.O. Niewig, D. van der Waaij, and H.F.K. van Saene.** 1980. Infection prevention in granulocytopenic patients by selective decontamination of the digestive tract. Eur. J. Cancer. **16:**859-869.
85. **Snoeyenbos, G.H., O.M. Weinack, and C.F. Smyser.** 1978. Protecting chicks and poults from *Salmonella* by oral administration of "normal" gut flora. Avian. Dis. **22:**273-287.
86. **Sugiyama, H., and D.C. Mills.** 1978. Intraintestinal toxin in infant mice challenged intragastrically with *Clostridium botulinum* spores. Infect. Immun. **21:**59-63.
87. **Surawicz, C.M., G.W. Elmer, P. Speelman, L.V. McFarland, J. Chinn, and G. van Belle.** 1989. Prevention of antibiotic-associated diarrhea by *Saccharomyces boulardii*: A prospective study. Gastroenterology **96:**981-988.
88. **Svanborg-Eden, C., T. Sandberg, K. Stenqvist, and S. Ahlstedt.** 1978. Decrease in adherence of *Escherichia coli* to human urinary tract epithelial cells *in vitro* by subinhibitory concentrations of ampicillin: a preliminary study. Infect. **6:**121-124.
89. **Tannock, G.W.** 1984. Control of gastrointestinal pathogens by normal flora, p. 374-382, *In* M.J. Khug and C.A. Reddy (ed.), Current Perspectives in Microbial Ecology. American Society for Microbiology, Washington, DC..
90. **Tannock, G.W., and D.C. Savage.** 1974. Influences of dietary and environmental stress on microbial populations in the murine gastrointestinal tract. Infect. Immun. **9:**591-598.
91. **Tempe, J.D., A.L. Steidel, H. Blehaut, M. Hasselmann, P. Lutun, and F. Maurier.** 1983.Prevention par *Saccharomyces boulardii* des diarrhees de l'alimentation enteral a debit continu. Sem. Hop. Paris. **51:**1409-1412.

92. **Toothaker, R.D.** 1984. Prevention of clindamycin-induced mortality in hamsters by *Saccharomyces boulardii*. Antimicrob. Agents. Chemother. **26**:552-556.
93. **van der Waaij, D.** 1982. Colonization resistance of the digestive tract: clinical consequences and implications. J. Antimicrob. Chemother. **10**:263-270.
94. **van der Waaij, D., J.M. Berghuis-de Vries, and J.E.C. Lekkerkerk-van der Wees.** 1971. Colonization resistance of the digestive tract in conventional and antibiotic-treated rice. J. Hyg. **69**:405-511.
95. **van Furth, R., and E.H. Nauta.** 1971. Principles of antibiotic treatment, p. 387-394. *In* F. Elkerbout, P. Thomas and A. Zwaveling (ed.), Cancer Chemotherapy. Leiden University Press, Leiden.
96. **Vosbeck, K., H. Handschin, E-B. Menge, and 0. Zak.** 1979. Effects of subminimal inhibitory concentrations of antibiotics on adhesiveness of *Escherichia coli in vitro*. Rev. Infect., Dis. **1**:845-851.
97. **Wade, J.C., C.A. de Jongh, K.A. Newman, J. Crowley, P.H. Wiernik, and S.C. Schimpff.** 1983. Selective antimicrobial modulation as prophylaxis against infection during granulocytopenia: trimethoprim-sulfamethoxazole vs. nalidixic acid. J. Infect. Dis. **147**:624-634.
98. **Wells, C.L., M.A. Maddaus, R.P. Jechorik, and R.L. Simmons.** 1988. Role of intestinal anaerobic bacteria in colonization resistance. Eur. J. Clin. Microbiol. Infect. Dis. **7**:107-113.
99. **Wilkins, T.D., and R.L. Van Tassell.** 1983. Production of intestinal mutagens, p. 265-288. *In* D.J. Hentges (ed.), Human Intestinal Microflora in Health and Disease. Academic Press, New York.
100. **Wilson, K.H.** 1981. Suppression of *Clostridium difficile* by norrnal hamster cecal flora and prevention of antibiotic-associated cecitis. Infect. Immun **34**:626-628.
101. **Wilson, K.H., M. Patel, P. Permoad, and L. Moore.** 1986. Ecologic succession-Use in development of synthetic microfloras. Microecol. Ther. **16**:181-189.
102. **Zilberberg, A., I. Ofek, and J. Goldhar.** 1984. Affinity of adherence *in vitro* and colonization of mice intestine by enterotoxigenic *Escherichia coli* (ETEC). FEMS. Microbiol. Lett. **23**:103-106.
103. **Zubrzycki, L., and E.H. Spaulding.** 1962. Studies on the stability of the normal human fecal flora. J. Bacteriol. **83**:968-974.

DISCUSSION

E. NURMI: I have to agree with Dr. Rolfe that it is a lot easier to study competitive exclusion in poultry than it is in humans.

N. STERN: You had made a statement that the diet contributes a great deal to change in the flora. We had a workshop before this symposium in which we really did have quite a bit of discussion, and there was disagreement as to what the ultimate floral profile is. Could you elaborate further on that point?

R. ROLFE: Diet has a tremendous influence on the development of the normal flora. For example, in humans if you compare breast-fed vs. formula-fed infants, the development of the intestinal flora differs tremendously. The final outcome is the same; the climax flora which is achieved is the same. However, the actual development differs tremendously. That is what I meant by the effect of diet. There have been several studies on the effect of diet on large bowel cancer. The effect of diet on a normal flora of adults has been examined. There is not much even when an individual switches over to a non-meat or all-meat diet; the intestinal flora is very similar. It normally takes very dramatic changes, for example starvation, to show a dramatic change.

S.STAVRIC: You didn't show any results of *Bifidobacterium*. The Japanese are claiming quite a beneficial effect of this organism on human health. Do you have any data on that?

R. ROLFE: Yes, there has been a lot of work on *Bifidobacterium*. However, I am not aware of any control blind studies that clearly show that *Bifidobacterium* has an effect. Primarily, they have examined its effect against antibiotic-associated diarrhea.

D. CORRIER: Were your studies with volatile fatty acids and *Clostridium difficile* in hamsters?

R. ROLFE: Yes, that was in hamsters.

D. CORRIER: You found VFA to be bacteriostatic. Did you manipulate the pH in your *in vitro* study to increase or decrease the amount of undissassociated VFA?

R. ROLFE: Yes, not only did we adjust the volative fatty acid concentrations at each time interval, but we also had to adjust the pH that was present. If we increased the pH, then there would have been a less inhibitory effect. We haven't looked at this *in vivo*. We have looked at breast milk vs. formula milk in humans, as well as in hamsters, and breast milk tends to decrease the pH t607ememdously, whereas formula milk has more of a buffering capacity. The pH and the anaerobic flora probably interact together to be one of the things that changes the normal flora.

G. MEAD: You mentioned that H_2S-producing organisms cound inhibit nutrient uptake under anaerobic conditions. Can you elaborate a little further on the types of organisms that will produce H_2S?

R. ROLFE: This is work that has primarily been done by Rolfe Freter. He uses continuous cultures and shows that the anaerobic flora produce H_2S. I can't give you the specific species.

M. OPITZ: You mentioned selective decontamination. Do you have any practical examples demonstrating that you could eliminate facultative bacterial flora and maintain the anaerobic bacterial flora?

R. ROLFE: Yes we do, and there have been several good studies. These papers have shown the effect that particular antimicrobial agents will have on the flora through very elaborate culturing techniques, and at the same time, examined the incidence of infection later.

EXPERIENCE WITH COMPETITIVE EXCLUSION IN THE NETHERLANDS

R.W.A.W. Mulder
N.M. Bolder

Spelderholt Centre for Poultry Research and Information
Services, Agricultural Research Service
7361 DA Beekbergen, The Netherlands

ABSTRACT

In recent years an increase in the reported incidence of foodborne diseases has become apparent in many countries in the so-called developed world. The presence of *Salmonella*, *Campylobacter*, *Listeria*, and other potentially pathogenic microorganisms on poultry and poultry products is therefore a serious threat to the marketing of these products. After 1973, when positive results from studies by Rantala and Nurmi (14) from Finland on the treatment of day-old chicks with a microflora which protects the young birds against colonization of *Salmonella* bacteria in the intestine were published, the same type of research was started in The Netherlands.

In The Netherlands an undefined microflora was prepared from adult caecal material. The procedure of preparation was described. Under laboratory conditions the microflora proved to be highly effective in preventing *Salmonella* from colonizing. Under the same conditions a similar protection against *Campylobacter* could not be demonstrated.

A technique for mass application to day-old chicks was developed, and a large field trial was carried out. In total, 284 flocks (143 treated with microflora, 141 untreated) totaling approximately eight million broilers were examined for the effect of the treatments on *Salmonella* colonization. The treatment reduced the number of *Salmonella*-positive flocks from 24% to 15% and the *Salmonella* contamination of live birds, as measured by sampling the caecal contents, from 3.5% positive samples in the non-treated to 0.9% positive samples in the treated group.

In another field study 58 flocks (29 treated, 29 untreated, involving 200,000 broilers) were examined for the effect of the treatment on *Campylobacter*. The number of *Campylobacter*-positive flocks was reduced from 50% to 31%. Also the number of *Campylobacter*-contaminated caeca and liver and gall samples was reduced. As with *Salmonella*, the incidence of *Campylobacter* within positive flocks was also considerably reduced.

Colonization Control of Human Bacterial Enteropathogens in Poultry

INTRODUCTION

During the past 30 years the production of poultry meat increased faster than that of most other animal products. Worldwide the production of poultry meat increased over that period by 6% (reaching 40 million tons in 1990), whereas the production of other products from animal origin increased by only 2 - 3%. As a percentage of the total meat consumption in the developed world, poultry meat consumption will increase from 17% in 1975 (total meat consumption: 35 million tons) to approximately 22% in 2000 (estimated total meat consumption: 82 million tons).

Poultry meat is perceived as a healthy food, despite all the genera of bacteria which have been isolated from poultry and poultry products and their role in human food poisoning. The presence of potentially pathogenic microorganisms such as *Salmonella, Campylobacter, Listeria monocytogenes*, and *Staphylococcus aureus* has been established. Table 1 summarizes the genera which have been reported in the literature. In The Netherlands, neither salmonellosis nor campylobacteriosis are reportable human diseases. Therefore, no exact information on the occurrence of these diseases is available. Data on human foodborne illnesses are incomplete. Beckers (1988) summarized the data from the years 1981 and 1982. In Table 2 the number of foodborne illness caused by microorganisms are summarized. Table 3 gives the details with respect to the commodities involved. In a recent report of the National Health Council in The Netherlands (Anon., 1988), mention is made of the increasing role of poultry and poultry products in the occurrence of human campylobacteriosis. It is clear that poultry and meat products are important sources of the most frequently named microorganisms.

The future of poultry meat products seems to be bright. Poultry meat has a lot of advantages over other meats, being low in saturated fat, high in protein content and not having religious barriers for large parts of the human population. Poultry meat products are also available in a large variety.

Nevertheless, a guaranteed market for poultry products in the future requires efficient and hygienic poultry processing resulting in a safe end product. In the near future, international trade of food products will increase considerably. International trade has a remarkable impact on outbreaks of human food poisoning. Foodborne illness is in general of worldwide concern, but especially in The Netherlands, where approximately 60% of the total poultry production is exported.

Because of the increased production of poultry with more animals per m^2 reared in climate controlled houses, day-old chicks will invariably be exposed to environments allowing microorganisms to invade the population. As the microbiological condition of the live animal influences the microbiology of the finished product, it is important to control the microorganisms of the live birds.

In this review paper, results of studies on competitive exclusion treatment of young broilers carried out in The Netherlands will be discussed. Competitive exclusion is the name for the treatment of day-old broilers with a microflora resulting in a colonization resistance towards potentially human pathogenic microorganisms. The effect of the treatment on *Salmonella* and *Campylobacter* bacteria was estimated under both laboratory and field conditions.

Table 1. Genera isolated from poultry products.

Achromobacter	*Corynebacterium*	*Neisseria*
Acinetobacter	*Cytophaga*	*Paracolobactrum*
Actinomyces	*Enterobacter*	*Peptostreptococcus*
Aeromonas	*Escherichia*	*Proteus*
Alcaligenes	*Eubacterium*	*Propionibacterium*
Arthrobacter	*Flavobacterium*	*Pseudomonas*
Bacillus	*Fusobacterium*	*Salmonella*
Bacteroides	*Gaffkya*	*Sarcina*
Bifidobacterium	*Haemophilus*	*Serratia*
Brevibacterium	*Klebsiella*	*Shigella*
Campylobacter	*Lactobacillus*	*Staphylococcus*
Chromobacterium	*Listeria*	*Streptococcus*
Citrobacter	*Microbacterium*	*Streptomyces*
Clostridium	*Micrococcus*	*Yersinia*

MATERIAL AND METHODS

In this section only the main procedures used in the described experiments and trials are given.

Preparation and Application of Intestinal Microflora under Laboratory and Field Conditions. The intestinal microflora used was an undefined flora originating from mature SPF birds and propagated in young birds kept in isolators. These day-old birds were inoculated orally; crops and intestines were harvested at three weeks of age. Homogenates of these organs were checked for avian pathogens (both *in vivo* and *in vitro*) and antibacterial drug-resistant bacteria, and were demonstrated to induce very significant protection against an oral challenge with up to 10^5- 10^6 CFU *Salmonella* organisms/chick under experimental conditions.

Table 2. Numbers of cases of foodborne diseases in The Netherlands in 1981 and 1982.

	1981	1982
Bacillus cereus	27	3
Campylobacter jejuni	1496	1728
Clostridium perfringens	13	22
Salmonella	7496	6795
Staphylococcus aureus	28	2
Yersinia enterocolitica	262	274

Beckers, RIVM, 1988

Table 3. Numbers of cases of foodborne diseases by foods involved (or suspected) 1981 and 1982.

	1981	1982
Meat and meat products	323	149
Fish and shellfish	406	56
Poultry	52	145
Egg and egg products	----	2
Dairy products	123	36
Chinese foods	486	414
Totals	3618	1376

Beckers, RIVM, 1988

Freeze-dried intestinal homogenates of SPF birds were resuspended in a suspension medium and given orally to day-old chicks which were reared in isolation chambers in the laboratory trials. For the field application the suspension was sprayed in the hatcher on the chicks and eggs by a Gloria knapsack spray apparatus using 10 ml of suspension per 100 eggs/chickens. The spray produced had a Median Mass Diameter of 130 μ and geometric s.d. of 2.2 by three bar (Dr. J.H.H. van Eck, Department of Poultry Diseases, University of Utrecht). The treatment took place on the 20th incubation day about 11 a.m. when about 30 - 40% of the chicks had been hatched.

***Salmonella* Isolation Procedure.** Pre-enrichment in buffered peptone water at 37 ° C for 18 hours was used only for the examination of feed samples (200 g each). All other samples were directly cultured in selenite-brilliant green-mannitol selective broth medium at 37 ° C for 24 and 48 hours. The weight ratio of sample to medium was 1:10. Thereafter, the samples were inoculated on brilliant green-phenol red agar (Oxoid) with 0.2% sodium desoxycholate (Merck) and cultured at 37 ° C for 24 hours. After biochemical and serological identification, *Salmonella* isolates were serotyped at the Dutch National Health Institute (RIVM, Bilthoven).

***Campylobacter* Isolation Procedure.** Enrichment (1:10) was in a *Campylobacter* enrichment medium containing per liter: 13 g nutrient broth (Oxoid), 2 g yeast extract, 3 g bile salts no. 3 (Oxoid), 1.5 g agar (Merck). After sterilizing the medium, 4 flasks of Oxoid SR84 growth supplement were added. Incubation was at 43 ° C under microaerophilic conditions by using BBL Campypak plus in anaerobic jars. Thereafter, the samples were inoculated on a *Campylobacter* blood medium containing per liter: 28 g Brucella broth (Difco), 17 g agar (Merck) antibitotic supplement (Oxoid), 0.15 ml cephaloridin (10%), 100 ml defibrinated sheep blood (Biotrading Benelux). Incubation was at 43 ° C for 48 hours under microaerophilic conditions. After biochemical identification, *Campylobacter* isolates were serotyped at the Dutch National Health Institute (RIVM, Bilthoven).

HISTORY OF CE RESEARCH IN THE NETHERLANDS

In The Netherlands, research on colonization resistance of the gastrointestinal tract was first reported by Van der Waay *et al.* (1971) with mice. Koopman and co-workers (1977, 1984) at the Central Animal Laboratory of the Catholic University of Nijmegen continued this type of research with mice, but in the late seventies they worked in close cooperation with Dr. E. Goren of the Poultry Health Institute in Doorn to incorporate their experience in the research projects related to the application in the poultry industry. This cooperation resulted in the development of a protective microflora, laboratory experiments and field trials. The laboratory experiments and field trials were also carried out by Bolder and Mulder (1985) at Spelderholt Centre for Poultry Research and Information Services at Beekbergen.

Previous research at Spelderholt had been focused on the treatment of chicks with colicins to prevent *Salmonella* colonization (Mulder, 1974). Colicins are compounds produced by some *Enterobacteriaceae*, e.g., *Escherichia*, and act antagonistically towards other members of the family of Enterobacteriaceae. Although laboratory experiments were successful in reducing the growth of *Salmonella* strains, experiments with broilers failed and the research stopped.

LABORATORY EXPERIMENTS

The positive results of the Finnish studies by Rantala and Nurmi (1973) on the treatment of day-old chicks with an undefined microflora to control *Salmonella infantis* infection prompted research in this area in several countries.

At the Poultry Health Institute in Doorn an undefined microflora was also produced by Goren and coworkers. The preparation of the microflora is described in the Materials and Methods section. The essential feature of this microflora is the use of homogenates of caeca and caecal contents. To protect the microflora, containing anaerobic and aerobic microorganisms, a special suspension medium was used (Koopman *et al.*, 1983). Table 4 gives the composition of this medium, in which cysteine and blood are the two compounds which protect against oxygen activity. It has been proven that the microflora given to day-old chicks housed in isolators protected the chicks from colonization with different *Salmonella* strains given orally at 10^5 or 10^6 CFU's. Figures 1 and 2 give some examples of results obtained after challenge with *Salmonella infantis* and *Salmonella typhimurium* (Goren *et al.*, 1984a; Goren, 1987).

Table 4. Suspension medium used for microflora (Koopman *et al.*, 1983).

Glucose	1.0 g
Starch	4.0 g
Tryptose	10.0 g
NaCl	5.0 g
N_2HPO_4	3.0 g
KH_2PO_4	0.5 g
$MgSO_4$	0.5 g
Cysteine-HCl	0.5 g
Water	1000 ml

5% sheep's blood is added after sterilization.

From Spelderholt Centre, Bolder *et al.* (1987) reported experiments, using the same undefined microflora, in which chicks were challenged at different times during their life time. This imitates the practice in which broilers may get infected several times during the rearing period, with different *Salmonella* serotypes. These experiments were

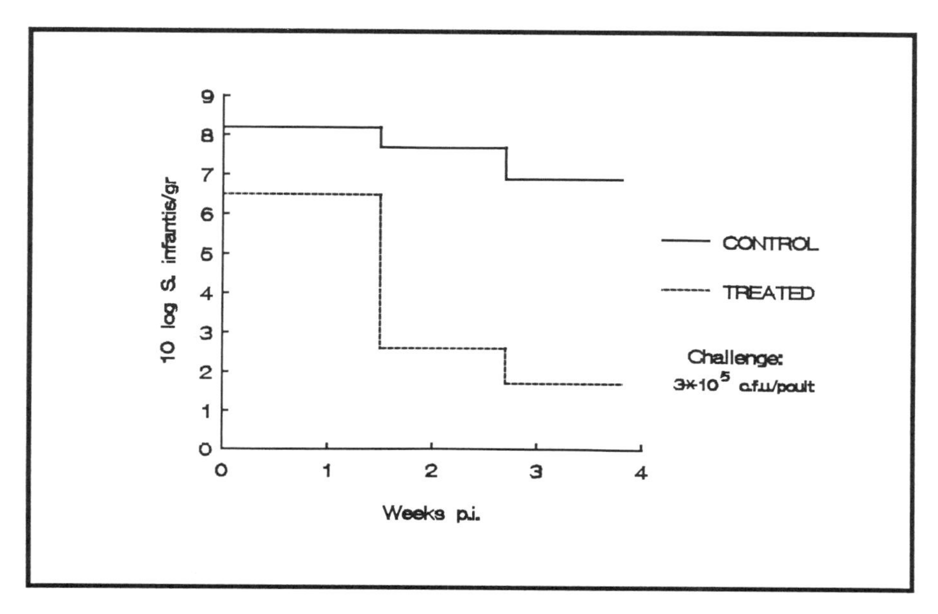

Fig. 1. *S. infantis* in caecal contents of chicks treated with microflora after challenge.

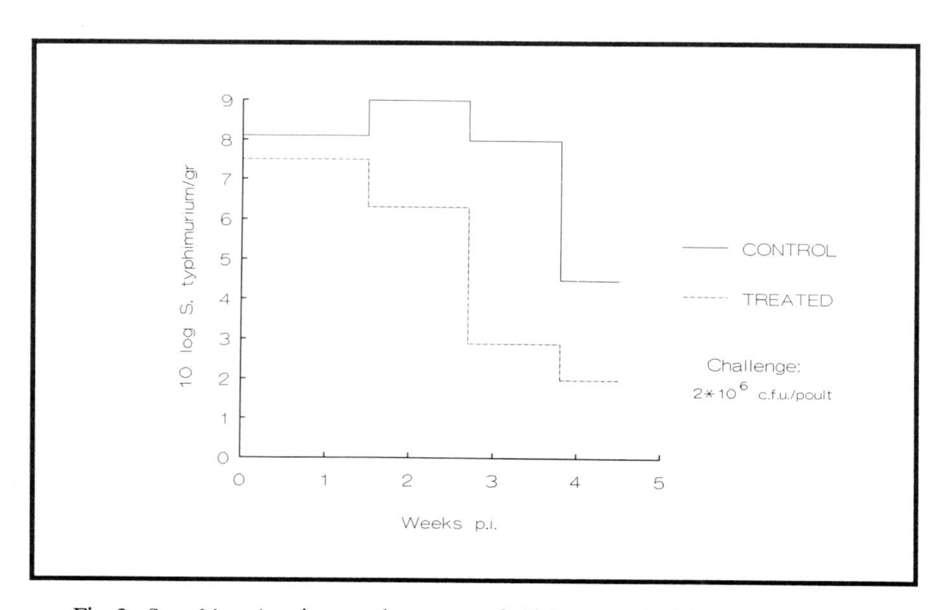

Fig. 2. *S. typhimurium* in caecal contents of chicks treated with microflora after challenge.

carried out with groups of broilers, which where (orally) treated with CE microflora and challenged with *Salmonella infantis* (10^7) one and 24 hours after CE treatment, and one group which was repeatedly challenged at one, two, and three weeks of age with *Salmonella virchow* (10^7), *S. blockley* (10^5) and *S. typhimurium* (10^5) respectively. Figure 3 shows some results of these experiments. As measured by the number of *Salmonella* organisms in the caeca, a significant reduction of *Salmonella* was observed in CE-treated groups in comparison to the reference group at two weeks of age.

At five weeks of age, *Salmonella* CFU's in the caeca of broilers from all experimental groups were 5.0 x 10^5 /g . *S. virchow* was not re-isolated. Especially after the challenge with *Salmonella typhimurium*, the protective effect was weakened. This finding was further confirmed in experiments, where the *S. typhimurium* challenge was given two weeks after CE treatment. After the protective effect of CE treatment was proven, the effect of antimicrobial or coccidial drugs, which have a widespread use in poultry husbandry practice, on the efficacy of the CE treatment was determined. Goren (1987) reported that the following drugs would interfere with the CE treatment: ampicillin, flavomycin, nicarbazin, TCN (a cocktail of tetracycline, chloramphenicol, and neomycin) and tylosin.

No effect on the CE treatment was observed for these drugs: amprolium, avoparcin, baytril, chloramphenicol, flumequine, furazolidone, lincospectin, monensin, oxytetracyclin, sulfadimidine, sulfaquinoxalin, sulfa + trimethoprim, virginiamycin and zinc-bacitracin. This demonstrates that, for implementation of CE treatment in actual practice, the industry has a good choice of drugs which can be used. Since 1986/1987, laboratory experiments at Spelderholt have involved the effect of CE treatment on *Campylobacter jejuni* as well as *Salmonella* challenge of day-old chicks. Under laboratory conditions no significant protective effect towards *Campylobacter* organisms (10^5 - 10^6) could be demonstrated. Later it was shown that the *Campylobacter* challenge was too high. In addition to the CE treatment, the effect of the addition of 4% D-mannose and 10% Germasil (a stabilized and enzyme-added germinated barley product) was tested. D-mannose added at a concentration of 4% to the feed protected against *Salmonella* and *Campylobacter* challenge. Because the mechanism of *Campylobacter* colonization and protection against colonization is not understood, but appears to differ from the mechanism of *Salmonella* colonization, additional research was started. Bolder and Mulder (1989) reported on the minimum effective dose for *Campylobacter* and the influence of the different (genetic origin) brands of broilers on colonization with *Campylobacter*.

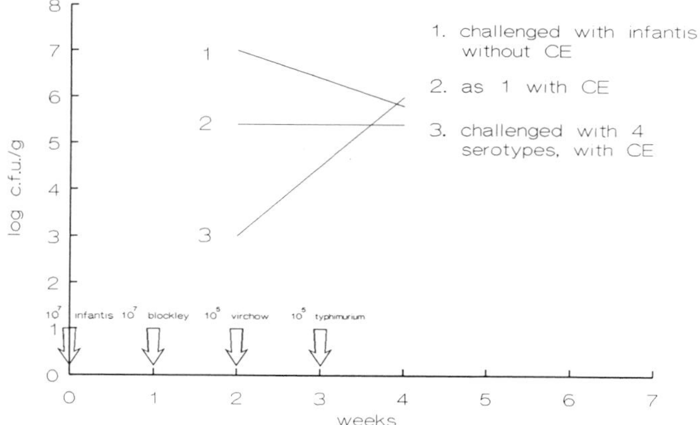

Fig. 3. Effect of CE-treatment on protection against several *Salmonella* serotypes.

FIELD TRIALS

After the laboratory experiments, the mass application of the microflora to day-old chicks was studied. As described in the Materials and Methods section, the CE is given to the chicks in the hatchery by spraying it on pipping embryo's on the hatcher trays the day before hatch. In cooperation with the Poultry Health Institute in Doorn, a large scale field trial was carried out in the southern part of The Netherlands. Special permission for using the undefined microflora under practical conditions was obtained from the Veterinary Service of the Ministry of Agriculture and Fisheries.

The field trial involved 46 commercial broiler farms supplied with day-old chicks by three hatcheries. The feed used was obtained from one feedmill, and the broilers were slaughtered in seven processing plants. Over a period of 18 months every farm received spray-treated flocks of broilers for four successive rearing periods, followed by four non-treated flocks or vice versa. In total, 284 flocks and approximately eight million broilers were involved. During rearing and at slaughter several samples were taken for examination for the presence of *Salmonella*. The method used for *Salmonella* isolation is described under the Materials and Methods section. Zootechnical data were also recorded. The results have been published by Goren *et al.* (1988). The most important results from the study will be highlighted. Table 5 shows the results of the examination of caecal contents for the presence of *Salmonella*. In untreated birds, 3.5% of the samples were *Salmonella*-positive, whereas after CE treatment only 0.9% of the caecal samples were *Salmonella*-positive. Also, the *Salmonella* incidence within *Salmonella*-positive flocks was reduced (on basis of the caecal samples) from 14.3% in the untreated flocks to 6.4% in the CE-treated flocks (Table 6). These results show that a CE treatment reduces the *Salmonella* contamination of live birds considerably, as well as the number of *Salmonella* organisms in caeca of positive birds (Bolder and Mulder, 1985). Thirty-eight different serotypes were isolated. Serotypes originating from the feed were not re-isolated from other material sampled.

Table 5. Effect of CE treatment on *Salmonella* contamination of caeca of broilers.

	Number of Samples	Positive
Non CE-Treated	14,099	486 (3.5%)
CE-Treated	14,400	134 (0.9%)

In a second smaller field trial, using the same protocol, 58 flocks, 29 CE-treated and 29 untreated involving 200,000 broilers, were examined for the effect of the CE treatment on the presence of *Campylobacter*. The caeca, liver, and gall were examined. The *Campylobacter* isolation method is described in the Materials and Methods section. Detailed results of the experiments are given by Mulder and Bolder (1989). The main results of this field trial are summarized in Tables 7 and 8. The number of *Campylobacter*-contaminated flocks was decreased by the treatment, as well as the number of *Campylobacter*-contaminated caeca, liver, and gall samples. As observed with *Salmonella*, the incidence of *Campylobacter* organisms within positive flocks was decreased considerably (Mulder and Bolder, 1989).

Table 6. *Salmonella* incidence in caecal contents from *Salmonella*- positive flocks.

	Number of Samples	Positive
Non CE-Treated	3399	486 (14.3%)
CE-Treated	2100	134 (6.4%)**

**P < 0.01

Table 7. Effect of CE treatment on *Campylobacter* contamination of broiler flocks.

	Total Number of Flocks	Positive
Non CE-Treated	29	18 (62%)
CE-Treated	29	12 (41%)

Table 8. Effect of CE treatment on *Campylobacter* contamination of caeca, liver, and gall samples.

	Total Number of Samples	Positive
Caeca		
Non CE-Treated	58	23 (40%)
CE-Treated	58	12 (21%)
Liver + Gall		
Non CE-Treated	286	79 (28%)
CE-Treated	290	35 (12%)

It was surprising, that under practical conditions a protective effect of the treatment of day-old chicks with a CE microflora could be demonstrated; whereas, under laboratory conditions a less protective effect was observed. In all these studies zootechnical data showed that the CE application had no adverse effects.

In the large field trial, the effect of the CE treatment on the *Salmonella* contamination of the skin of processed carcasses was also estimated. No differences between CE-treated and non CE-treated flocks were found. It proved, by comparison of the isolated *Salmonella* serotypes, that cross-contamination occurred during transport and in the early phases of processing. Efficiency of crate washing equipment and improvement of scalding and plucking techniques has since been studied by Spelderholt Centre and the results will be published in the near future.

COSTS OF THE TREATMENT, PRACTICAL IMPLEMENTATION, AND FUTURE RESEARCH

CE treatment in The Netherlands has been used in field trials using an undefined microflora produced by the Poultry Health Institute in Doorn. It has been calculated that production of this microflora and spraying the suspended flora on the day-old chicks will cost Dfl. 0.025 per chicken (1 Dfl. = approx. $0.5026 dollars). The year after our field trial a large broiler operation tried the CE treatment in further studies. It was found that nine out of 51 lots of CE-treated day-old chicks were *Salmonella*-positive. The serotypes involved were highly correlated to the serotypes found on the breeding farms. The hatchery thus proved not to be the so-called "hygienic" barrier which everyone thought it was. This fact, together with the cost of the CE treatment prompted a temporary halt to the CE treatment. Hygiene programs for breeding farms are now tested, and after improvement the CE treatment will be tried again.

Undefined microflora, although effective and harmless (free of pathogens) has the disadvantage that legislators cannot accept the implementation of CE treatment easily. The main disadvantage of the use of defined microflora is the problem in maintaining viability and effectiveness over a long period of time. Therefore our future research will also be focused on stabilizing defined microflora by using coating techniques. As coating techniques are very expensive, the selected defined microflora should not contain more than 10 different microorganisms. The only known defined microflora with this number of organisms, having protective properties against *Salmonella* colonization, has been described by Stavric *et al.* (1985).

CONCLUSION

CE treatment of day-old chicks may contribute significantly to the reduction of the *Salmonella* and *Campylobacter* contamination of live broilers and the reduction of the incidence of these organisms in positive flocks. No adverse effects of the treatment on zootechnical data, including mortality, were observed. Not all questions on the mechanism of colonization or protection from colonization have been answered. Research should focus on understanding these mechanisms.

To maintain the beneficial effect of the CE treatment to the consumer ready end-product, improvements in the hygiene of transportation and slaughtering are necessary. Separation of negative from positive flocks in the slaughterhouses by a microbiological monitoring system is necessary too. This microbiological monitoring system will become mandatory in the upcoming years in EEC countries. To overcome the problems with the increased costs of production of CE-treated broilers, governments should be more active in encouraging industry to use this type of treatment. They can then produce microbiologically safe products to offer to the consumers.

REFERENCES

1. **Anon.** 1988. Report of the Health Council of The Netherlands on *Campylobacter jejuni* infections in The Netherlands. The Hague. Oct., 1988.
2. **Beckers, H.J.** 1988. Incidence of Foodborne Diseases in The Netherlands. Annual Summary 1982 and an overview from 1979-1982. Journal of Food Protection **51:** 327-334.
3. **Bolder, N.M., and R.W.A.W. Mulder.** 1985. Administration of a microflora to protect chicks against *Salmonella* infection. Proc. 7th. European Symposium on Poultry Meat Quality (ed. T. Ambrosen), Vejle, Danmark, p. 284.
4. **Bolder, N.M., and R.W.A.W. Mulder.** 1989. Minimum infective number of *Camopylobacter* bacteria for broilers. Proceedings of An International Symposium on Colonization Control of Human Bacterial Enteropathogens in Poultry, Atlanta, GA.
5. **Bolder, N.M., R.W.A.W. Mulder, M.C. van der Hulst, and J.L. Meijer.** 1987. Some observations on the Nurmi-concept in laboratory studies. Proceedings of the International Workshop on Competitive Exclusion of *Salmonella* from Poultry, ed. by G.C. Mead, Bristol, UK.

6. **Goren, E.** 1987. Colonization resistance to *Salmonella* infection. Poultry 3(no.5):37-39.
7. **Goren, E., W.A. de Jong, R. Doornenbal, J.P. Roopman, and H.M. Kennis.** 1984a. Protection of chicks against *Salmonella infantis* infection induced by strict anaerobically cultured intestinal microflora. The Veterinary Quarterly **6:**22-26.
8. **Goren, E., W.A. de Jong, P. Doornenbal, J.P. Roopman, and H.M. Kennis.** 1984b. Protection of chicks against *Salmonella* infection induced by spray application of intestinal microflora in the hatchery. The Veterinary Quarterly **6:**73-79.
9. **Goren, E., W.A. de Jong, P. Doornenbal, N.N. Bolder, R.W.A.W. Mulder, and A. Jansen.** 1988. Reduction of *Salmonella* infection of broilers by spray application of intestinal microflora: a longitudinal study. The Veterinary Quarterly **10:**249-255.
10. **Koopman, J.P., and H.M. Kennis.** 1977. Two methods to assess the gastrointestinal transit time in mice. Zeitschrift Versuchstierkunde **21:**72-77.
11. **Koopman, J.P., R.A. Prins, J.W.M.A. Mullink, G.W. Welling, H.M. Kennis, and M.P.C. Hectors.** 1983. Association of germ-free mice with bacteria isolated from the intestinal tract of normal mice. Zeitschrift Versuchstierkunde **25:**57-62.
12. **Mulder, R.W.A.W.** 1974. Antagonistische aktiviteiten. Spelderholt report 4272.
13. **Mulder, R.W.A.W., and N.M. Bolder.** 1989. Reduction of *Campylobacter* infections of broilers by competitive exclusion treatment of day-old chicks. Proceedings of An International Symposium on Colonization Control of Human Bacterial Enteropathogens in Poultry, Atlanta, GA.
14. **Rantala, M., and E. Nurmi.** 1973. Prevention of the growth of *Salmonella infantis* in chicks by the flora of the alimentary tract of chickens. Br. Poult. Sci. **14:**627-630.
15. **Stavric, S., T.M. Gleeson, B. Blanchfield, and H. Pivnick.** 1985. Competitive exclusion of *Salmonella* from newly hatched chicks by mixtures of pure bacterial cultures isolated from faecal and caecal contents of adult birds. Journal of Food Protection **48:**778-782.
16. **Waay, D. van der, J.M. Berghuis-De Vries, and J.E.C. Lekkerkerk-Van der Wees.** 1971. Colonization resistance in the digestive tract in conventional and antibiotic treated mice. J. Hyg. Camb. **69:**405-411.

DISCUSSION

G. MEAD: I'm intrigued on one particular point, I think I can understand why it is that Dr. Mulder would prefer to use cecal homogenates because obviously thse retain the maximum protective capability. However, the point that intrigues me is how he manages to treat all those millions of birds using that approach because surely this would have involved a very large number of donor birds. Can I just ask how many donor birds were used in treating the commercial flock?

R. MULDER: Thirty thousand.

G. SNOEYENSBOS: I hope we have all noticed what a low incidence of *Salmonella* that you secured without treatment which really is very good. Very good compared to many countries as well as the United States. It shows what can be done by more basic approaches.

R. MULDER: This study was done in the southern part of the country. The infection route through the feet was more or less closed there. I think the publication of 1988 described studies in the same area approximately one half a year later. The estimate of contamination in our country will be approximately 20 - 25%.

DEVELOPMENTS IN COMPETITIVE EXCLUSION TO CONTROL *SALMONELLA* CARRIAGE IN POULTRY

G. C. Mead

Agricultural and Food Research Council
Institute of Food Research
Bristol Laboratory, Langford
Bristol BS18 7DY, United Kingdom

ABSTRACT

Because of known limitations in the use of defined mixtures of caecal bacteria to protect chicks against *Salmonella* colonization by competitive exclusion, attention has been given to safety requirements for the application of undefined caecal cultures to commercial poultry flocks. In the United Kingdom, such material was used successfully following antibiotic therapy in 20/22 field trials involving *ca*. 250,000 adult breeder birds with a history of salmonellae infection.

Although the protective mechanism is still uncertain, early administration of appropriate caecal cultures readily prevented the growth of salmonellae entering the crop or caeca of chicks challenged experimentally, and reduced the spread of *Salmonella enteritidis* phage type 4 from infected 'seeders' to contact chicks in delivery boxes, with no significant effect of transportation. Under laboratory conditions, caecal-culture treatment was also used in combination with acid disinfection of feed, and was the major factor in protecting chicks against an environmental salmonellae challenge.

Further work is needed to determine conditions of husbandry and farm hygiene that willl maximize the benefit obtained from competitive exclusion treatment, when used commercially.

Colonization Control of Human Bacterial Enteropathogens in Poultry

INTRODUCTION

The indigenous microbiota of man and animals has long been appreciated as a major factor in protecting the host against colonization by 'foreign' microorganisms, and is considered more effective in this respect than protection mediated by the classical immune system (6). However, in spite of widespread interest in the topic in relation to disease control (e.g., 4, 31, 35), most attempts to manipulate the natural microbiota in a particular part of the host, and prevent the establishment of hazardous microorganisms, have yet to progress beyond the experimental stage. As Savage (26) points out, the main barrier to progress is present lack of detailed knowledge concerning those ecological factors that control the composition and activities of indigenous microbiotas and their interactions with host tissues.

Since salmonellosis in food animals is a zoonotic disease which poses significant health problems for man throughout the developed world (36), there is a perceived need to find better means of preventing intestinal colonization by the causative organisms in live animals. This is particularly so for poultry, an important source of human foodborne disease, where commercial production utilizes large-scale, intensive systems for rearing and processing that facilitate transmission of any pathogen present. It is in this area that research on colonization control in the live bird, known usually as 'competitive exclusion' (CE) or the Nurmi Concept, has shown particular promise, and offers the prospect of successful use on a wide scale.

This paper describes some of the more recent work at the author's laboratory and highlights certain practical aspects of applying CE to commercial poultry production.

PROGRESS WITH DEFINED TREATMENT PREPARATIONS

In early work at the UK Institute of Food Research and elsewhere, it was recognized that development of a defined bacterial mixture to protect chicks against *Salmonella* colonization would avoid including any avian or human pathogens in the treatment material. Other prospective benefits were the possibility of limiting the mixture to only a small number of strains and being able to ensure constant composition and activity of the preparation for commercial use.

It was soon established that a defined mixture could be used in place of undefined treatment material, and the 48-organism mixture described by Impey *et al.* (13) was just as effective against a low-dose challenge (*ca.* 10^3 cells/chick) with *S. kedougou, S. typhimurium*, or *S. virchow*, as undefined suspensions or cultures of caecal content from an adult bird (17). Subsequently, several other bacterial mixtures of comparable activity were described (20, 32). With these, the numbers of component strains varied from 10 to 50.

Although the above-mentioned preparations were defined in the sense that each individual strain had been characterized, there were difficulties in identifying many of the nonsporing anaerobes. In taxonomic studies of chicken caecal bacteria, reviewed by Mead (16), only *ca.* 25% of the anaerobes isolated could be identified with known species. Complete strain identification may be unimportant provided that suitable tests for pathogenicity can be carried out and any hazardous organisms excluded from the final treatment mixture.

At present, there appears to be no defined bacterial preparation that is available commercially and has been specifically designed to prevent chicks from becoming symptomless carriers of salmonellae. The practical difficulties in developing such a product include the following:

1. In laboratory trials, the most effective mixtures, giving consistent protection to treated chicks, have been too large for economically feasible product development. The necessary inclusion of various obligate anaerobes is also problematical because of their sensitivity to oxygen, tendency for poor growth in ordinary culture media and loss of viability during storage.

2. Protection provided by defined mixtures is more host-specific and less able to withstand challenge levels of $>10^4$ salmonellae/chick than is the case with suspensions or crude cultures of gut content (14, 32). This suggests the absence of key organisms from the defined mixtures and/or some change in properties of the component strains in artificial culture.

3. Loss of protective activity during subculture in laboratory media and cold storage is a common problem and was experienced, e.g., with a strain of *Streptococcus (Enterococcus) faecalis* (5). In this case, the activity was restored by animal passage.

Experiments with the 48-organism mixture (13) showed that omitting certain species or groups of bacteria led to a reduction in protective capability for chicks, and hence an increase in the level of salmonellae carriage following challenge. The most active organisms in protecting chicks were the lactobacilli (*L. acidophilus, L. fermentum*, and *L. salivarius*) and certain Gram-positive anaerobic cocci (18). Other obligate anaerobes present at high levels in the caeca of adult birds, (e.g., *Gemmiger formicilis*) appeared to play little or no part in the protective process (14). Experience suggests, however, that interactions between bacterial groups are important for protection, particularly those between facultative and obligate anaerobes (25, 9), and can even occur among strains of the same species, such as *Escherichia coli* (2).

Further development of defined treatment preparations is hampered by the lack of information on the exclusion mechanism and on the essential properties of key organisms. Also, in relation to isolating potentially protective strains, selective media presently available appear to be inadequate for anaerobes present at levels of one per cent or less of the total population, while evidence suggests that such organisms may be involved in the protective process (3).

BEHAVIOUR OF SALMONELLAE IN RESPONSE TO PRIOR CE TREATMENT

In relation to commercial poultrymeat production, the main purposes of CE treatment are (a) to prevent salmonellae shedding in the live-bird environment, thus reducing the opportunities for cross-infection in the flock during rearing, and (b) to avoid subsequent transfer of salmonellae to carcasses via faecal contamination during processing. Clearly, the extent to which salmonellae are carried in the contents of the lower intestine of the live bird has important implications for the ultimate contamination of carcasses.

Since the crop and caeca of the bird are the main sites where salmonellae colonization can occur in poultry (30), attention has been given to the behaviour of the organisms at these particular sites, using chicks challenged orally after pretreatment with an anaerobic culture of cecal content from a salmonellae-free donor bird (12). The effect of protective treatment on the persistence of *Salmonella kedougou* in the crop and caeca are shown in Table 1, which combines the results of three experiments. The table shows the extent to which the caeca in particular can become colonized with *Salmonella*, the proportion of positive birds increasing with time. As expected, there was no marked caecal colonization by *Salmonella* in the treated chicks, whereas in the untreated (control) birds, *Salmonella* had increased in number after 24 hours, reaching $\log_{10}$ 6.8/g at 48 hours. The slow disappearance of *Salmonella* from the caeca of chicks given CE treatment suggests a mainly bacteriostatic effect.

Table 1. Behaviour of *Salmonella* in the crop and caeca of control chicks and chicks pre-treated with a protective caecal culture.

	*No. of Positive Samples and **$\log_{10}$ *Salmonella*/g () in Each Case									
	TIME AFTER CHALLENGE (H)									
	1	2	4	6	8	10	12	24	36	48
Control Chicks										
crop	16 (2.4)	14 (2.3)	13 (1.9)	9 (3.5)	17 (3.2)	13 (1.6)	12 (2.3)	5 (1.7)	3 (1.4)	4 (2.8)
caeca	1 (2.7)	8 (2.7)	5 (2.8)	12 (2.1)	13 (2.7)	16 (3.0)	22 (3.3)	25 (3.5)	28 (5.0)	29 (6.8)
Treated Chicks										
crop	2 (2.1)	4 (1.5)	2 (2.4)	2 (1.2)	2 (<1.1)	0 (<1.1)	0 (1.9)	1 (<1.1)	0 (1.5)	3 (1.2)
caeca	2 (2.3)	1 (1.9)	4 (2.4)	3 (2.1)	3 (2.8)	3 (2.7)	1 (3.2)	3 (2.3)	6 (2.2)	10 (<1.6)

*Data from 30 chicks at each sampling time.
**Separate experiment (six chicks at each sampling time).

The behaviour of *Salmonella* in the crop was studied further in a feed-slurry model where conditions resembled those of moistened crop contents (7). The feed was a chick starter mash, without antimicrobial additives, that had been sterilized by gamma-irradiation. The *Salmonella* (*ca*. 10^4/g) were used in all tests and, where required, a protective cecal culture was also incorporated at *ca*. 10^7 organisms/g. The inoculated slurries were mixed continuously by gentle rotation during aerobic incubation at 37 ° C. Samples were taken immediately after inoculation and at hourly intervals up to seven hours. Tests were made for total anaerobic counts, coliforms, faecal streptococci, lactobacilli, and salmonellae.

The feed slurry (pH 7.1) supported rapid growth of the *Salmonella* in the absence of other microorganisms, but not with the addition of caecal culture. In this case (Fig. 1), *Salmonella* levels showed little change over the first four hours but then numbers declined, as did those of coliforms and faecal streptococci. This phenomenon coincided with the growth of lactobacilli to *ca*. 10^9/g and a pH value of 5.5. During incubation, counts of lactobacilli followed closely the total anaerobic counts (not shown), suggesting that lactobacilli had become the dominant organisms.

The bacteriostatic and bactericidal effects of the caecal culture on *Salmonella* added to the feed slurry were similar to those reported previously (7) for the inhibition of *Escherichia coli*, and were attributed mainly to the fall in pH value resulting from lactic acid production. Colonization of the crop by lactobacilli, whether by natural means or by administration of CE cultures, may be important in preventing this organ from becoming a salmonellae reservoir.

Although the fate of *Salmonella* administered orally to chicks pretreated with CE cultures has been determined, the mechanism of protection in the complex environment of the caecum, is far from clear. Possible factors include pH and oxidation-reduction potential (Eh); inhibitory substances such as H_2S, bacteriocins, fatty acids, and deconjugated bile acids; competition for nutrients and receptor sites; and local immunity (6, 26). However, because these factors are interdependent *in vivo*, it is difficult to elucidate any of their effects without the use of appropriate *in vitro* models (6).

One aspect which has received particular attention is the apparent competition between protective microorganisms and salmonellae for specific receptor sites in the caeca. Whether this involves an intimate association with the surface of the caecal mucosa or merely entrapment in the overlying 'blanket' of mucus is unclear, and the nature of the association appears to vary from one type of bacterium to another (26). A practical consequence of protective bacteria becoming 'attached' to the mucosal surface is that they are not readily removed in the laboratory by standard washing procedures. Thus, there would appear to be an advantage in analyzing washed sections of the caecal wall to increase the chances of isolating the effective organisms (33). In developing a treatment preparation for preventing salmonellae colonization of poultry, it has been considered desirable to select bacterial strains with ability to adhere to epithelial cells obtained from the alimentary tract of chicks (22). On the other hand, the results of such an *in vitro* test (23) did not predict whether a *Lactobacillus* strain would associate with epithelium in the alimentary tract of inoculated piglets, although epithelial association *in vivo* appeared to be an important factor in maintaining populations of lactobacilli in the tract.

The emphasis on protective bacteria from administered cultures being effective through their ability to associate with the caecal wall arose largely from the rapidity with which protection first became evident following oral inoculation of chicks with treatment material. The first indications of developing resistance to salmonellae colonization have been observed within approximately one hour of treatment (28, 33), although full protection was not apparent until 24-48 hours after treatment (29, 30). Nevertheless, even after 6-8 hours, all types of bacteria necessary for initiating protection appeared to be adhering to the mucosa (33).

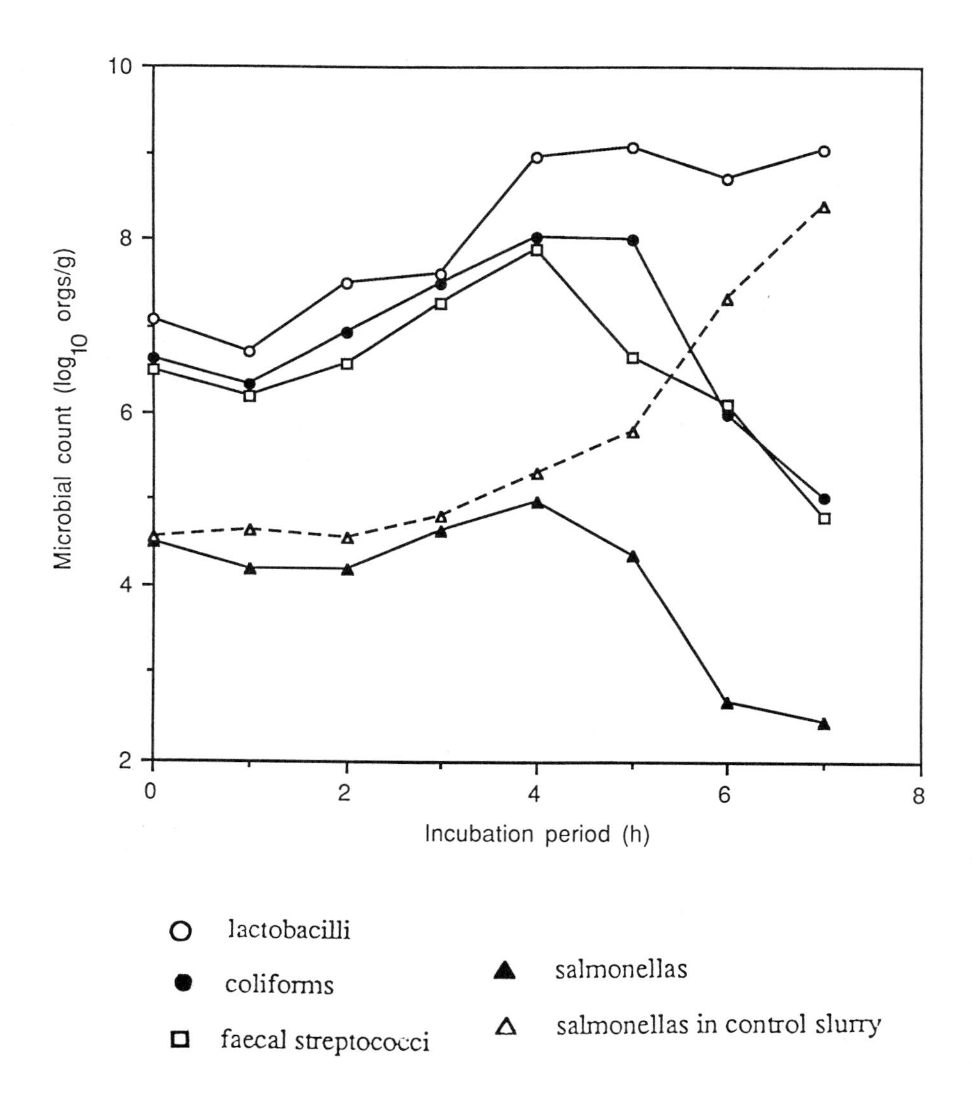

Fig. 1. Behaviour of *Salmonella kedougou* in feed slurry inoculated with mature caecal culture (12).

PRACTICAL ASPECTS OF CE APPLICATION

The time necessary for protective bacterial populations to become established in the caeca could be an unavoidable practical constraint in using CE treatment to its fullest advantage under field conditions. When chicks reared under laboratory conditions were challenged with *Salmonella* incorporated in the feed, it was found that protection was more evident in chicks exposed to the inoculated feed 24 or 48 hours after treatment than in birds which had received the *Salmonella* challenge at 0 or 5 hours (15). The advantage in applying CE at the earliest opportunity led Dutch workers (8) to develop and apply a method of spray-inoculating chicks in the hatchery, as they emerged from the eggs.

Although full protection of chicks by CE treatment is not immediate, the treatment is still likely to be advantageous in controlling the spread of salmonellae during long distance distribution of stock, e.g., for export purposes. In trials carried out at Bristol, in conjunction with a local hatchery, newly hatched chicks were challenged with *S. enteritidis* phage type 4, using the 'seeder' bird principle, with each seeder receiving an oral

Table 2. Protection of chicks by CE in (a) standard assay and (b) delivery boxes against *Salmonella enteritidis* phage type 4.

Type of Experiment	Chick Group	Transported (T) or Static (S)	Proportion of Chicks Pos. (%)	*Geometric Mean Count from Caeca
Standard Assay	Untreated	NA**	100	7.5
	Treated	NA	17	2.0
Delivery Boxes	Untreated 'Seeders'	S	100	7.3
		T	100	8.3
	Untreated 'Contacts'	S	97	6.4
		T	90	5.7
	Treated 'Seeders'	S	100	7.4
		T	100	8.0
	Treated 'Contacts'	S	40	3.6
		T	63	4.2

*Log_{10} *Salmonella*/g wet weight of caecal content.
**NA - not applicable.
Standard assay: Mead *et al.* (19).
Delivery boxes: 30 contact chicks and two seeders group.
Transportation - 4 h by road; total holding period - 72 h.
In all cases, results are a mean of three replicate trials.

dose of *ca*. 10^4 *Salmonella*. After inoculation, the seeders were placed in cardboard delivery boxes, in contact with other chicks, in the ratio 2:30. Each chick then received 0.5 ml of an adult caecal culture grown in VL medium (1) and administered orally. Equivalent numbers of similarly boxed seeders and control chicks were each given 0.5 ml of sterile water instead of the caecal culture. The space allocation in all boxes was 20 cm^2/chick.

Batches of treated and untreated (control) chicks were subjected to one of two subsequent procedures. Either the birds were kept at ambient temperature for 72 hours under static conditions in a separate room, or they were first transported by road for four hours. Results of the trials are shown in Table 2. Also included in Table 2 are results obtained from a standard assay for assessing treatment efficacy (19), in which the interval between treatment and challenge was 24 hours.

By comparison with tests involving an appreciable interval between treatment and challenge, protection was less evident in birds held in delivery boxes and exposed immediately to a developing seeder challenge. Nevertheless, both the proportion of positive contact chicks and the levels of *Salmonella* carriage were reduced by protective treatment. Although transportation could have been a stress factor, leading to reduced protection, analysis of variance showed no significant differences between static and transported birds in either treated or untreated groups.

Because of the limitations of defined culture treatments discussed previously, interest has moved towards the use of undefined preparations and the requirements for safe application to commercial poultry flocks. The undefined material used in the author's laboratory is invariably a caecal culture prepared from a healthy, salmonellae-free adult of the same species as the birds to be treated, and applied via the drinking water or by spraying. Although the exact composition of the culture will be unknown, there is detailed information on the main types of bacteria present in both cultures and starting material (27, 16), and sufficient experience of field use to show no adverse effect of the treatment on chick mortality, weight gain or feed conversion. In Sweden, this type of preparation was used over a five year period for the treatment of 2.86 million chickens (34), and in the Netherlands for trials involving eight million broilers (8). Both studies showed benefits in reducing flock infection with salmonellae, and no undesirable side effects were observed.

Even the process of culturing the caecal material provides a safeguard against some microbial hazards (24). For example, viruses and protozoa are unable to multiply in bacteriological culture media and, if present initially, would be eliminated by subculturing. Mycoplasmas are known to compete poorly in mixed cultures, as does *Listeria monocytogenes* under the conditions used (Mead and ter Boo, unpublished observations). No growth of *Campylobacter jejuni* has been observed (10) and, again, the organism would be eliminated by subculturing several times before the material is used to treat poultry flocks.

In addition, treatment cultures may be screened directly for any of the known agents of human foodborne disease, and indirectly as possible vehicles of particular avian pathogens following oral inoculation of Specific Pathogen Free (SPF) chicks and monitoring the birds for development of relevant antibodies. Culture material used by the author and his colleagues to treat chicken breeder flocks in the UK was checked by these means to ensure absence of the agents listed in Table 3. The list is not exhaustive, however, and could be extended, if required.

Table 3. Screening of undefined treatment cultures by bacteriological tests and *following oral inoculation of SPF chicks.

Campylobacter spp.	Newcastle disease
Clostridium perfringens	Reovirus
Salmonella spp.	Reticuloendotheliosis virus
Staphylococcus aureus	Marek's disease
Mycoplasma gallisepticum	Adenovirus
M. synoviae	Infectious bursal disease
Egg drop syndrome	Lymphoid leukosis A, B, C
Infectious bronchitis	

*In collaboration with Dr. P. A. Barrow, AFRC Institute for Animal Health - Houghton Laboratory.

In the case of turkeys and ducks, SPF flocks were not generally available in the UK, and the policy adopted in all cases (including trials on chickens) was to restrict the use of treatment material from a particular company's stock to recipient birds belonging to the same company. This precaution was taken to avoid the possibility of transmitting any unsuspected pathogens between companies.

Having tested adult birds supplied by each producer to identify suitable donors of salmonellae-free cecal material, cultures were tested for protective activity using the standard assay (19) and a *Salmonella* challenge of 10^4 and 10^6 organisms/chick respectively. Only cultures giving protection against 10^6 *Salmonella*/chick were used in the trials.

In all except the one trial with ducks, the birds were moved at about 19 weeks of age from the rearing site to a laying farm. All the flocks involved in these trials had a previous history of salmonellae infection and, during the 12 days prior to transfer, appropriate antibiotic medication was given by each company's veterinarian. Chickens received via the feed (g/t): furazolidone 400 + chlortetetracycline 200; turkeys were given neomycin 223 in place of the furazolidone, and ducks neomycin 220. Following the combined treatment, fresh litter was added to the duck pens.

The antibiotics were withdrawn on the day prior to CE treatment. When required, a fresh caecal culture was diluted up to 1:20 in the drinking water and made available after withholding water from the birds for several hours to induce thirst. Then, the birds rapidly consumed large quantities of liquid. The flocks were monitored for salmonellae before and after the combined treatment by culturing cloacal swabs and liner samples taken monthly from each pen.

Table 4. Field trials on breeder birds involving flocks with a history of *Salmonella* infection.

	Chickens	Turkeys	Ducks	Totals
No. of trials	8	13	1	22
Birds treated	187,100	37,600	120	224,820
Pens treated	80	165	1	246
Pens positive for salmonellae	1	4	0	5
Serotypes present initially:	*S. enteritidis* *S. typhimurium*	*S. enteritidis* *S. heidelberg* *S. senftenberg* *S. shubra* *S.virchow*	*S. typhimurium*	
Serotypes present finally:	*S. enteritidis* *S. typhimurium* (1 pen)	*S. agona* (4 pens)		

In total, 22 trials were carried out between 1986 and 1988, involving approximately 250,000 chickens, turkeys and ducks (Table 4). Results taken as indicative of successful protection were three consecutive monthly occasions when all samples were salmonellae-negative. According to this criterion, 20 of the trials were completely successful, and only in two cases were any positive pens identified. Surprisingly few serotypes were encountered, and the majority of isolations were either *S. enteritidis* or *S. typhimurium*, the former coming to prominence as a cause of human food poisoning in England and Wales shortly after completion of the trials (11). No tests were made to determine the possibility of ovarian or oviduct infection in the birds. The four positive turkey pens in one trial all yielded *S. agona*, a serotype not detected in the birds prior to treatment, and presumably acquired on the laying farm.

With chickens kept in cages, it has been found that successful use of antibiotic medication combined with CE treatment depended on the birds being moved to a second farm (21). In the UK, there is interest in controlling salmonellae both by CE treatment and acid disinfection of the feed, now widely used for breeder diets. The main advantages in incorporating certain organic acids in feed, e.g., formic and propionic acids, are that any salmonellae present will be destroyed and the feed protected against subsequent contamination.

To determine whether CE treatment and feed disinfection can be used as a combined strategy for salmonellae control, some preliminary experiments were carried out in collaboration with Dr. Michael Hinton, Bristol University School of Veterinary Science. For each of three replicate trials, four groups of ten chicks were kept on litter. On arrival from the hatchery, two of the groups were given CE treatment, using a caecal culture of proven efficacy. Feed was withheld for *ca.* six hours. Then, one treated and one untreated group were given feed supplemented with 0.68% (w/w) of a formic acid product, while the remaining chicks, both treated and untreated, were given feed without the acid supplement. All birds were exposed to a *Salmonella* challenge for four hours at one week of age, using a culture of *S. enteritidis* phage type 4 added to the drinking water. The estimated dose/chick was *ca.* 10^5 CFU. Examination of chicks at two and three weeks of age showed no effect of dietary acid on the incidence of *Salmonella* colonization in the caeca, and the *S. enteritidis* strain was found in 82% of 118 chicks that had received either acid alone or no acid. By contrast, only 8% of 120 chicks given CE treatment, with and without dietary acid, yielded *Salmonella*. The corresponding figures for liver isolations were 52% and 12.5% respectively, thus showing a lower tendency for systemic infection in CE treated chicks. Table 5 shows the incidence of *Salmonella* in chicks of each experimental group.

Table 5. Effects of competitive exclusion (CE) treatment and acid disinfection of feed on infection of chicks via the drinking water with *Salmonella enteritidis*.

			No. of Birds at Each Level of Carriage							
Type of Sample	CE Treatment	Acid in Feed	2 weeks*				3 weeks			
			0**	1-2	3-5	6-8	0	1-2	3-5	6-8
Caecal Contents	-	-	5	0	8	17	4	2	19	5
	-	+	5	1	6	18	7	0	5	16
	+	-	28	2	0	0	25	2	2	1
	+	+	29	0	0	1	28	2	0	0
Liver Parenchyma	-	-	10	18	2	0	18	12	0	0
	-	+	10	19	1	0	19	6	3	0
	+	-	27	3	0	0	26	3	1	0
	+	+	26	4	0	0	26	4	0	0

*Bird age; **Log_{10} *Salmonella*/g (0, not found); + treatment used; - treatment not used

The trials also indicated that CE treatment was effective against an environmental challenge (i.e., from the drinking water), while dietary acid was without influence on this mode of infection. Withholding the acidified feed for *ca.* six hours was apparently sufficient to avoid any interference with CE-mediated protection, and there could be some benefit in controlling both feedborne and environmental salmonellae infection by the combined treatments, to maximize protection.

CONCLUSION

In the context of the CE principle, the establishment of protective populations of intestinal bacteria in poultry is most likely to occur when the administered organisms reflect the complexity of the natural microbiota, thus avoiding an ecological imbalance in the gut. For this reason, protection against salmonellae colonization is more readily achieved through the use of undefined caecal cultures, of known complexity, than with defined bacterial mixtures that may lack key components, are inherently less protective and more prone to losing their protective activity. Further progress in developing defined treatment mixtures will be difficult in the foreseeable future because of our limited knowledge of the protective mechanism and hence the necessary basis for selecting the required strains.

Experience of using undefined treatment material on a commercial scale has shown no adverse effects on bird health or growth performance (8, 34). However, any risk of spreading unsuspected avian or human pathogens to recipient birds can be dealt with in one of two ways. Either the material is rigorously screened for organisms of concern, using standard bacteriological tests and SPF birds that can be given the CE treatment and monitored for antibody development or, as in the UK trials on breeder birds, use of any one source of caecal cultures is restricted to flocks belonging to the company that supplied the donor material. The latter approach is obviously more laborious and requires that caecal material from each donor bird be monitored for: (a) absence of salmonellae, and (b) high protective capability in a laboratory assay.

Available field data referred to in this paper and in a previous review (24) suggest that CE treatment can usefully reduce salmonellae infection in commercial poultry flocks, both with chicks and adult breeders. It now remains to evaluate the treatment on a wider scale and detemiine the necessary conditions of husbandry and farm hygiene that will maximize the benefits obtained.

REFERENCES

1. **Barnes, E.M., and C.S. Impey.** 1970. The isolation and properties of the predominant anaerobic bacteria in the caeca of chickens and turkeys. Br. Poult. Sci. **11:**467-481.
2. **Barrow, P.A., and J.F. Tucker.** 1986. Inhibition of colonization of the chicken caecum with *Salmonella typhimurium* by pre-treatment with strains of *Escherichia coli.* J. Hyg. **96:**161-169.
3. **Blanchfield, B., S. Stavric, T. Gleeson, and H. Pivnick.** 1984. Minimum intestinal inoculum for Nurmi cultures and a new method for determining competitive exclusion of *Salmonella* in chicks. J. Food Prot. **47:**542-545.

4. **Bruce, A.W., and G. Reid.** 1988. Intravaginal instillation of lactobacilli for prevention of recurrent urinary tract infections. Can. J. Microbiol. **34:**339-343.
5. **Cumming, R.B.** 1983. Further studies on *Streptococcus faecalis* in competitive exclusion, p. 301-302. *In* Australian Veterinary Poultry Association and International Union of Immunological Societies, Proceedings No. 66: Disease Prevention and Control in Poultry Production, Sydney, Australia.
6. **Freter, R.** 1984. Interdependence of mechanisms that control bacterial colonization of the large intestine. Microbiol. Therap. **14:**89-96.
7. **Fuller, R.** 1977. The importance of lactobacilli in maintaining normal microbial balance in the crop. Br. Poult. Sci. **18:**85-94.
8. **Goren, E., W.A. de Jong, P. Doornenbal, N.M. Bolder, R.W.A.W. Mulder, and A. Jansen.** 1988. Reduction of salmonella infection of broilers by spray application of intestinal microflora: a longitudinal study. Vet. Q. **10:**249-255.
9. **Goren, E., W.A. de Jong, P. Doornenbal, J.P. Koopman, and H.M. Kennis.** 1984. Protection of chicks against *Salmonellla infantis* infection induced by strict anaerobically cultured intestinal microflora. Vet. Q. **6:**22-26.
10. **Hudson, W.R., C.S. Impey, and G.C. Mead.** 1987. 'Nurmi concept' and possible transmission of campylobacters to poultry. Vet. Rec. **120:**439.
11. **Humphrey, T.J., G.C. Mead, and B. Rowe.** 1988. Poultry meat as a source of human salmonellosis in England and Wales. Epidemiol. Infect. **100:**175-184.
12. **Impey, C.S., and G.C.Mead.** 1989. Fate of salmonellae in the alimentary tract of chicks pre-treated with a mature caecal microflora to increase colonization resistance. J. Appl. Bacteriol. **66:**469-475.
13. **Impey, C.S., G.C. Mead, and S.M. George.** 1982. Competitive exclusion of salmonellae from the chick caecum using a defined mixture of bacterial isolates from the caecal microflora of an adult bird. J. Hyg. **89:**479-490.
14. **Impey, C.S., G.C. Mead, and S.M. George.** 1984. Evaluation of treatment with defined and undefined mixtures of gut microorganisms for preventing *Salmonella* colonization in chicks and turkey poults. Food Microbiol. **1:**143-147.
15. **Impey, C.S., G.C. Mead, and M. Hinton.** 1987. Influence of continuous challenge via the feed on competitive exclusion of salmonellas from broiler chicks. J. Appl. Bacteriol. **63:**139-146.
16. **Mead, G.C.** 1989. Microbes of the avian caecum: types present and substrates utilized. J. Exp. Zool. Suppl. **3:**48-54.
17. **Mead, G.C., and C.S. Impey.** 1985. Control of salmonella colonization in poultry flocks by defined gut-flora treatment, p. 72-79. *In* Proceedings of the Intenational Symposium on *Salmonella*, New Orleans, USA. American Association of Avian Pathologists, University of Pennsylvania, Kennett Square, PA.
18. **Mead, G.C., and C.S. Impey.** 1987. The present status of the Nurmi Concept for reducing carriage of food-poisoning salmonellae and other pathogens in live poultry, p. 57-77. *In* F.J.M. Smulders (ed.) Elimination of Pathogenic Organisms from Meat and Poultry. Elsevier Science Publishers, Amsterdam.
19. **Mead, G.C., P.A. Barrow, M.H. Hinton, F. Humbert, C.S. Impey, C. Lahellec, R.W.A.W. Mulder, S. Stavric, and N.J. Stern.** 1989. Recommended assay for treatment of chicks to prevent *Salmonella* colonization by 'competitive exclusion'. J. Food Prot. **52, No. 7:**500-502.
20. **Nurmi, E.** 1985. Use of competitive exclusion in prevention of salmonellae and other enteropathogenic bacteria infections in poultry, p. 64-71. *In* G.H. Snoeyenbos (ed.) Proc. Int. Symp. on *Salmonella*, New Orleans, LA. Amer. Assoc. of Avian Pathologists, University of Pennsylvania, Kennett Square, PA.

21. **Nurmi, E., J. Hirn, L. Oivanen, and M. Kauppi.** 1988. The combined use of antibiotic treatment and competitive exclusion (CE) in eradication of *Salmonella* and *Campylobacter* in poultry, p. 698. *In* Proceedings of the 6th World Conference on Animal Production, Helsinki.
22. **Nurmi, E.V., C.E. Schneitz, and P.H. Makela.** 1981. Process for the production of a bacterial preparation for the prophylaxis of intestinal disturbances in poultry. European Patent Application No. 0 033 584A2.
23. **Pederson, K., and G.W. Tannock.** 1989. Colonization of the porcine gastrointestinal tract by lactobacilli. Appl. Environ. Microbiol *55:*279-283.
24. **Pivnick, H., and E. Nurmi.** 1982. The Nurmi Concept and its role in the control of salmonellae in poultry, p. 41-70. *In* R. Davies (ed.) Developments in Food Microbiology-1. Applied Science Publishers, London.
25. **Rantala, M., and E. Nurmi.** 1973. Prevention of the growth of *Salmonella infantis* in chicks by the flora of the alimentary tract of chickens. Br. Poult. Sci. **14:**627-630.
26. **Savage, D.C.** 1987. Factors influencing biocontrol of bacterial pathogens in the intestine. Food Technol. **41(7):**82-87.
27. **Schneitz, C.E., E. Seuna, and A. Rizzo.** 1981. The anaerobically cultured caecal flora of adult fowls that protect chickens from salmonella infections. Acta Pathol. Microbiol. Scand. Sec. B **89:**109-116.
28. **Seuna, E.** 1979. Sensitivity of young chickens to *Salmonella typhimurium* var. *copenhagen* and *S. infantis* infection and the preventive effect of cultured intestinal microflora. Avian Dis. **23:**392-400.
29. **Soerjadi, A.S., R. Rufner, G.H. Snoeyenbos, and O.M. Weinack.** 1982. Adherence of salmonellae and native gut microflora to the gastrointestinal mucosa of chicks. Avian Dis. **26:**576-584.
30. **Soerjadi, A.S., S.M. Stehman, G.H. Snoeyenbos, O.M. Weinack, and C.F. Smyser.** 1981. Some measurements of protection against paratyphoid *Salmonella* and *Escherichia coli* by competitive exclusion in chickens. Avian Dis. **25:**706-712.
31. **Sprung, K., and G. Leidy.** 1988. The use of bacterial interference to prevent infection. Can. J. Microbiol. **34:**332-338.
32. **Stavric, S., T.M. Gleeson, B. Blanchfield, and H. Pivnick.** 1985. Competitive exclusion of *Salmonella* from newly hatched chicks by mixtures of pure bacterial cultures isolated from fecal and cecal contents of adult birds. J. Food Prot. **48:**778-782.
33. **Stavric, S., T.M. Gleeson, B. Blanchfield, and H. Pivnick.** 1987. Role of adhering microflora in competitive exclusion of *Salmonella* from young chicks. J. Food Prot. **50:**928-932.
34. **Wierup, M., M. Wold-Troell, E. Nurmi, and M. Hakkinen.** 1988. Epidemiological evaluation of the salmonella-controlling effect of a nationwide use of a competitive exclusion culture in poultry. Poult. Sci. **67:**1026-1033.
35. **Wilson, K., L. Moore, M. Patel, and P. Permoad.** 1988. Suppression of potential pathogens by a defined colonic microflora. Microb. Ecol. Hlth. Dis. **1:**237-243.
36. **World Health Organization.** 1988. Salmonellosis control: the role of animal and product hygiene. Report of a WHO Expert Committee. Technical Report Series No. 774. World Health Organization, Geneva.

ALTERNATIVE ADMINISTRATION OF COMPETITIVE EXCLUSION TREATMENT

N. A. Cox
J. S. Bailey
L.C. Blankenship

Poultry Microbiological Safety Research Unit
USDA, ARS, Russell Research Center
Athens, Georgia 30613

ABSTRACT

Chicks were treated orally (OR) or intracloacally (IC) with competitive exclusion (CE) cultures (native intestinal microflora). Two days after treatment, all birds were orally challenged with *S. typhimurium* and no differences in the degree of protection were observed between OR and IC administration of CE. All treated birds were shown to be protected. Hatchery samples (egg fragments, swabs of belting material, paper pads, fluff) from three integrated broiler hatcheries, three primary breeder hatcheries, and one great, great, grandparent hatchery were tested for the presence of salmonellae. Salmonellae organisms were detected in 75, 91, and 67% of the samples from the integrated broiler hatcheries I, II, and III, respectively; 1.3, 5.0, and 22.5% of the samples from the primary breeder hatcheries I, II, and III, respectively; and 36% of the samples from the great, great, grandparent hatchery. Salmonellae contamination of the hatchery and hatchery environment limits the effectiveness of CE. Therefore, in an attempt to apply CE treatment before exposure to salmonellae, we introduced the treatment *in ovo* to the unhatched embryo. Undefined, anaerobically-cultured CE, diluted 1:1,000 or 1:1,000,000 was inoculated (0.1 ml) onto the air cell membrane or beneath the air cell membrane of an 18-day-old incubating hatching egg. Early experiments produced day-of-hatch chicks resistant to a 10^5 oral challenge of *S. typhimurium,* whereas untreated chicks were readily colonized. Subsequent experiments have produced variable and inconsistent results. At present we are refining the procedure of applying undefined CE *in ovo* to achieve significant and consistent protection of the new hatchling. Ultimately, we hope to be able to administer a limited number of defined organisms *in ovo* to effectively protect the chick from environmental sources of salmonellae without adversely affecting hatchability.

INTRODUCTION

Research studies in Finland in the early 1970's demonstrated that susceptibility of the newly hatched broiler chick to salmonellae colonization was probably due to the delayed establishment of normal gut microflora in chicks reared according to modern mass production methods (Nurmi and Rantala, 1973). They also demonstrated that salmonellae infections could be prevented by feeding anaerobic cultures of normal, intestinal, adult fowl flora to newly hatched chicks (Nurmi and Rantala, 1973; Rantala and Nurmi, 1973). This concept is known as competitive exclusion (CE) or the Nurmi concept. Since the early 1970's, the efficacy of the CE concept has been demonstrated in laboratories around the world (Snoeyenbos *et al.*, 1978; Barnes *et al.*, 1980; Pivnick *et al.*, 1981; Bailey *et al.*, 1988).

Although the efficacy of CE has been clearly demonstrated in the laboratory, the limited number of large-scale field trials which have been conducted have given mixed results (Huttner *et al.*, 1981; Weirup *et al.*, 1987; Goren *et al.*, 1988). At present, the CE approach has demonstrated the most potential compared to other intervention attempts to impact on salmonellae colonization of chickens. The objective of our present study was to explore some alternative means of applying CE to improve its effectiveness (e.g., intracloacal and *in ovo*).

COMPETITIVE EXCLUSION FLORA

Fecal and cecal droppings were collected from a one-year-old, salmonellae-free, caged layer and 0.5 g of these droppings were inoculated into 10 ml of VL broth anaerobically-incubated at 35° C for 48 hours, then 0.2 ml of the broth culture was transferred to another 10 ml tube of VL broth. After an additional 48 hours of anaerobic incubation at 35° C, glycerol was added to the broth at a final concentration of 15%. The mixture was divided into 1.0 ml aliquots and frozen at -70° C until needed. For each experiment, a vial of the frozen culture was thawed, and 0.2 ml of the culture inoculated into 10 ml VL broth and incubated for 48 hours at 35° C.

SALMONELLA CHALLENGE

Chicks were challenged at the appropriate time with a strain of *Salmonella typhimurium* that had previously been induced to be resistant to nalidixic acid.

Experiment 1. In a previous study (Cox *et al.*, 1990a), we demonstrated that approximately 100-fold fewer *S. typhimurium* cells were required to colonize young chicks by the intracloacal (IC) route as compared to oral (OR) gavage (Table I). We hypothesized that the low pH of the upper gastrointestinal tract contributes to the higher levels of *Salmonella* required to colonize chicks via the OR route. Therefore, the objective of this experiment was to determine if the hostile acidity would also adversely affect the protective microorganisms of the CE mixture.

Table I. Cecal colonization of one-day-old broiler chicks challenged orally (OR) or intracloacally (IC) with varying levels of *S. typhimurium*.

Route of Inoculation	Colonization Dose-50%
OR	$10^{2.4}$
IC	$10^{0.6}$

Broiler chicks were obtained from a local commercial hatchery on the day of hatch and transported within one hour to our rearing facility. Chicks were placed ten per cage in a wire-floored battery brooder (Petersime, Gettysburg, OH) equipped with heating elements and watering facilities. Unmedicated, crumbled, broiler starter feed and water were available *ad libitum*. Fecal droppings of each lot of birds were sampled shortly after the birds were placed in the brooders to ensure the absence of indigenous salmonellae. Within two hours after placing the chicks in the brooders, 0.2 ml of the CE was introduced by gavaging directly into the crop (OR) or cloacum (IC) using a sterile syringe and a 22-gauge needle with a small tygon catheter tube attached to the end of the needle.

Two days after treatment, the chicks were orally challenged by gavage with varying levels of cells of a nalidixic-acid-resistant strain of *S. typhimurium*. Five days post challenge, the birds were killed by CO_2 asphyxiation and soaked in 70% ethyl alcohol for one minute to reduce external bacterial contamination. Ceca were then aseptically excised from the chick and sampled according to a swab-plate method described by Bailey *et al.* (1988) for recovery of *S. typhimurium*. No differences were observed in the protectiveness of the CE introduced OR or IC.

Experiment 2. In a recent study (Cox *et al.*, 1990b), the majority of the samples obtained from several integrated broiler hatcheries were found to be contaminated with salmonellae (Table II). In addition, several primary breeder hatcheries (Table III) were also found to be contaminated and one great, great, grandparent hatch produced 36% salmonellae-positive samples (data not shown). Also, 40 randomly-selected and semiquantitatively analyzed hatchery samples were shown to contain greater than 10^3 salmonellae per sample. In a large-scale field trial study in Holland (Goren *et al.*, 1988) involving more than 8 million broilers, 7.6% of the 263 flocks were shown to be contaminated with salmonellae at the hatchery. We know that hatchery contamination limits the effectiveness of CE. Consequently, if a hatchery influence to negate the beneficial effects of a treatment can be demonstrated with only 2% of the hatchery samples being positive in the European study (Goren *et al.*, 1988), it seems reasonable to assume that CE would indeed have limited effectiveness in the United States in view of the present hatchery situation.

Experiment 3. In an attempt to circumvent the hatchery contamination and apply treatment prior to exposure of the newly hatched chick to salmonellae, a novel approach for introducing undefined competitive exclusion mixtures of bacteria to the incubating embryo inside of the egg was developed. The method involves the administration of CE, and dilutions thereof, to the air cell or beneath the air cell membrane of an 18-day-old incubating embryo by injecting 0.1 ml quantities of undefined CE culture (diluted either 1:1000 or 1:1,000,000) on Day 18 of incubation. Three days later, day-of-hatch chicks (treated and untreated) were orally challenged with *S. typhimurium*. Seven days after challenge, the birds were sacrificed and tested for the marker organism as described previously in Experiment 1.

Table II. Presence of salmonellae contamination in the integrated broiler hatchery.

Hatchery	Egg Fragments	Belting Material	Paperpads	Total
1	14/20[a]	17/20	9/13	40/53
2	18/20	18/20	4/4	40/44
3	18/30	21/30	13/18	52/78
Total	50/70	56/70	26/35	132/175

[a]Number of salmonellae-positive samples/number of samples tested.

Table III. Presence of salmonellae contamination in the primary breeder hatchery.

Hatchery	Egg Fragments	Fluff	Paperpads	Total
1	0/35[a]	0/20	1/20	1/75
2	3/20	0/20	0/20	3/60
3	3/20	1/20	14/40	18/80
TOTAL	6/75	1/60	15/80	22/215

[a]Number of salmonellae-positive samples/number of samples tested.

Table IV. Resistance of day-of-hatch chicks to a 10^6 oral challenge of *S. typhimurium* following *in ovo* injection of CE into the air cell and beneath the air cell membrane on the 18th day of incubation.

Treatment	Number of Chicks Colonized/Challenged
Untreated	11/12
AC(T)[1]	0/4
AC(M)[2]	4/5
AM(M)[3]	2/2

[1] Air cell was the site of injection and the 48 hr VL broth was diluted 1:1,000.
[2] Air cell was the site of injection and the 48 hr VL broth was diluted 1:1,000,000.
[3] The site of injection and the 48 hr VL broth was diluted 1:1,000,000.

Table V. Resistance of day-of-hatch to varying levels orof orally administered *S. typhimurium* following *in ovo* injection into the air cell of CE on the 18th day of incubation.

Treatment	Challenge Level			
	10^1	10^3	10^5	10^7
Untreated[3]	6/8[3]	7/8	4/4	6/6
AC(T)[1]	1/8	0/8	0/7	4/8
AC(M)[2]	0/8	4/6	7/8	7/8

[1] Air cell was the site of injection and the 48 hr VL broth was diluted 1:1,000.
[2] Air cell was the site of injection and the 48 hr VL broth was diluted 1:1,000,000.
[3] Number of chicks colonized/numberof chicks challenged.

In the first trial (Table IV) on a small number of chicks and with one challenge level of 10^6 *S. typhimurium* on day-of-hatch, the birds receiving the treatment of CE diluted 1:1,000 and injected into the air cell were resistant to colonization. In a second trial with a larger number of birds and varying challenge levels from 10^1 to 10^7, similar results were observed (Table V). The birds receiving CE diluted 1:1,000 injected into the air cell were all resistant to challenges of 10^3 and 10^5 and only 50% were colonized at 10^7. Three-fourths of the untreated birds were colonized with only ten (*S. typhimurium*) cells. Table VI shows the hatchability data when CE culture was administered *in ovo.* Only when CE was diluted 1:1,000 and 1:1,000,000 and injected into the air cell was the hatchability similar to commercial hatchability figures. No differences in livability and body weight were observed with CE diluted 1:1,000 or 1:1,000,000 and injected into the air cell (data not shown). In a third trial (Table VII), where only the air cell was treated with CE diluted 1:1,000 and 1:1,000,000, the data was less encouraging than previously observed.

Several more trials produced data similar to that shown in Table VII. At this point, we decided to study what was responsible for the inconsistency of the process. To begin with, we wanted to know the effect of atmospheric oxygen on our undefined CE culture, since the inoculum into the air cell may be exposed to oxygen many hours before ingestion by the embryo. The presence of oxygen was shown to have no effect on the effectiveness of the anaerobically-grown CE (Tables VIII and IX). We also wanted to know where the CE organisms injected into the air cell and beneath the air cell membrane would localize in the embryo and how consistently this would occur.

Table VI. Effect of *in ovo* CE culture administration on hatchability.

In ovo Treatment	Hatchability
Noninjected	96[3]
AC[1]-Undiluted	56
AC(T)	81
AC(M)	78
AM[2]-Undiluted	0
AM(T)	0
AM(M)	48

[1]AC-Injection site was onto the air cell membrane.
[2]AM-Injection site was beneath the air cell membrane
[3]Out of 100 fertile eggs set.

Table VII. Resistance of day-of-hatch chicks to varying levels of orally administered *S. typhimurium* following *in ovo* injection of CE into the air cell on the 18th day.

Treatment	Challenge Level		
	10^3	10^5	10^7
Untreated	6/8[3]	6/8	7/7
AC(T)[1]	4/8	4/6	5/6
AC(M)[2]	4/7	2/8	6/6

[1]Air cell injection and the 48 hr VL broth was diluted 1:1,000.
[2]Air cell injection and the 48 hr broth was diluted 1:1,000.000.
[3]Number of chicks colonized/number of chicks challenged.

Table VIII. The effect of atmospheric oxygen on the performance of anareobically cultured CE (undefined) administered by gavage to live chicks.

Treatment	Challenge Level	
	10^5	10^7
CE[1]	4/10[4]	3/10
CE + 24[2]	2/10	3/10
CE + 72[3]	1/10	7/10

[1]Anaerobically grown CE in VL broth at 35 °C for 48 hr.
[2]Same as #1 but exposed to air for 24 hr after anaerobic incubation.
[3]Same as #1 except that exposure time was 72 hr.
[4]Number of chicks colonized/number of chicks challenged.

Table IX. Effect of atmospheric oxygen during and following the incubation of undefined CE administered by gavage to live chicks.

Treatment	Number of Chicks Colonized/Challenged
48 AN[1]	0/10
48A[2]	9/10
48 AN-48A[3]	1/10

[1]CE anaerobically grown for 48 hr at 35° C in VL broth
[2]CE aerobically grown for 48 hr at 35° C in VL broth.
[3]CE anaerobically grown for 48 hr at 35° C in VL broth, then exposed to atmospheric oxygen for 48 hr before administering to chicks.

Table X. Level of *S. typhimurium* recovered from the intestinal tract of unhatched embryos 48 hours after *in ovo* administration into the air cell or beneath the air cell membrane of 18-day-old incubating eggs.

Treatment Site	No. of *S. typhimurium* Introduced	CQ[1]
AC	10^2	2.5
AC	10^4	5.2
AM	10^2	8.7
AM	10^4	8.7

[1]Colonization Quotient = Log_{10} per gram of cecal material of the geometric means of the colony froming units of the organism within that experimental group.

Table XI. Level of organisms recovered from the intestinal tract of unhatched embryos 24 hours after *in ovo* administration of undefined CE (diluted 1:1,000) into the air cell or beneath the air cell membrane of 18-day-old incubating eggs.

Treatment	Level Recovered				
	$< 10^2$	10^2	10^4	10^6	$> 10^6$
Control	4[1]	2	0	0	0
AC	2	6	5	2	1
AM	4	1	0	1	3

[1]Number of embryos from which this level of organisms were recovered from the intestinal tract.

Table XII. Level of organisms recovered from the intestinal tract of unhatched embryos 24 hours after *in ovo* administration of defined CE (diluted 1:1,000) into the air cell or beneath the air cell membrane of 18-day-old incubating eggs.

Treatment	Level Recovered				
	$< 10^2$	10^2	10^4	10^6	$> 10^6$
AC	3[1]	0	2	0	0
AM	0	0	0	0	5

[1]Number of embryos from which this level of organisma were recovere from the intestinal tract.

The first experiment was conducted with the marker strain of *S. typhimurium* for ease of recovery. We injected either 10^2 or 10^4 cells of *S. typhimurium* into the air cell or beneath the air cell membrane of 18-day-old incubating embryos. The presence and level of the marker in the intestinal tract of the unhatched embryo was determined 48 hours later on Day 20 (Table X). The organism reached the highest levels in the intestinal tracts of embryos receiving the injection beneath the air cell membrane. This experiment was then repeated using undefined CE, diluted 1:1,000 and injected into the air cells and beneath the air cell membrane of 18-day-old incubating embryos.

The results from these experiments are shown in Tables XI and XII. Air cell administration produces inconsistent attainment of adequate numbers of microorganisms in the intestinal tracts of the unhatched embryo. With administration beneath the air cell membrane, adequate numbers are consistently achieved; however, hatchability from these treatments has been somewhat disappointing.

SUMMARY

1.Intracloacal introduction of CE was not more effective than the oral route.

2. In commercial broiler industries, all levels of hatcheries in the breeding operation are reservoirs of salmonellae contamination and should be considered as main targets for an intervention method.

3. To circumvent hatchery contamination, *in ovo* introduction of CE to the 18-day-old incubating embryo was demonstrated to be a feasible route of administration.

4. Refinements are underway to improve consistency of air cell CE treatments and hatchability of treatments beneath the air cell membrane, so that adequate field trials can be conducted.

5. Research has been initiated to establish necessary chemical treatment protocols to prevent horizontal transmission of salmonellae from breeder hen to progeny.

ACKNOWLEDGMENTS

The authors wish thank D. Posey and L. Tanner for their technical assistance and J. Bandler and S. Smith for their secretarial assistance.

REFERENCES

1. **Bailey, J.S., L.C. Blankenship, N.J. Stern, N.A. Cox, and F. McHan.** 1988. Effect of anticoccidial and antimicrobial feed additives on prevention of *Salmonella* colonization of chicks treated with anaerobic cultures of chicken feces. Avian Dis. **32:**324-329.
2. **Barnes, E.M., C.S. Impey, and D.M. Cooper.** 1980. Competitive exclusion of salmonellae from newly hatched chicks. Vet. Rec. **106(3):**61.

3. **Cox, N.A., J.S. Bailey, L.C.Blankenship, R.J. Meinersmann, N.J. Stern, and F. McHan.** 1990a. Colonization dose 50% values for *Salmonella* administered orally and introcloacally to young broiler chicks. Poultry Sci. (In Press).
4. **Cox, N.A., J.S. Bailey, J.M. Mauldin, and L.C. Blankenship.** 1990b. Presence and impact of salmonellae contamination in the integrated broiler hatchery. Poultry Sci. (In Press).
5. **Goren, E., W.A. deJong, P. Doornenbal, N.M. Bolder, R.W.A.W. Mulder, and A. Jansen.** 1988. Reduction of *Salmonella* infection of broilers by spray application of intestinal microflora: a longitudinal study. Vet Quarterly **10(4)**:249-255.
6. **Huttner, B., H. Landgraf, and E. Vielitz.** 1981. Kontrolle der Salmonelleninfektionen in Mastelterntier-Bestanden durch Verabreichung von SPF-Darmflora on Eintagskuken. Dtsch. Tieraerztl. Wochenschr. **88**:527-532.
7. **Nurmi, E., and M. Rantala.** 1973. New aspects of *Salmonella* infection in broiler production. Nature **241**:210-211.
8. **Pivnick, H., B. Blanchfield, and J.Y. D'Aoust.** 1981. Prevention of *Salmonella* infection in chicks by treatment with fecal cultures from mature chickens (Nurmi cultures). J. Food Prot. **44**:909-916.
9. **Rantala, M., and E. Nurmi.** 1973. Prevention of the growth of *Salmonella infantis* in chickens by flora of the alimentary tract of chickens. Br. Poultry Sci. **14**:627-630.
10. **Snoeyenbos, G.H., O.M. Weinak, and C.F. Smyser.** 1978. Protecting chicks and poultry from salmonellae by oral administration of "normal" gut microflora. Avian Dis. **22**:273-287.
11. **Wierup, M., M. Wold-Troell, E. Nurmi, and M. Hakkinen.** 1987. Epidemiological evaluation of the *Salmonella* controlling effect of a nation-wide use of a "competitive exclusion" culture in poultry, p. 7. *In* Proceedings of an International Workshop on Competitive Exclusion of *Salmonella* from Poultry. Bristol Laboratory, Langford, UK.

DISCUSSION

N. STERN: Nelson,you had presented some data showing that the hatcheries are a primary source of salmonellosis and that these were not novel sources of salmonellae. You are proposing *in ovo* as a means to intervene in salmonellae colonization of the chick. The salmonellae infection of the egg may occur before Day 18 of hatching. What are some of the philosophies that would still propel this sort of research if some of the infections are occurring before the bird is hatched?

N. COX: The salmonellae can penetrate almost immediately after the hen lays the egg, and in those cases, penetrates, hangs around a while, and then proliferates around nutritious fluids surrounding the embryo. The chick can live in there and hatch out or be colonized with many salmonellae before our treatment would ever get there on Day 18. Obviously, no one thing is going to eliminate salmonellae. You could have a rapid chemical treatment of the eggs, and the CE *in ovo* might get the treatment there early enough to be more protective against environmental sources of salmonellae. Exposure can still occur even though you apply treatment *in ovo.*

W. JOCHLE: Dr. Mead, I'm referring to the undefined protective cecal flora preparation you have been using in your studies. How similar or dissimilar are they to those Nurmi used in Sweden and reported on last year in *Poultry Science*. This study involved several million chickens with competitive exclusion preparation giving similar results.

G. MEAD: Our donor material was selected on the basis of a standard assay which has now been published for efficacy against salmonellae colonization in chicks. We actually take it a little bit further than that because we introduce a higher challenge dose than is necessary for the assay. We only use donor material which will protect against a challenge of 10^6 *Salmonella*/chick. So we are deliberately choosing what we think is the best material that we can find according to that assay system. The product, Broilact, I can't comment on, but essentially, our material is different in the sense that for the experiments that I've described, in the field trials we would use donor birds provided by each company to protect their own stock. So, in that sense, we kept changing the material because that was necessary for those particular trials. I suppose you could say that generically the two types of approaches are the same in that undefined material was used without a mixture of defined organisms.

W. JOCHLE: Professor Nurmi, the paper that you and your co-workers published last year in *Poultry Science* refers to Broilact, which is now industrially produced in several European markets and marketed specifically in your home country. Can you give us an idea of how successful this is when used on a large scale as we have heard? Specifically, does it have an effect on *Salmonella* only or does it affect the *Campylobacter* picture as well?

E. NURMI: As you know, the product has been used in Finland for several years. The product today is different than what we started with in the early eighties and middle seventies. It is manipulated more and better controlled. For instance, one company in Finland used it during the period 1986-1989. They treated about 400 flocks with approximately four million birds, and these groups had an infection contamination of about 70-80%, compared to 25% in the untreated group. Everything was similar except, of course, they chose the flocks that had been infected earlier and the control flocks which had never had *Salmonella.* We have also done a lot of experiments on *Campylobacter*, and we have very promising results against *Campylobacter*. That flora is more specific and defined and is not the same as Broilact. The flora in Broilact is not very protective against *Campylobacter*. You will need a different flora. We have been following the use of Broilact by these companies for three years. About 70% of Finnish companies use Broilact today. In Sweden, if you have salmonellae in your flocks, it is obligatory to use Broilact after infection is detected. Sweden is the only country that says they do not have salmonellae in their broilers.

R. GUSTAFSON: Geoff, you mentioned that you had tried probiotic products with negative results. Earlier today, Dr. Rolfe mentioned that he had seen optimistic results with the SF68 strain of *Streptococcus faecium*, and I wondered if you had tried that particular organism.

G. MEAD: No we haven't tried that specific organism. We've done some preliminary trials with commercially available products according to the assay system that has just been published. Of course, we were following the manufacturer's instructions. This was a short term trial with an assay designed to select the best protective material. We don't follow the use of these products thoughout the life of the flock. So we have only tried out these products in our assay system, but we didn't find any commercial products that we tried to be useful in this respect. But, we have not tried the specific organism that you mentioned.

R. GUSTAFSON: You showed a slide on the effect of the organism used on CFU's/g of cecum. Was that geometric mean of batch samples from a group of chickens or one chicken?

G. MEAD: I'm sorry if I didn't make that clear. We do express our results as geometric means from groups of chickens, usually at least five birds.

A. MORRIS-HOOKE: Earlier you mentioned challenge organisms growing in the crop, and you mentioned it was growing like a weed. I was really impressed by the growth. There seemed to be a rather long lag period and then the challenge organism took off. It looked like there was about a three log increase in about two hours, which translates into a mean generation time of about 15 minutes. I wondered if you wanted to comment. Do you have any ideas on what's going on during the long lag period and are there any bactericidal mechanisms in the crop?

G. MEAD: Yes, it was a slight surprise as to how good microbial growth was in the feed, and I think more than the salmonellae it was the lactobacilli which, after all, are more demanding organisms that grew extremely well, and I think, very clearly, are the key to controlling salmonellae in the crop. I can't comment much further on what we did because it was a very simple model and physiologically the crop is relatively simple. I felt that Fuller's crop model was relevant to look at the behavior of salmonellae and the interaction with the protective flora. It does seem, as you say, that there is a fairly lengthy lag and then the salmonellae progressively decline in numbers. By analogy with Fuller's findings, this would appear to be principally a pH affect. There is a fermentation induced by lactobacilli which lowers the pH to a point where salmonellae do not manage to survive very well. There are conceivably other effects going on as well, and I don't think we got as far as sorting all that out.

G. SNOEYENBOS: Dr. Mead, would you care to comment on the criteria used for selection of the antibiotic to be used for the first 12 days prior to movement and treatment with CE? Another related question stems from the lack of uniformity among different results reported in disruption of protective microflora by different antibacterials, which has suggested that there may be differences among the protective microflora in their antibiotic sensitivity.

G. MEAD: We, of course, have to take the advice of the company veterinarian in each case. The actual control of the health status of the flocks is the responsibility of that person and he decided the broad spectrum antibiotic treatment that would be used. The antibiotics that were employed were chlorotetracycline, neomycin, and furazolidone. Very clearly you can't use furazolidone for turkeys, so in that particular case neomycin and chlorotetracycline in combination were used.

G. MEAD: The literature does suggest that a fairly wide range of antimicrobials can be used in feed without disrupting the protective effect, and yet we know that some of them have quite a broad spectrum of antimicrobial activity. For example, if you take monensin, which in reported experiments has a broad spectrum of activity, you can use it with no problems in the feed used in these sorts of trials. I don't think I really quite understand why there is this difference between what you can see *in vitro* and what appears to happen when you use the products for trial purposes. It does at least suggest that this needn't cause us any problem. We can easily choose antimicrobial strategies for growth promotion, control of coccidia and so on, which will not interfere with the treatment. So, I feel fairly confident myself that antimicrobials shouldn't present a problem, but clearly it is necessary to be aware of the particular antimicrobials being used.

E. NURMI: I think we don't know enough about what the anitmicrobials do in the gut. They may be beneficial for, or they may have a bad influence upon the colonization of *Campylobacter*. We lack knowledge about the effect of antimicrobials in feed on the colonization by pathogens.

M. OPITZ: When we want to use CE for protection of laying or breeder flocks we need a very long lasting effect. How stable is the colonization of the competitive flora, especially in view of viral infections which frequently occur, such as rotavirus or infectious bursal disease, or other enteritis-causing factors such as feed changing, stress, etc.?

G. MEAD: There are some cases in which we have kept birds for relatively long periods, not perhaps as long as a laying fowl would be kept under commercial circumstances. I have observed protection to remain for at least six weeks and others have made similar observations. One of the things we didn't appreciate at the time of trials on breeder birds was the behavior of *S. enteritidis* with respect to laying fowl and what we know now about the ability of this particular organism. The birds can suddenly stop shedding and organisms can become invasive and you can't find *S. enteritidis* unless you kill the birds and examine them fully. There could be complicating factors of that kind, but my feeling is that if you can stop the colonization at a time when birds are most sensitive (the young stage), it has to have some benefit in preventing the spread of salmonellae.

S. STAVRIC: Dr. Mead, have you used defined mixtures in combination with antibiotics?

G. MEAD: No, we haven't done that.

S. STAVRIC: Dr. Cox, could you elaborate more on the defined treatment that you used *in ovo*?

N. COX: If someone could come up with a unique way to screen, it might make the production of a defined culture easier. We cannot give any details of the defined culture right now, because we are trying to determine if there are any patent rights.

S. STAVRIC: Did you use same mixture in treating chicks by gavage?

N. COX: No, not yet.

S. STAVRIC: Was it a mixture of many organisms?

N. COX: No, not a large number.

E. NURMI: We will bring to a close the competitive exclusion session. There was a WHO meeting in Geneva this year. They stated an objective of salmonellae-free poultry production. It might be quite difficult, but not impossible. The goal is to reduce contamination by salmonellae and *Campylobacter*. The report called for grandparents (primary breeders) free of salmonellae within one year. This surprises me, because it means that you accept the presence of salmonellae in great grandparents (pedigree) flocks. Then it would be possible to give them salmonellae-free feed and to keep them free of salmonellae in that way. This approach works in Scandinavia and in grandparent flocks in some countries. The next objective was to have parent flocks (multiplier breeders) free of salmonellae, because we have to work with the whole chain to reduce the salmonellae that are present today. In conclusion, we have much research to do to achieve these goals.

SESSION III: MECHANISMS OF COLONIZATION

Convener: *Norman J. Stern*

USDA, ARS, Russell Research Center
Athens, Georgia

Mechanisms of colonization may mean different things to different people. I divide the concept into two parts for your consideration. The first part includes the array of tools which assist the bacterium in exploiting a specific niche within the host animal. Such an array of determinants can include specific biomolecules enabling the organism to adhere to gut epithelial tissue, specialized means for motility, adaptive utilization of substrates within the GI tract, unique morphology evolved for the GI tract niche, chemotactic response toward specific GI tract constituents, invasiveness, entertoxin production, and the ability to resist the cidal host immune response-such as intracellular survival within phagocytic cells.

The second part of my definition for mechanisms for colonization considers the host portion of the host-bacterium relation. This can vary among hosts, and could lead to identifying specific research thrusts providing resistance to colonization by human bacterial enteropathogens. Variation could exist in numbers of phagocytes, killing ability of the phagocyte, immune status of the host relative to cell mediated and secretory immunoglobulin expressions, expression of macromolecules on the epithelium, expression of macromolecules on immunologically involved cells and within the mucin layer, the gastric acidity, toxicity of the bile acids, the makeup of the intestinal flora, and the nutrient availability for the bacterium.

One example where the host controls the potentials for colonization can be seen with pigs and colonization by *Escherichia coli*, which is dependent on the organism colonizing by means of a fimbria adhering to a receptor on the surface the enterocytes. A single gene within the pig encodes for this receptor, and therefore, pigs resistant to *E. coli* can be bred. Another example where we have observed success, has been in breeding chickens resistant to Marek's disease. The immunological status of the animals will be covered in the following session.

COLONIZATION OF CHICKS BY *CAMPYLOBACTER JEJUNI*

M. P. Doyle

Food Research Institute
University of Wisconsin-Madison
Madison, Wisconsin 53706

ABSTRACT

Poultry is a common reservoir of *C. jejuni* and a leading vehicle of human cases of *Campylobacter* enteritis. The organism colonizes primarily the lower gastrointestinal tract of chicks, principally the ceca, large intestine, and cloaca where densely packed cells of *C. jejuni* localize in mucus within crypts. Based on electron microscopic examination, the campylobacters appear to freely pervade the lumina of crypts without attachment to crypt microvilli. Studies of the chemotactic behavior of *C. jejuni* revealed the organism is chemoattracted to mucin and L-fucose, an important constituent of mucin. Additionally, *C. jejuni* can utilize mucin as a sole substrate for growth. It is suggested that the chemoattraction of *C. jejuni* to mucin may attract the organism to mucus, in which it moves by its highly active flagellum to mucus-filled crypts where the organism establishes itself. Campylobacters likely remain in the crypt because of their attraction to and metabolism of mucin.

Colonization Control of Human Bacterial Enteropathogens in Poultry

INTRODUCTION

A wide variety of animals both wild and domestic have been identified as reservoirs of *Campylobacter jejuni* (40). Typically, these animals appear healthy with no symptoms of illness even though large numbers (10^4 to 10^7 CFU/g) of *C. jejuni* are often excreted in their feces (1, 11, 23, 31, 42). Epidemiologic investigations of outbreaks and sporadic cases of *Campylobacter* enteritis among humans have identified poultry as a principal vehicle of illness (5, 8, 12, 14, 15, 17, 27, 29, 34, 36) and as an important reservoir of the serotypes of *C. jejuni* that are often involved in human infection (2, 18, 19, 30, 33). Many surveys have been done to determine the rate of intestinal carriage of *C. jejuni* by poultry; most studies revealed a carriage rate between 30% and 100% (11, 13, 31, 32, 35, 42). A major step toward reducing the incidence of *Campylobacter* enteritis is to reduce the level and prevalence of *C. jejuni* carriage by poultry. Defining the mechanism of *Campylobacter* colonization of poultry may provide critical information for the development of practical means to eliminate *C. jejuni* from poultry.

Sites of *C. jejuni* Localization in Poultry. Peroral inoculation studies by several different investigators have revealed that *C. jejuni* can readily colonize chicks (3, 20, 35, 38, 39, 41). Several factors influence the organism's ability to colonize chicks, including the number of cells and strain of *C. jejuni,* and the lineage and age of chicks (20, 39, 41). Microbiological analysis of chick tissue specimens at different times postinoculation revealed the ceca and cloaca are the principal sites of *Campylobacter* colonization (3, 20, 38). Results of a study by Beery *et al.* (3) illustrate the preferential localization of *C. jejuni* in the chick's lower gastrointestinal tract, especially in the ceca, where the organism was often detected at levels of 10^4 to 10^7 cells/g. The organism was recovered most frequently from the ceca (20 of 36 chicks, 55.6%), distal small intestine (14 of 36 chicks, 38.9%), and large intestine plus cloaca (5 of 14 chicks, 35.7%), but only twice from the spleen, once from the gallbladder and blood (obtained by cardiac puncture), and not from the liver. By histologic examination the investigators detected *C. jejuni* in 17 cecal and 11 cloacal plus large intestinal specimens from 18 chicks inoculated with *C. jejuni* and examined seven days postinoculation.

A detailed histologic study of the ceca and cloaca of chicks perorally inoculated with *C. jejuni* revealed campylobacters were localized in the lumen of mucus-filled crypts (Figs. 1, 2). *C. jejuni* appeared to be confined to the lumen of the crypt and there was no evidence of pathological change.

Between four and 35% of cecal crypts were filled with densely packed *C. jejuni* cells. Campylobacters generally filled the crypts from the proximal end to the distal end but were most densely concentrated at the distal end.

Examination of *C. jejuni*-colonized cecal crypts by electron microscopy revealed that the campylobacters pervaded the lumina of crypts and were occasionally in close opposition to the glandular microvilli, but were never in direct contact with the microvillus outer membrane (Fig. 2). It appeared that campylobacters colonized crypt mucus without attaching to crypt microvilli. These observations are in general agreement with those made by Lee *et a*l. (21) who studied the colonization of gnotobiotic mice by *C. jejuni.* They reported that *C. jejuni* colonizes mucus on the outer surface and deep within the intestinal crypts, with the cecal crypts being preferentially colonized. They observed that campylobacters do not adhere to the intestinal surface but are highly motile and track rapidly along intestinal mucus.

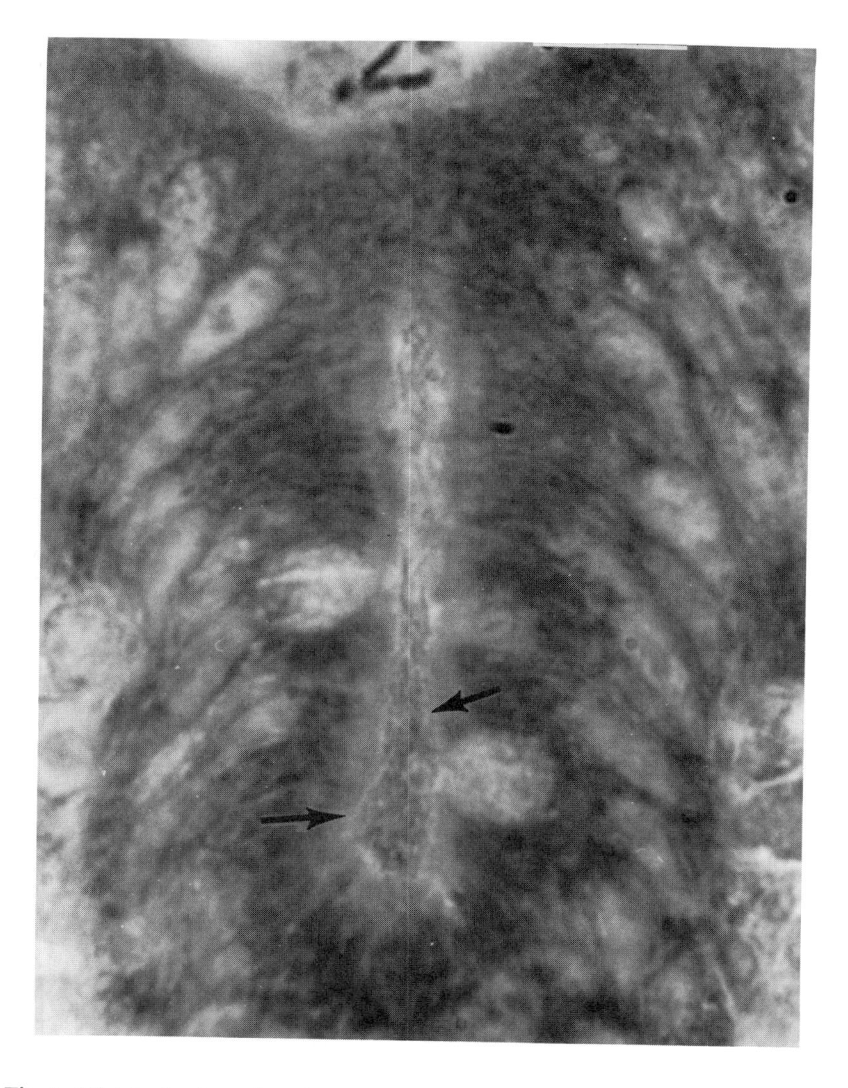

Fig. 1. Histologic view of section of proximal cecum from eight-day-old chick perorally inoculated with *C. jejuni*. Cecal gland contains densely packed cells of *C. jejuni* in the deep lumen of the crypt (arrows). Bar, 10 μm. (Ref. 3).

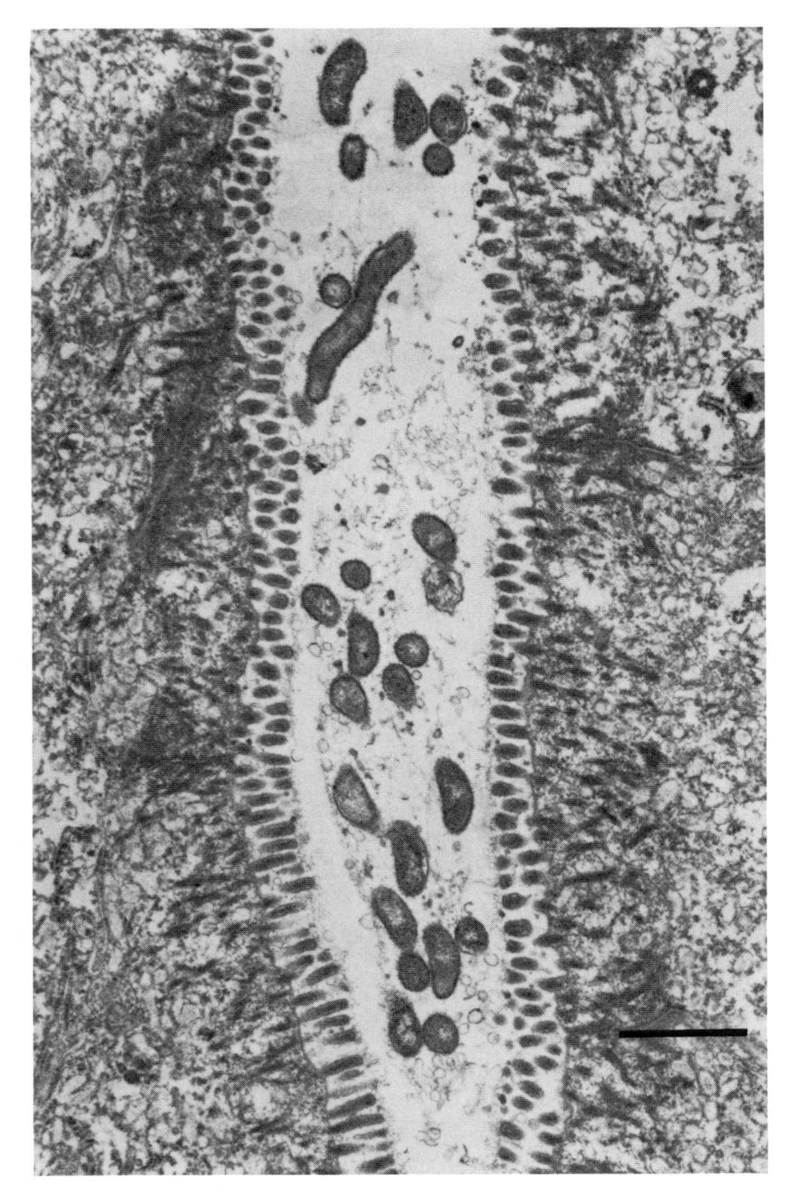

Fig. 2. Electron micrograph of *C. jejuni*-dominated proximal cecal crypt from an eight-day-old chick perorally inoculated with *C. jejuni*. The campylobacters appear to be freely associated and do not attach to cecal glandular microvilli or their glycocalyx. Bar, 2μm. (Reprinted from Beery *et al.*, Appl. Environ. Microbiol. **54**:2365-2370, 1988).

Mechanism of Intestinal Colonization by *C. jejuni*. Several approaches, both *in vitro* and *in vivo,* have been taken to define the mechanism by which *C. jejuni* colonizes the intestine. For example, many investigators have used tissue culture cells to determine if *C. jejuni* has adhesive properties and can attach to cells originally derived from intestinal tissue. Several studies have demonstrated adherence by the organism to INT 407 human intestinal epithelial cells (7, 24, 25).

Although *C. jejuni* appears to lack fimbriae, which are important colonization factors for many bacteria including enterotoxigenic *Escherichia coli,* some investigators have suggested that the organism may possess other adhesins. Studies with INT 407 cells revealed that adherence of *C. jejuni* is inhibited by L-fucose and certain other carbohydrates such mannose (7, 25). These findings led Cinco *et al.* (7) to suggest that L-fucose may be a receptor for binding a *C. jejuni* adhesin.

Newell *et al.* (28) studied the role of flagella of *C. jejuni* in the organism's ability to colonize infant mice. The investigators were able to obtain two variants of a flagellated, motile clinical isolate of *C. jejuni.* They included: (a) a flagellate but nonmotile strain and (b) an aflagellate strain. Mice were inoculated either intragastrically or orally with the organisms and, after about one week postinoculation, sections of intestine were homogenized and plated to enumerate campylobacters. They found that the flagellate, nonmotile variant colonized the gastrointestinal tract as successfully as the wild-type strain; however, the aflagellate variant poorly colonized the mice. The investigators further determined that the flagellate, nonmotile strain attached more efficiently to human epithelial tissue culture cells than either the wild-type strain or the aflagellate strain. They concluded that these differences in attachment and colonization suggest there is an adhesin intimately associated with the flagella of *C. jejuni,* and that flagella, whether active or inactive, are necessary for the efficient colonization of the gastrointestinal tract of infant mice by *C. jejuni.*

McSweegan and Walker (25) also determined that flagella were an important adhesin of *C. jejuni.* Removing flagella from *C. jejuni* reduced the organism's ability to adhere to INT 407 cells or intestinal mucus, whereas immobilizing the flagella with KCN increased adhesion. Purified flagella was also observed to adhere to INT 407 cells and mucus. They also identified lipopolysaccharide (LPS) as a second adhesin of *C. jejuni.* LPS specifically bound to INT 407 cells and mucus but not after the LPS was treated with periodate. Using an experimental approach similar to the infant mouse model used by Newell *et al.* (28), Morooka *et al.* (26) examined intestinal colonization of suckling mice by a wild-type strain of *C. jejuni* and several nonmotile mutant strains. They found that the wild-type strain colonized the intestinal tract two days after inoculation, whereas all of the nonmotile strains (with or without flagella) were cleared from the intestinal tract within two days postinoculation. The investigators concluded that motility of *C. jejuni* is an important factor in the colonization of the intestinal tract of mice.

Its spiral shape and corkscrew motility confer an advantage to *C. jejuni* in viscous environments such as mucus. Ferrero and Lee (9) observed that compared with conventional rod-shaped bacteria such as *Salmonella* or *Escherichia coli, C. jejuni* is better suited for mobility in viscous solutions (such as mucus). They suggest that *C. jejuni* has a motility suited to movement in a viscous environment, and that this ability may provide the organism with an ecological advantage when in intestinal mucus.

The significance of an adhesin having a role in the colonization of the intestinal tract by *C. jejuni* was brought into question by Lee *et al.* (21) after an elegant study of the colonization of germfree mice by *C. jejuni.* The investigators intragastrically introduced bicarbonate solution and human isolates of *C. jejuni* into germfree adult mice. They observed that the organism colonized the mucus deep within the intestinal crypts of the animals. The cecal crypts were preferentially colonized and, by using a combination of microscopic techniques, including videotaping of fresh wet preparations of intestinal mucosae, they determined that campylobacters do not adhere to the surface of intestinal tissue. Rather, the campylobacters were seen to be highly motile in the intestinal tract, rapidly tracking along the intestinal mucus. Hence, it was concluded that the ability to colonize intestinal mucus is a major determinant of pathogenicity in intestinal *Campylobacter* infections. The investigators further suggested that the possession of specific adhesins is not likely to be a significant determinant of pathogenicity. If campylobacters indeed colonize mucus but do not adhere to the surface of crypts as proposed by Lee *et al.* (21) and as observed in chicks by Beery *et al.* (3), then what mechanism promotes the attraction and retention of campylobacters in the mucus? Freter *et al.* (10), in studying *Vibrio cholerae,* determined that chemotaxis, particularly toward mucus, is an important mechanism in the association of this pathogen with mammalian hosts. Following this observation, Hugdahl *et al.* (16) studied the chemotactic behavior (i.e., the movement of an organism toward or away from a chemical stimulus) of *C. jejuni* to determine if chemotaxis might be a factor in the colonization potential of the organism.

The chemotactic behavior of two strains of *C. jejuni* was determined in the presence of different carbohydrates, amino acids, and preparations and constituents of mucin and bile. More than twenty carbohydrates, representing a variety of pentoses, hexoses, amino sugars, and deoxy sugars, were assayed and, of these, only L-fucose was a chemoattractant. Interestingly, neither D-fucose nor glactose, which is a fucose analogue, were chemoattractants. Of sixteen amino acids and analogues evaluated, only four, i.e., L-cystiene, L-glutamate, L-aspartate, and L-serine, were chemoattractants.

Of relevance to mucus colonization, *C. jejuni* was chemoattracted to mucin. Mucin, which is a glycoprotein of high molecular weight, is the principal constituent of mucus. Mucins, which are excreted from epithelial cells of intestinal, gastric, and gallbladder tissue, have a similar structure consisting of an extended protein core with attached oligosaccharides. The oligosaccharides attach to the protein core via serine and threonine ether linkages. The carbohydrate constituents are sialic acid, i.e., substituted neuraminic acid, or L-fucose or both attached to N-acetylglucosamine, N-acetylgalactosamine, and galactose. Mucin is principally composed of carbohydrate molecules, and L-fucose is invariably a terminal sugar.

The protein core of mucin primarily contains serine, threonine, and proline, characteristically in large amounts. However, the protein core is primarily sheathed by the oligosaccharide units. Chemotaxis toward serine or other amino acid components of mucin would be less likely than chemotaxis toward the peripheral L-fucose. Therefore, it is likely that the L-fucose moiety is the principal chemoattractant in mucin.

Some investigators have reported that *C. jejuni* may localize in the gallbladder and survive for long periods of time in the bile of healthy hosts (4, 6, 22, 37). Hence, studies were done to determine if *C. jejuni* was chemoattracted to bile. Oxgall, fresh chick bile, and fresh beef bile were all determined to be chemoattractants; however, many of the individual components of bile were chemorepellents, including cholic acid, deoxycholic acid, taurocholic acid, glycocholic acid, and the other acids that are prominent in bile. The only component of bile determined to be a chemoattractant was mucin. Beef bile was then fractionated by column chromatography and the mucin component purified. The purified mucin and the pooled, concentrated fractions of bile constituents without mucin were assayed for chemotactic activity. Mucin was determined to be a chemoattractant, whereas the pooled fractions of bile constituents without mucin were chemorepellents. When mucin was separated from bile, the chemoattractive activity of bile was lost. Hence, it appears that the chemoattractant activity of the mucin component of bile prevails over the chemorepellent activity of the remaining bile constituents, indicating that the mucin component is the influential chemoattractant in bile. In addition to its chemoattractant activity, mucin can also serve as a sole substrate for growth of *C. jejuni* (3, 16). Considering the observations described above, Beery *et al.* (3) proposed that *C. jejuni* colonization of chicks occurs primarily in the lower gastrointestinal tract where the organism localizes principally in cecal and cloacal crypts. They further hypothesized that the chemoattraction of *C. jejuni* to mucin attracts the organism to mucus, in which it moves by a highly active flagellum to mucus-filled crypts where the organism establishes itself. Within the crypts, *C. jejuni* probably grows by using mucin as a substrate. The organism is likely to remain established in the crypt because of its attraction to and metabolism of mucin. *C. jejuni* does not appear to attach to the crypt; adherence to the crypt is likely not important for colonization of the ceca and cloaca.

REFERENCES

1. **Altmeyer, V.M., P. Krabisch, and P. Dorn.** 1985. Zum Vorkommen und zur Verbreitung von *Campylobacter jejuni/coli* in der Jungmastgeflugel Produktion. 1. Mitt. Dfsch. Tierarztl. Wsehr. **92:**456-459.
2. **Banffer, J.R.J.** 1985. Biotypes and serotypes of *Campylobacter jejuni* and *Campylobacter coli* strains isolated from patients, pigs and chickens in the region of Rotterdam. J. Infect. **10:**277-281.
3. **Beery, J.T., M.B. Hugdahl, and M.P. Doyle.** 1988. Colonization of gastrointestinal tracts of chicks by *Campylobacter jejuni.* Appl. Environ. Microbiol. **54:**2365-2370.
4. **Blaser, M.J., H.L. Hardesty, B. Powers, and W.-L.L. Wang.** 1980. Survival of *Campylobacter fetus* subsp. *jejuni* in biological milieus. J. Clin. Microbiol. **11:**309-313.
5. **Brouwer, R., M.J. Mertens, T.H. Siem, and J. Katchaki.** 1979. An explosive outbreak of *Campylobacter* enteritis in soldiers. Antonie van Leeuwenhoek. J. Microbiol. **45:**517-519.
6. **Bryner, J. H., P.A. O'Brien, C. Estes, and J.W. Foley.** 1972. Studies of vibrios from gallblader of market sheep and cattle. Am. J. Vet. Res. **33:**1439-1444.
7. **Cinco, M., E. Banfi, E. Ruaro, P. Crevatin, and D. Crotti.** 1984. Evidence for L-fucose (6-deoxy-L-galactopyranose)-mediated adherence of *Campylobacter* spp. to epithelial cells. FEMS Microbiol. Lett. **21:**347-351.
8. **Deming, M.S., R.V. Tauxe, P.A. Blake, S.E. Dixon, B.S. Fowler, T.S. Jones, E.A. Lockamy, C.M. Patton, and R.0. Sikes.** 1987. *Campylobacter* enteritis at a university: transmission from eating chicken and from cats. Am. J. Epidemiol. **126:**526-534.

9. **Ferrero, R.L., and A. Lee.** 1988. Motility of *Campylobacter jejuni* in a viscous environment: comparison with conventional rod-shaped bacteria. J. Gen. Microbiol. **134:**53-59.
10. **Freter, R., P.C.M. O'Brien, and M.S. Macsai.** 1979. Effect of chemotaxis on the interaction of cholera vibrios with intestinal mucosa. Am. J. Clin. Nutr. **32:**128-132.
11. **Grant, I.H., N.J. Richardson, and V.D. Bokkenheuser.** 1980. Broiler chickens as a potential source of *Campylobacter* infections in humans. J. Clin. Microbiol. **11:**500-510.
12. **Harris, N.V., N.S. Weiss, and C.M. Nolan.** 1986. The role of poultry and meats in the etiology of *Campylobacter jejuni/coli* enteritis. Am. J. Public Health **76:**407-411.
13. **Hartog, B.J., G.J.A. deWilde, and E. deBoer.** 1983. Poultry as a source of *Campylobacter jejuni.* Arch. Lebensmittelhyg. **34:**109-122.
14. **Hopkins, R.S., and A.S. Scott.** 1983. Handling raw chicken as a source forsporadic *Campylobacter jejuni* infections. J. Infect. Dis. **148:**770.
15. **Hopkins, R.S., R. Olmsted, and G.R. Istre.** 1984. Endemic *Campylobacter jejuni* infection in Colorado: identified risk factors. Am. J. Public Health **74:**249-250.
16. **Hugdahl, M.B., J.T. Beery, and M.P. Doyle.** 1988. Chemotactic behavior of *Campylobacter jejuni.* Infect. Immun. **56:**1560-1566.
17. **Istre, G.R., M.J. Blaser, P. Shillam, and R.S. Hopkins.** 1984. *Campylobacter* enteritis associated with undercooked barbecued chicken. Am. J. Public Health **74:**1265-1267.
18. **Jones, D.M., J.D. Abbott, M.J. Painter, and E.M. Sutcliffe.** 1984. A comparison of biotypes and serotypes of *Campylobacter* sp.isolated from patients with enteritis and from animal and environmental sources. J. Infect. **9:**51-58.
19. **Juven, B.J., and M. Rogol.** 1986. Incidence of *Campylobacter jejuni* and *Campylobacter coli* serotypes in a chicken processing factory. J. Food Prot. **49:**290-292.
20. **Kai no, K., H. Hayashi dani, K. Kaneko, and M .Ogawa.** 1988. Intestinal colonization of *Campylobacter jejuni* in chickens. Jpn. J. Vet. Sci. **50:**489-494.
21. **Lee, A., J.L. O'Rourke, P.J. Barrington, and T.J. Trust.** 1986. Mucus colonization as a determinant of pathogenicity in intestinal infection by *Campylobacter jejuni:* a mouse cecal model. Infect. Immun. **51:**536-546.
22. **Luechtefeld, N.W., and W.-L.L. Wang.** 1982. Animal reservoirs of *C.jejuni,* p.249-252. *In* D.G. Newell (ed.), Campylobacter - epidemiology, pathogenesis and biochemistry. MTP Press Ltd., Lancaster, England.
23. **Luechtefeld, N.W., W.-L.L. Wang, M.J. Blaser, and L.B. Reller.** 1981. Evaluation of transport and storage techniques for isolation of *Campylobacter fetus* subsp. *jejuni* from turkey cecal specimens. J. Clin. Microbiol. **13:**438-443.
24. **McBride, H., and D.G. Newell.** 1983. *In vitro* models of adhesion for *Campylobacter jejuni,* p.110. *In* A.D. Pearson, M.B. Skirrow, B. Rowe, J.R. Davies, and D.M. Jones (ed.), Campylobacter II. Public Health Laboratory Service, London.
25. **McSweegan, E., and R.I. Walker.** 1986. Identification and characterization of two *Campylobacter jejuni* adhesins for cellular and mucus substrates. Infect. Immun. **53:**141-148 .
26. **Morooka, T., A. Umeda, and K. Amako.** 1985. Motility as an intestinal colonization factor for *Campylobacter jejuni.* J. Gen. Microbiol. **131:**1973-1980.

27. **Mouton, R.P., J.J. Veltkamp, S. Lauwers, and J.P. Butzler.** 1982. Analysis of a small outbreak of campylobacter infections with high morbidity, p. 129-134. *In* D.G.Newell (ed.), Campylobacter - epidemiology, pathogenesis and biochemistry. MTP Press Ltd., Lancaster, England.
28. **Newell, D.G., H. McBride, and J.M. Dolby.** 1985. Investigations on the role of flagella in the colonization of infant mice with *Campylobacter jejuni* and attachment of *Campylobacter jejuni* to human epithelial cell lines. J. Hyg. **95:**217-227.
29. **Norkrans, G., and A. Svedhem.** 1982. Epidemiological aspects of *Campylobacter jejuni* enteritis. J. Hyg. **89:**163-170.
30. **Oosterom, J., J.R.J. Banffer, S. Lauwers, and A.E. Busschbach.** 1985. Serotyping of and hippurate hydrolysis by *Campylobacter jejuni* isolates from human patients, poultry and pigs in the Netherlands. Antonie van Leeuwenhoek J. Microbiol. **51:**65-70.
31. **Oosterom, J., S. Notermans, H. Karman, and G.B. Engels.** 1983. Origin and prevalence of *Campylobacter jejuni* in poultry processing. J. Food Prot. **46:**339-344.
32. **Prescott, J.F., and C.W. Bruin-Mosch.** 1981. Carriage of *Campylobacter jejuni* in healthy and diarrheic animals. Am. J. Vet. Res. **42:**164-165.
33. **Rogol, M., I. Sechter, Z. Greenberg, R. Mizrachi, Y. Shtark, and S. Alfi.** 1985. Contamination of chicken meat and environment with various serogroups of *Campylobacter jejuni/coli.* Int. J. Food Microbiol. **1:**271-276.
34. **Rosenfield, J.A., G.J. Arnold, G.R. Savey, R.S. Archer, and W.H. Woods.** 1985. Serotyping of *Campylobacter jejuni* from an outbreak of enteritis implicating chicken. J. Infect. **11:**159-165.
35. **Shanker, S., A. Lee, and T.C. Sorrell.** 1986. *Campylobacter jejuni* in broilers: the role of vertical transmission. J. Hyg. **96:**153-159.
36. **Skirrow, M.B., R.G. Fidoe, and D.M. Hones.** 1981. An outbreak of presumptive food-borne campylobacter enteritis. J. Infect. **3:**234-236.
37. **Smibert, R.M.** 1969. *Vibrio fetus* var. *intestinalis* isolated from the intestinal content of birds. Am. J. Vet. Res. **30:**1437-1442.
38. **Soerjadi, A.S., G.H. Snoeyenbos, and 0.M. Weinack.** 1982. Intestinal colonization and competitive exclusion of *Campylobacter fetus* subsp. *jejuni* in young chicks. Avian. Dis. **26:**520-524.
39. **Soerjadi-Liem, A.S., G.H. Snoeyenbos, and 0.M. Weinack.** 1984. Comparative studies on competitive exclusion of three isolates of *Campylobacter fetus* subsp. *jejuni* in chickens by native gut microflora. Avian. Dis. **28:**139-146.
40. **Stern, N.J., and S.U. Kazmi.** 1989. *Campylobacter.* p.71-110. *In* M.P. Doyle (ed.), Foodborne bacterial pathogens. Marcel Dekker, Inc., New York.
41. **Stern, N.J., J.S. Bailey, and L.C. Blankenship.** 1988. Colonization characteristics of *Campylobacter jejuni* in chick ceca. Avian Dis. **32:**330-334.
42. **Wempe, J.M., C.A. Genigeorgis, T.B. Farver, and H.I. Yusufri.** 1983. Prevalence of *Campylobacter jejuni* in two California chicken processing plants. Appl. Environ. Microbiol. **45:**355-359.

DISCUSSION

C. GYLES: It wasn't clear to me how you distinguished between chemotaxis and inhibition of growth by the various substances?

M. DOYLE: We took these studies one step further and I didn't explain the details of our tethering studies. Mucin was the only substrate we tested that was chemoattractive. We did tethering studies using specific antibodies for the bacterium. The *C. jejuni* were attached to a glass surface and then we monitored the organism microscopically by observing the rate at which the flagella twirled.

L. BLANKENSHIP: What is the approximate percentage of fucose and serene in mucin?

M. DOYLE: Fucose is approximately .3% of the mucin.

L. BLANKENSHIP: What is the relationship between the chemoreplusion for the bile salts in relationship to the natural occurrence of bile salts in the GI tract?

M. DOYLE: The concentrations of the bile salts were close to those that we tested. By purifying the bile salts in the mucin, we showed that these were chemorepellent.

L. BLANKENSHIP: Did the bile salts produce an *in vitro* effect on the morphology or motility of *Campylobacter jejuni*?

M. DOYLE: We didn't address that specifically, but Mary Hugdahl did observe that the organisms, when exposed to the bile, did not appear to be as healthy as your typical *Campylobacter*.

D. ROLLINS: The entire intestinal tract would be mucin-lined. Is there any idea what attracts these bacteria to the crypts?

M. DOYLE: Dr. Adrian Lee did observe the *Campylobacter* moving along the mucin within the intestine itself. I don't know why it actually ends up in the cecum. Perhaps because the cecum is more of a dead end space and it doesn't have the physically clearing peristaltic activity that occurs in the gut.

L. KIRKEGAARD: Why do you suppose that some chickens shed *Campylobacter* and some do not, when they all have mucin?

M. DOYLE: I think, based on our recent experiences, that if you use a sensitive enough procedure you will find that almost all chickens do carry *Campylobacter.* We have been doing studies for the last two or more years trying to identify *Campylobacter*-negative birds (laying hens). We have looked at three different flocks and thousands of birds, and we have come up with eight that did not carry *Campylobacter*. That first screening by direct plating and then going to an enrichment procedure, which is quite sensitive, to screen these birds to rule out that they don't shed *Campylobacter*. In studies where *Campylobacter* was not found, I think that one has to look critically at the methods that were used. In these instances perhaps, if a more sensitive procedure had been used *Campylobacter* might have been found. Perhaps seasonal variation, changes in feed, etc., stimulate the excretion of larger numbers of *Campylobacter*.

J. HASSAN: Is there anything in the immune response that controls the colonization in mucin?

M. DOYLE: There very well could be, but our work did not address that at all. We did not specifically look at the duodenum when we actually did our initial studies. We determined where *Campylobacter* localizes by taking the entire large intestine, but we did not segregate it into different segments. We didn't find that the small intestine had unusually large numbers of *Campylobacter*. I would guess that it might vary from bird to bird. We did find a few birds that had *Campylobacter* in the gall bladder, but not in all birds. In fact, gall bladder colonization by *Campylobacter* seems to be less frequent than more frequent. Even though the organism is chemoattracted to bile, it is not as chemoattracted to bile as is is to mucus because mucin is a smaller component of bile that it is in mucus.

J. HASSAN: Is there an interaction of motility and colonization?

M.DOYLE: There have been studies specifically addressing motility and looking at flagellated versus aflagellated strains of *Campylobacter*. It appears that motility is a very important factor in the colonization process.

D. WALTMAN: Is L-fucose a common chemotaxin among bacteria?

M. DOYLE: I can't say that is is. *E. coli* is attracted to L-fucose, but beyond that organism I can't answer your question.

H. LILLEHOJ: This is pure speculation, and I think that the hypothesis has to be tested. From my experience with different macrophage distribution in the gut, lymphocytes isolated from the ceca had the highest amount of macrophages, and from literature, most of the enteric pathogens first encounter macrophages. If this is correct, maybe macrophages carry bacteria and go back to the ceca. This is where they come from. This may be one way that the bacteria can be spread throughout the body.

M. DOYLE: It is an interesting idea that deserves further research.

ASPECTS OF VIRULENCE AND INTESTINAL COLONIZATION BY *SALMONELLA*

C. L. Gyles
C. Poppe
R. C. Clarke

Department of Veterinary Microbiology and Immunology
Veterinary College, University of Guelph
Guelph, Ontario, Canada N1G 2W1

ABSTRACT

Avian *Salmonella* isolates representing 38 serovars were examined for association of large plasmids with serovar, serum resistance, and virulence. Six serovars possessed typical, large virulence plasmids. Small plasmids were associated with virulence in two serovars. Serum resistance was not related to serovar or virulence. DNA probe studies indicated that plasmid-borne genes for virulence were related or identical in plasmids from different serovars.

Wild and mutant strains of *S. typhimurium* were administered orally to calves or injected into ligated ileal segments. Invasion of the intestine was monitored by light microscopy, transmission electron microscopy, scanning electron microscopy, and by immunoperoxidase methods. The ileum was most severely affected and jejunum, colon, and cecum were affected to a lesser extent although similar numbers of *Salmonella* were present in these tissues. Degree of invasion of intestinal mucosa was related to numbers of organisms injected into ligated segments of ileum and to virulence of the strains.

Colonization Control of Human Bacterial Enteropathogens in Poultry

INTRODUCTION

This presentation is in two distinct parts. The first part deals with virulence plasmids in avian isolates of *Salmonella*; the second deals with colonization of the gut by *Salmonella*. Plasmid-borne genes that mediate virulence have been reported in a limited number of serovars of *Salmonella* (1, 4, 14, 15, 17, 21, 23, 29) and resistance to the bactericidal activity of serum has been associated with this plasmid-related virulence (15, 22). The plasmids that have been identified in this role are typically 30-60 megadaltons and are characteristic for a serovar, as shown in Table 1. We were interested in asking the following questions with respect to *Salmonella* isolates from poultry: 1) Are there virulence plasmids in other serovars of *Salmonella*? 2) Can serum resistance be used as a screen for virulence plasmids? 3) How does virulence in mice relate to virulence in chicks? 4) Are "plasmid virulence genes" in one serovar the same as those in others? 5) Are "plasmid virulence genes" sometimes located in the chromosome?

Table 1. Reported association of plasmids with virulence in serovars of *Salmonella*.

Serovar	Virulence Plasmid (Mdal)
S. typhimurium	60
S. gallinarum	60
S. pullorum	60
S. dublin	50
S. enteritidis	36
S. choleraesuis	32

The second study was an investigation of virulence of wild and mutant strains of *Salmonella typhimurium* in the calf. Although there are numerous studies of experimental salmonellosis, most of these studies have involved *S. typhimurium* in mice and therefore reflect typhoidal disease. Interest in the control of foodborne salmonellosis through live vaccines requires that we develop methods of screening for virulence in suitable models. Live *Salmonella* vaccines for poultry, for example, need to be shown to be avirulent not only for poultry but also for mammals. We chose to study virulence of *S. typhimurium* in the calf because the calf is a highly susceptible natural host for non-typhoidal salmonellosis. We used the wild virulent *S. typhimurium* strain 3860 and mutants derived from this strain (5). Initially, the virulence and the effects on the intestine were established for strain 3860 in two-week-old colostrum-fed calves that were infected by oral inoculation. Subsequently, the virulence of *gal*E mutant of *S. typhimurium* LT2 was assessed in orally infected calves, and finally virulence of wild and several mutant strains was evaluated in ligated ileal loops in calves.

MATERIALS AND METHODS

The following materials and procedures were used for the study of virulence plasmids.

Bacteria. We examined 230 avian isolates of *Salmonella*, most of which had been isolated in Ontario (25). These isolates represented 38 serovars. We also investigated five *S. choleraesuis* and five *S. dublin* isolates that were not of avian origin since these serovars had been reported to contain virulence plasmids.

Determination of Plasmid Profiles. Plasmid DNA was isolated by the method of Portnoy and White as cited by Crosa and Falkow (8), then subjected to electrophoresis in 0.7% agarose. The molecular masses of the plasmids were calculated, based on the mobility of plasmids of known molecular masses (19). On the basis of plasmid profiles, we identified plasmids of a particular molecular mass that were present in most or all strains of a serovar.

Serum Bactericidal Test. The protocol for the serum bactericidal test was that described by Helmuth and coworkers (15) in their report on virulence plasmids in *Salmonella*. These workers had observed an association of virulence plasmids with serum resistance. Our test used pooled guinea pig serum or pooled germfree chicken serum, both of which were free of agglutinins to *Salmonella*. The test measured the extent to which *Salmonella* grew or were killed during three hours of incubation at 37 ° C in 80% serum. Serum resistance was the extent of growth in serum in the presence of complement: the number of organisms at 3 h was expressed as the percentage of the number of organisms at time 0.

Virulence Tests. Virulence studies were carried out for 29 strains in orally and intraperitoneally inoculated Balb/c mice and day-old *Salmonella* leghorn chicks. These 29 strains included ten strains which were resistant to guinea pig serum, nine strains which were pairs or triplets of strains of the same serovar with different plasmid profiles, and ten strains selected at random.

Implication of Plasmids in Virulence. In subsequent experiments we investigated whether plasmids in *S. typhimurium*, *S. enteritidis*, *S. pullorum*, *S. gallinarum*, *S. heidelberg*, and *S. agona* could be implicated in virulence. The approach was as follows: We attempted to label the plasmid of interest with transposon Tn*1*, to eliminate labelled or unlabelled plasmid by chemicals or by heat (26), and to attempt to reintroduce the plasmid or a cloned fragment of the plasmid, and to assess the virulence of strains with and without the plasmid of interest.

Hybridization Studies. Hybridization studies were carried out with a 3.75 Kb fragment of the 60 Mdal virulence plasmid of *S. typhimurium*. This probe was made available to us through the generosity of Paul Gulig and Roy Curtiss (13). Plasmid DNA and chromosomal DNA were isolated (20) from 68 strains belonging to 18 serovars. The plasmid DNA was blotted onto nitrocellulose and hybridized with the radio-labelled probe DNA. Hybridization with total DNA employed a slot blot method.

For the second study, the following materials and methods were used:

Strains of *Salmonella*. *Salmonella typhimurium* strain 3860 is a wild virulent strain originally recovered from a pig. A nalidixic-acid-resistant mutant of the organism was used to facilitate recovery and enumeration of challenge organisms from the intestine. *S. typhimurium* strain G30D was a *gal*E mutant of *S. typhimurium* strain LT2, kindly supplied to us by Dr. C. Wray, Weybridge, England (31, 32). The parent strain LT2 was also tested in ligated segments of calf intestine. Mutants of strain 3860 that were selected and tested were a *gal*E mutant, a *dap* mutant, an *aro*A mutant, and a *dap*, *aro*A mutant. These mutants were made by transposon Tn*10* insertion into the gene of interest, followed by transduction of the mutant gene into strain 3860 and subsequent selection on Bochner medium (3) of isolates that had lost the transposon Tn*10* (18).

Calves. Two-week-old calves were colostrum-fed and were maintained in isolation facilities. Repeated checks were made to ensure that the calves were not shedding *Salmonella*.

Infection of Calves. In the first experiment each of eight calves was given 10^{10} organisms of strain 3860 in milk replacer. Two calves were used as uninoculated control animals. Two of the infected calves were euthanized at 6 h, 12 h, 24 h, and 34 h post infection. The calves were monitored for development of illness. In the second experiment eight calves were infected orally with the *gal*E mutant G30D: four received 10^6 organisms and four received 10^{10}. These calves were monitored for signs of illness.

Sampling Tissues from Calves. At time 0 for the uninfected calves and just prior to the assigned times for euthanasia of the infected calves, samples were removed under anaesthesia. The samples that were taken were abomasum, duodenum, ileum, jejunum, colon, rectum, mesenteric lymph nodes, liver, and spleen. Samples were processed to determine numbers of salmonellae per gram of tissue, and for examination by light, transmission electron, and scanning electron microscopy (5). Samples processed for light microscopy were also subjected to an indirect immunoperoxidase test with antiserum specific for *S. typhimurium* (5).

Ileal Loop Studies. Ligated segments were created (5) in the ileum of six calves and were inoculated with graded doses of the wild and mutant strains of *S. typhimurium*. The calves were euthanized 16-18 h after inoculation of the intestine and samples of intestine were immediately removed for processing as described previously. Volumes of fluid and lengths of ileal loop were recorded.

RESULTS

Plasmid Profiles, Serum Resistance, and Virulence. In the first phase of the study, plasmid profiles were determined for 230 isolates of *Salmonella*. Serovar-associated plasmids were identified for seven serovars (Table 2). In the case of *S. typhimurium*, a 2.3 Mdal plasmid was more frequently associated with strains than was the 60 Mdal plasmid. A small plasmid was also associated with the serovars *heidelberg* and *gallinarum*. Digestion of 36 Mdal plasmids from three of the six *S. enteritidis* strains with three restriction endonucleases gave identical restriction fragments. Plasmid profile was sometimes associated with source of isolates. For example, four isolates of *S. johannesburg*, which had identical plasmid profiles were traced to a common source.

Table 2. Association of plasmids with serovars of *Salmonella* examined in this study.

Serovar	Plasmids (Mdal)	No. /Total
S. typhimurium	60	11/24
	2.3	17/24
S. enteritidis	36	4/4
S. heidelberg	2.2	20/24
S. pullorum	60	6/7
S. gallinarum	60	2/2
	1.5	2/2
S. choleraesuis	32	5/5
S. dublin	50	5/5

Table 3. Resistance to guinea pig serum among serovars for which three or more strains were tested.

Serovar	No. resistant / No. tested	Serovar	No. resistant / No. tested
S. anatum	3/9	*S. saintpaul*	1/5
S. enteritidis	3/3	*S. schwarzengrund*	1/5
S. hadar	7/9	*S. senftenberg*	6/22
S. heidelberg	6/17	*S. typhimurium*	6/7

Resistance to guinea pig serum was determined for 117 isolates, chosen so that each plasmid profile was represented. Forty-three strains grew in the presence of guinea pig serum and serum resistance of these isolates was generally not a serovar related phenomenon. This is illustrated for those strains for which three or more had been tested (Table 3). Of the 43 serum resistant isolates, 25 were tested for their ability to grow in the presence of serum from germfree chickens. The results of these tests are illustrated in Table 4. When incubated in chicken serum, all isolates had reduced numbers at 3 h compared with the numbers at time 0 (all less than 100%). The three *S. enteritidis* strains were the least susceptible to the bactericidal action of chicken serum. There was no relationship between resistance to guinea pig serum and resistance to chicken serum.

Table 4. Resistance to growth in chicken serum of 29 isolates of *Salmonella* resistant to growth in guinea pig serum.

Strain	Serovar	Plasmids (Mdal)	Percent Resistance	
			GP Serum	Chicken Serum
118	*S. typhimurium*	2.3	1163	1
137	"	60	909	21
165	"	60,2	241	1
166	"	84,60,42,30	344	1
167	*S. senftenberg*	120,48,42,4.1	279	0
294	"	120,48,24,16	325	3
299	"	3.2	250	0
402	"	120	1802	20
139	*S. enteritidis*	36	760	28
183	"	36	601	43
271	"	36	1030	34

The main features of the findings in intraperitoneally injected Balb/c mice and day-old leghorn chicks are illustrated by the data presented in Table 5.

LD_{50} values in mice and in chicks were quite different for the same strain and there were several strains for which there were low LD_{50} values in chicks but high values in mice. Resistance to guinea pig serum could not be related to virulence for mice or chickens. Strains of *S. typhimurium* which lacked a 60 megadalton plasmid but possessed a 2.3 megadalton plasmid could be highly virulent for chicks.

Table 5. Relation of LD_{50} values in mice and chickens to plasmid profiles and resistance to guinea pig serum.

Strain	Serovar	Plasmids (Mdal)	% Serum Resistance	LD_{50} in Mice (i.p.)	LD_{50} in Chicks (i.p.)
139	*S. enteritidis*	36	760	10^4	10^4
137	*S. typhimurium*	60	909	10^5	10^3
118	"	2.3	1163	10^5	10^1
8	*S. heidelberg*	120, 70	561	10^6	10^1
144	"	2.2	198	10^4	10^0
126	*S. agona*	120	64	10^6	10^2
110	"	none	158	10^6	10^1

Virulence of Strains with and without Plasmids. Labelling was obtained for 11 of 30 strains. Elimination of plasmids was not always readily achieved. The use of chemical agents resulted in loss of plasmids from six of 17 strains that were treated. Incubation of cultures at elevated temperature (45 ° C) was much more effective and eliminated plasmids in 18 of 25 strains. This latter procedure was sufficiently efficient in that loss of unlabelled as well as labelled plasmids was detectable. The virulence of strains with and without plasmids is illustrated in Tables 6-8. Data for strains of *S. enteritidis* and *S. typhimurium* are shown in Table 6. These were the only serovars which were sufficiently virulent in orally infected mice to show effects in this model. Chicks were also injected intramuscularly with *S. gallinarum* to permit comparison with the work by Barrow *et al.* (2) (Table 7).

There was only partial restoration of virulence when the 1.5 Mdal plasmid, cloned in plasmid pUC18, was reintroduced into *S. gallinarum* and when the 2.3 Mdal plasmid, cloned in the same vector, was reintroduced into *S. heidelberg*.

Hybridization with a DNA Probe from the *S. typhimurium* Plasmid Virulence Genes. The results of this study are as follows: The only plasmids which hybridized with the probe were 60 Mdal plasmids in *S. typhimurium*, *S. pullorum*, and *S. gallinarum* and 32, 36, and 50 Mdal plasmids in *S. choleraesuis*, *S. enteritidis*, and *S. dublin*, respectively. None of the chromosomal DNA preparations, none of the other plasmids in strains of the six serovars listed above, and none of the plasmids in strains of 12 other serovars hybridized with the probe.

Probing of Southern blots of restriction endonuclease-digested plasmid DNA showed that a 3.7 Kb *Bgl*I fragment and 2.5 and 1.5 Kb *Pvu*II fragments were common to the plasmids which were probe positive.

Table 6. Effect on virulence in orally inoculated Balb/c mice of elimination of plasmids from *S. enteritidis* and *S. typhimurium*.

Organism	Plasmid	Log LD_{50}	
		with plasmid	without plasmid
S. enteritidis	36[a]	5	>8
	36,30	<4	>8
	30	>8	>7
S. typhimurium	60	5	>8

[a]Underlined indicates the plasmid that was eliminated.

Virulence in calves of the wild strain *S. typhimurium* 3860 and of *gal*E mutant G30D. Strain 3860 proved to be highly virulent in calves and produced disease in all eight calves that were challenged. Tables 9-11 summarize the extent of association of the organism with different areas of the intestine, the degree of invasion of various areas of the intestine, and the extent of mucosal damage produced. In mild damage (+), villi were observed to be slightly contracted, with some separation of basement membranes from epithelial cells at the tips of villi, and there were cells and some fluid in sub-epithelial bullae. Moderately affected intestine (+ +) was characterized by short, blunt villi, exfoliation of enterocytes from the tips of villi, the presence of neutrophils and epithelial cells in the lumen, and increased inflammatory exudation in the lamina propria. Severe damage (+ + +) resulted in loss of the upper portions of most villi, with large numbers of neutrophils and rounded epithelial cells on the surface; there was thrombosis of venules and a heavy, mainly neutrophilic inflammatory infiltrate in the lamina propria.

Invasion was readily observed in sections stained by the immunoperoxidase method. In mild invasion (+), the occasional cluster of *Salmonella* was observed the lamina propria. In heavy invasion (+ +), each villus had many clusters of *Salmonella* in the epithelial cells and in the lamina propria. Heavy invasion was observed not only in enterocytes overlying Peyer's patches but also in enterocytes in areas not associated with Peyer's patches.

The *gal*E mutant G30D administered orally at a dose of 10^{10} to three calves and at a dose of 10^{6} to four calves was virulent and produced illness in all calves. The signs of illness included fever, diarrhea, and inappetence. Severity of disease was dose-related. The virulence of this strain may have been due to some degree of reversion to galactose resistance (5).

Virulence of Wild and Mutant Strains of *S. typhimurium* in Calves. Stable, non-reverting mutants of 3860 were developed by transduction of Tn*10*-labelled mutant *dap*, *aro*A, and *gal*E genes. The effects of the wild parent, mutants derived from strain 3860, and the *gal*E strain G30D were then tested in ligated segments of calf ileum. The wild strain 3860 induced fluid accummulation at doses greater than 10^7 and mucosal damage at doses greater than 10^8. The laboratory strain LT2 and all mutants induced low volumes of fluid at a dose of 10^9.

Mucosal damage parallelled mucosal invasion. All mutants did invade the mucosa and produced mild damage but only when the numbers were of the order of 10^9 organisms.

Table 7. Effect on virulence in chicks associated with elimination of 60 and 1.5 megadalton plasmids.

Organism	Plasmids (Mdal)	Log LD_{50}	
		with plasmid	without plasmid
I.P. route			
S. gallinarum	60, 1.5[a]	<1	7
	60	>8	>7
I.M. route			
S. gallinarum	60, 1.5	<3	>9
	60	>9	>9

[a]Underline indicates plasmid that was eliminated.

DISCUSSION

In a previous study of large numbers of *S. typhimurium* (15) more than 90% of isolates possessed a 60 Mdal plasmid, whereas in this study only 41% of the 27 strains had such a plasmid. The difference may relate to different sources of the isolates. A small plasmid, 2.3 Mdal, was more commonly associated with *S. typhimurium* in this study, being present in 63% of strains. It would be interesting to investigate whether the 2.3 Mdal plasmid plays a role in virulence of *S. typhimurium*, especially since strains which possessed this plasmid but lacked the 60 Mdal plasmid were highly virulent. Another small plasmid, a 2.2 Mdal plasmid, was highly associated with *S. heidelberg* and was associated with virulence of this serovar. Studies to investigate relationships between these two small plasmids might be useful.

Table 8. Effect of elimination of plasmids on virulence in chicks[a] of strain of *S. heidelberg*, *S. pullorum*, and *S. schwarzengrund*.

Organism	Plasmids (Mdal)	Log LD_{50}	
		with plasmid	without plasmid
S. heidelberg	28,23,2.3[b],1.0	<1	4
	31,23,1.0	4	5
	23,2.3,1.0	<1	5
S. pullorum	60,2.8,1.6,1.5	<4	>8
S. schwarzengrund	2.3	3	8

a Day-old chicks were inoculated intraperitoneally.
[b]Underlined indicated plasmid that was eliminated.

It was very clear that the phenomenon of a large molecular weight serovar-associated virulence plasmid was limited to only six serovars. If one considers that although *S. typhimurium* causes disease in a wide variety of hosts it is host-adapted to mice, then all six serovars are host-adapted *Salmonella*. Our results with *S. gallinarum* differ from those reported by Barrow and colleagues (2) in that elimination of the 1.5 Mdal plasmid had a marked effect on virulence for chicks in our study. That at least some of the serovar-associated plasmids clearly enhance virulence has now been demonstrated by several workers (1, 13, 30). Based on hybridizations and Southern blot analysis following restriction enzyme digestions, the virulence genes in all six serovars are highly related and may be identical. However, these plasmids may contain unique DNA sequences which reflect their association with different hosts. If such sequences are identified they could be exploited in rapid detection of low numbers of organisms, such as *S. enteritidis* in poultry, eggs, and the environment of poultry.

Differences in virulence in mice compared with chicks emphasizes the importance of the host species in determination of virulence. The test species should be selected in relation to the animal host in which virulence needs to be assessed.

Failure to completely restore virulence to strains of *S. heidelberg* and *S. gallinarum* from which the small plasmids had been eliminated may have been due to damage to the organisms during curing or to alteration in gene expression resulting from cloning into a vector.

Table 9. Numbers of *Salmonella* in different areas of the intestine in calves euthanized at various times after challenge with *S. typhimurium* 3860.

		Log *Salmonella*/gram					
Calf	Time	ABO[a]	DUO	JEJ	IL	CE	CO
1	6	4	2	7	7	8	8
2	6	3	3	6	7	8	8
3	12	3	2	7	7	7	6
4	12	6	7	8	7	9	9
5	24	5	4	6	6	7	7
6	24	5	4	6	6	7	7
7	34	6	5	8	8	8	8
8	34	6	6	8	8	8	7

[a]abomasum, duodenum, jejunum, ileum, cecum, colon

Serum resistance was generally not related to plasmids or virulence in this study, but *S. typhimurium* strains with a 60 Mdal plasmid and *S. enteritidis* strains with a 36 Mdal plasmid were serum resistant. Differences among researchers with respect to serum resistance and serovar-associated plasmids in *Salmonella* (13, 15, 17) may be due to differences in the way the serum bactericidal tests were carried out and/or to differences in the strains of *Salmonella* studied. The strong bactericidal effect of serum from chickens (27), coupled with the intestinal barrier (24), may be important in protecting chickens against disease due to *Salmonella*. It is noteworthy that *S. enteritidis* strains were the least affected by chicken serum.

Reduction in virulence was associated with elimination of a serovar-associated plasmid from strains of *S. enteritidis*, *S. gallinarum*, *S. heidelberg*, *S. pullorum*, *S. schwarzengrung*, and *S. typhimurium*. The *S. dublin* and *S. choleraesuis* strains which had serovar-associated plasmids were not studied for the effects of elimination of plasmids.

The studies in calves provided evidence of the susceptibility of calves to oral infection and suggested that ligated segments of calf ileum may be a useful screen for mutants of *Salmonella* developed as avirulent organisms for vaccines (5, 6, 7, 9, 10, 11, 12, 16). The method permits testing of several organisms and of control organisms in the same animal. Effective vaccine strains will almost certainly be invasive and detection of invasion in ligated ileal loops is no indication of unsuitability of such organisms as vaccine organisms. Some degree of mucosal damage, however, was noted with all mutants tested, indicating that these mutants were of reduced virulence but not avirulent.

The *gal*E mutant of 3860 produced not only microscopic lesions but also grossly visible lesions. The organism in which mutations are produced clearly influence the virulence. For example, damage in ligated ileal loops of calves produced by a *gal*E mutant of *S. typhimurium* 3860 was more severe than that produced by a similar mutant of *S. typhimurium* LT2 (5).

Table 10. Extent of mucosal damage produced in different areas of the intestine in calves euthanized at various times after challenge with *S. typhimurium* 3860.

		Mucosal Damage					
Calf	Time (Hours)	ABO[a]	DUO	JEJ	IL	CE	CO
1	6	-	-	-	+	-	-
2	6	-	-	-	+	+	-
3	12	-	-	+ +	+ + +	-	
4	12	-	-	-	+ +	+ +	+
5	24	-	-	+ +	+ + +	+ + +	+
6	24	-	-	+	+ + +	+ + +	-
7	34	-	-	+ +	+ + +	+ + +	+ +
8	34	-	-	+ + +	+ + +	+ + +	+ +

[a]abomasum, duodenum, jejunum, ileum, cecum, colon

- = no damage, + = mild damage, + + = moderate, + + + = severe

Recently, Curtiss and colleagues (9, 13) have developed mutants which show much promise as potential live *Salmonella* vaccine organisms. It would be interesting to evaluate the behaviour of such mutants in ligated ileal segments in calves, which are highly susceptible hosts that may die as a result of intestinal lesions in the absence of significant systemic invasion (5).

Table 11. Extent of mucosal invasion in different areas of the intestine in calves euthanized at various times after challenge with *S. typhimurium* 3860.

		Mucosal Invasion					
Calf	Time (Hours)	ABO[a]	DUO	JEJ	IL	CE	CO
1	6	-	-	-	+ +	-	-
2	6	-	-	-	+ +	-	-
3	12	-	-	+	+ +	+	-
4	12	-	-	-	+ +	+ +	+
5	24	-	-	+	+ +	+ +	+
6	24	-	-	+	+ +	+ +	-
7	34	-	-	+ +	+ +	+ +	+ +
8	34	-	-	+ +	+ +	+ +	+ +

[a]abomasum, duodenum, jejunum, ileum, cecum, colon

- = no *Salmonella* were observed; + = mild invasion, only occasional organism was seen in the villi; + + = severe invasion, most villi were invaded by large numbers of *Salmonella*

ACKNOWLEDGEMENTS

The authors wish to acknowledge that this research was supported by Agriculture Canada and by the Ontario Ministry of Agriculture and Food.

REFERENCES

1. **Baird, G.D., B.J. Manning, and P.W. Jones.** 1985. Evidence for related virulence sequences in plasmids of *Salmonella dublin* and *Salmonella typhimurium*. J. Gen. Microbiol. **131**:1815-1823.
2. **Barrow, P.A., J.M. Simpson, J.A. Lovell, and M. Binns.** 1987. Contribution of *Salmonella gallinarum* large plasmid toward virulence in fowl typhoid. Infect. Immun. **55**:388-392.
3. **Bochner, E.R., H. Huang, J.L. Schieven, and B.N. Ames.** 1980. Positive selection for loss of tetracycline resistance. J. Bacteriol. **143**:926-933.
4. **Chikami, G. K., J. Fierer, and D. G. Guiney.** 1985. Plasmid mediated virulence in *Salmonella dublin* demonstrated by use of a Tn*5*-*ori*T construct. Infect. Immun. **50**:420-424.
5. **Clarke, R.C.** 1985. Virulence of wild and mutant strains of *Salmonella typhimurium* in the calf. Ph.D. Thesis. Univ. of Guelph, Guelph, Canada.
6. **Clarke, R.C., and C.L. Gyles.** 1986. Galactose epimeraseless (*gal*E) mutants of *Salmonella typhimurium* as live vaccines for calves. Can. J. Vet. Res. **50**:165-173.
7. **Clarke, R.C., and C.L. Gyles.** 1987. Virulence of wild and mutant strains of *Salmonella typhimurium* in ligated intestinal segments. Am. J. Vet. Res. **48**:504-510.
8. **Crosa J. H., and S. Falkow.** 1981. Plasmids. *In* Manual of Methods for General Bacteriology. P. Gerhardt (Ed.), American Society for Microbiology, Washington, D.C. pp. 267-268.
9. **Curtiss, R., S.M. Kelly, P.A. Gulig, C.R. Gentry-Weeks, and G.E. Galàn.** 1988. Avirulent salmonellae expressing bacterial virulence antigens from other pathogens for use as orally administered vaccines. *In* Virulence Mechanisms of Bacterial Pathogens, J.A. Roth (Ed.), American Society for Microbiology, Washington, D.C. pp. 311-328.
10. **Germanier, R.** 1970. Immunity in experimental salmonellosis. I. Protection induced by rough mutants of *Salmonella typhimurium*. Infect. Immun. **2**:309-315.
11. **Germanier, R.** 1971. Immunity in experimental salmonellosis. III. Comparative immunization with viable and heat-inactivated cells of *Salmonella typhimurium*. Infect. Immun. **5**:792-797.
12. **Germanier, R., and E. Fuerer.** 1971. Immunity in experimental salmonellosis. II. Basis for avirulence and protective capacity for *gal*E mutants of *Salmonella typhimurium*. Infect. Immun. **4**:663-673.
13. **Gulig, P.A., and R. Curtiss.** 1988. Cloning and transposon insertion mutagenesis of virulence genes of the l00-kilobase plasmid of *Salmonella typhimurium*. Infect. Immun. **56**:3262-3271.
14. **Hackett, J., I. Rotlarski, V. Mathan, K. Francki, and D. Rowley.** 1986. The colonization of Peyer's patches by a strain of *Salmonella typhimurium* cured of the cryptic plasmid. J. Infect. Dis. **153**:1119-1125.
15. **Helmuth, R., R. Stephan, C. Bunge, B. Hoog, A. Steinbeck, and E. Bulling.** 1985. Epidemiology of virulence-associated plasmids and outer membrane protein pattern within seven common *Salmonella* serotypes. Infect. Immun. **48**:175-182.
16. **Hoiseth, S.K., and B.A.D. Stocker.** 1981. Aromatic-dependent *Salmonella typhimurium* are non-virulent and effective as live vaccines. Nature (Lond.) **291**:238-239.

17. **Jones, G. W., D. K. Rabert, D. M. Svinarich, and H. J. Whitfield.** Association of adhesive, invasive, and virulent phenotypes of *Salmonella typhimurium* with autonomous 60-megadalton plasmids. Infect. Immun. **38:**476-486.
18. **Kleckner. N., K. Reichorde, and D. Botstein.** 1979. Inversions and deletions of the *Salmonella* chromosome generated by the translocatable element Tn*10*. J. Mol. Biol. **127:**89-115.
19. **Macrina, F. L., D. J. Kopecko, K. R. Jones, D. J. Ayers, and S. M. McCowen.** 1978. A multiple plasmid-containing Escherichia coli strain: convenient source of size reference plasmid molecules. Plasmid **1:**417-420.
20. **Maniatis, T., E. F. Fritsch, and J. Sambrook.** 1982. Molecular Cloning. A Laboratory Manual. Cold Spring Harbor Laboratory, Cold Spring Harbor, New York. pp. 97-106.
21. **Manning, E. J., G. D. Baird, and P. W. Jones.** 1984. Possible plasmid involvement in the virulence of *Salmonella dublin*. Biochem. Soc. Trans. **12:**847-848.
22. **Manning, E. J., G. D. Baird, and P. W. Jones.** 1986. The role of plasmid genes in the pathogenicity of *Salmonella dublin*. J. Med. Microbiol. **21:**239-243.
23. **Nakamura, M., S. Sato, T. Ohya, S. 8uzuki, and S. Ikeda.** 1985. Possible relationship of a 36-megadalton *Salmonella enteritidis* plasmid to virulence in mice. Infect. Immun. **47:**831-833.
24. **Popiel, I., and P. C. B. Turnbull.** 1985. Passage of *Salmonella enteritidis* and *Salmonella thompson* through chick ileocecal mucosa. Infect. Immun. **47:**786-792.
25. **Poppe, C.** 1988. Virulence-associated Plasmids in *Salmonella*. Ph.D. Thesis. Univ. of Guelph, Guelph, Canada.
26. **Poppe, C. and C.L. Gyles.** 1988. Tagging and elimination of plasmids in *Salmonella* of avian origin. Vet. Microbiol. **18:**73-87.
27. **Schwab, G. E., and P. R. Reeves.** 1966. Comparison of the bactericidal activity of different vertebrate sera. J. Bacteriol. **91:**106-112.
28. **Stocker, B.A.D., S.H. Hoiseth, and B.P. Smith.** 1983. Aromatic-dependent *Salmonella* sp. as live vaccine in mice and calves. Develop. Biol. Standard. **53:**47-54.
29. **Terakado, N., T. Sekizaki, R. Hashimoto, and S. Naitoh.** 1983. Correlation between the presence of a fifty-megadalton plasmid in *Salmonella dublin* and virulence for mice. Infect. Immun. **41:**443-444.
30. **Williamson, C.M., G.D. Baird, and E.J. Manning.** 1988. A common virulence region on plasmids from eleven serotypes of *Salmonella*. J. Gen. Microbiol. **134:**975-982.
31. **Wray, C., W.J. Sojka, J.A. Morris, and W.J.B. Morgan.** 1977. The immunization of mice and calves with *gal*E mutants of *Salmonella typhimurium*. J. Hyg. **79:**17-24.
32. **Wray, C., W.J. Sojka, D.G. Pritchard, and J.A. Morris.** 1983. Immunization of animals with *gal*E mutants of *Salmonella typhimurium*. Develop. Biol. Standard. **53:**41-46.

DISCUSSION

N. STERN: Why, if we are trying to get rid of *Salmonella* from any domestic animals, would we want to use a *Salmonella* to infect our animals with a *Salmonella* mutant?

C. GYLES: The literature indicates that live vaccine have been superior to killed vaccine in effectiveness. We know of the understandable aversion to the use of live vaccine because of the history of attuated organisms reverting to become virulent. Nonetheless, we now are more confident and still hope to generate non-revertable mutants and can evaluate their safety. Therefore, we should not rule out live vaccines before thoroughly investigating all prospects. In the long run, a vaccine organism producing a desired antigen would need to replicate in the animal. A single vaccine organism, carrying a wide range of antigen for a wide variety of pathogens, could be achieved through such a live vaccine.

R. MEINERSMANN: Could you please tell me about the inflammatory response in reference to your histology studies as well as the kinetics of that response?

C. GYLES: We looked at the samples within six hours and by then the inflammatory response was in full flight. Initially, we observed huge, massive exudation of neutrophils into the lumen of the gut and, we could also see mononuclear cells, although they tended to come on a little later on. The initial phase was a neutrophilic response. In the Peyer's patches, we saw evidence of damage to the lymphocytes, there was central depletion of lymphocytes at that stage, and pyknosis of the lymphocytes. Over time, what happened in some of the villi, especially in the colon, one could see replacement of the epithelial covering of the colon. In terms of the kinetics of the response, what we saw was an initial, very rapid, neutrophilic response, likely due to chemotaxic factors associated with the organism, and over time we started to see more mononuclear cells. We saw evidence of thrombi within blood vessels as the infection progressed, and we believe this must have been in response to the endotoxins.

SPEAKER X: What method do you use to cure the plasmid?

C. GYLES: We use both chemical methods, such as acridine orange, SDS, methylene blue, and combinations of those which are not very effective methods. We then tried incubating at 45 degrees, and this approach was very efficient.

R. CURTISS: With avirulent mutants, at what point in dosing do you get a situation where you wouldn't have significant inflammation, while maintaining protection?

C. GYLES: That is a difficult question. The effective live *Salmonella* vaccine strains are going to invade or be taken up. In fact, that is one of the things that we want to happen. One would have to determine how many organisms should be necessary to provide an effective antigenic response without interferring with the normal metabolisism or health of the host. Screening studies would tell which strains are pathogenic. I think use of isolates which invade to a very mild degree is desirable. Then you would have to assess these isolates in a full battery of various parameters to see whether there is any effect on the health of those animals.

VIRULENCE PROPERTIES OF *SALMONELLA ENTERITIDIS* ISOLATES

C.E. Benson
R.J. Eckroade

Department of Clinical Studies
University of Pennsylvania
Kennett Square, Pennsylvania 19348

ABSTRACT

Delineation of the virulence characteristics of a facultative intracellular pathogen such as *Salmonella enteritidis* is dependent upon the development of appropriate assay systems. The HeLa cell invasion model has been adapted to monitoring the stages of pathogenesis, i.e., attachment, penetration of the host cell, intracellular growth, and dissemination. An infection value was calculated for each strain evaluated by the *in vitro* cell culture system. Individual isolates have values in a consistent range, however, the values among different isolated varied. Certain *S. enteritidis* isolates have distinctly high infection patterns (50%) which compared favorably with the animal studies performed by other investigators. The pattern of attachment was both diffuse and localized and it did not vary significantly when the cell cultures were examined at 1 hr and 3 hr post inoculation. The exception to this observation was the yolk isolate (Y8-P2 or Benson strain), which had decreased cell-association after 3 hr incubation. Mannose did not alter the percent of cells with associated bacteria nor the pattern of attachment. Thirty-four isolates of *S. enteritidis* acquired from a variety of different sources and different epidemics were negative for enterotoxin production. No distinctive antibiogram patterns were noted between those strains, and plasmid fingerprints fit the characteristic pattern for *S. enteritidis*. Fingerprints of genomic DNA have been acquired with 16 different endonucleases and Southern Blots of these preparations have been probed with radiolabeled ribosomal RNA. No significant deviation between strains has been observed.

Colonization Control of Human Bacterial Enteropathogens in Poultry

INTRODUCTION

The best introduction to *Salmonella enteritidis*, serotype *enteritidis* (hereafter referred to as *S. enteritidis*), is via a brief historical presentation of our data relative to salmonellae and chickens.

During the summer of 1987 we were invited to assist the USDA in a microbiological evaluation of ovaries obtained from chickens implicated in an outbreak of salmonellosis in humans in a New York hospital. A total of 555 individually collected ovary specimens were processed through a selenite enrichment protocol and 69% (383 specimens) were productive for *S. enteritidis*. The specimens could have been contaminated by the collection process in the processing plant. However, other studies in our laboratory, utilizing aseptically collected tissues of field-infected chickens indicates that *S. enteritidis* can be associated with the ovary and not be present in the intestines.

The story does not stop here. We were asked to evaluate some eggs. Little did we appreciate that "some eggs" would turn out to be six cases of eggs! Over 1,200 eggs were sampled and *S. enteritidis* was found to be associated only with the yolk fractions.

All spoiled and cracked eggs were not cultured. To date we have examined over 6,000 eggs from serology positive flocks and we have successfully isolated *S. enteritidis* in only 15 specimens. Thus, the incident of salmonellae associated with eggs is probably much lower in a randomized sample. The microbiological techniques for surveillance of eggs are not efficient and some positive specimens may have been missed.

As this data was being reported to the interested agencies and institutions, a common speculation was that these isolates must be different than those *S. enteritidis* of years ago. These comments were accepted as a challenge to investigate. Our approach to a comparative study of *S. enteritidis* isolates has been based upon our research experiences with the *S. typhimurium*. Thus a brief historical discussion of those activities may help explain the approach and interpretation.

Seven or so years ago, I was studying the pathogenic characteristics of *Salmonella typhimurium*. The HeLa cell invasion system described by investigators in Dr. Formal's laboratory at Walter Reed had been shown to correlate fairly well with animal-based evaluations and could serve as an indicator of the pathogenic nature of salmonellae. We modified that model to include several additional manipulations in an effort to increase the sensitivity of the assay (Table 1). In this modification there is an incubation period to permit the initial attachment of the pathogen, and that was followed by an incubation to facilitate invasion by those bacteria specifically associated with the HeLa cell. Coverslips can be removed throughout the incubation to facilitate a progressive assessment of the attachment:invasion:intracellular process. In all instances the coverslips were removed and rinsed in sterile PBS, fixed in methanol and stained in the Diff-Quik preparation for one minute in each solution, followed by a water rinse. The coverslips were air dried prior to mounting onto a slide. Visual evaluation of the 24 hr specimen revealed that some isolates were able to grow in the intracellular environment and had spread to uninfected cells, presumably via the cytoplasmic bridges. Other strains grew quite slowly within the cells but caused extrinsic formation of vacuoles and a decrease in viable HeLa cells, while a third category of isolates did not survive in, or cause significant decrease in, the HeLa cell population. Quantitation of these dynamics has been partially achieved via the use of an infection index system (Table 2). With this calculation, we compared the impact of a decrease in the HeLa cell population (since the dead cells detach and are lost in to the culture fluid) with a quantitation of the intracellular

Table 1. HeLa cell infection protocol - *Salmonella typhimurium.*

TIME (HRS)	STEP	MANIPULATION
-24	1.	HeLa cells innoculated into petri dishes containing coverslips immobilized with sterile vaseline. Use antibiotic-free media (ABF-DMEM).
	2.	Bacteria inoculated into 5 ml Brain Heart Infusion broth (BHI), aerated at 37 ° C.
-1.5	3.	Collect bacteria by centrifuge, wash 3X with sterile PBS, resuspend to 1×10^9 cfu/ml.
	4.	Inoculate 0.1 ml bacterial suspension into 10 ml ABF:DMEM.
0	5.	Remove coverslip. Replace media in HeLa cell cultures with preparation of step 4.
1	6.	Remove media. Wash 1X with sterile PBS. Add fresh pre-warmed ABF:DMEM. Remove coverslip.
4	7.	Wash cultures as in step 6. Remove coverslip. Add DMEM containing 100 μg gentamicin per ml media.
24	8.	Remove coverslip. Transfer 0.1 ml culture media to BHI.

Table 2. Calculation of infection index.

A. $$\text{\% HeLa Cell Survival (24 hrs)} = \frac{\text{\#cells (24 hrs)/ x fields}}{\text{200 cells (0 hrs)/ x field}} \times 100$$

B. $$\text{\% Infection (24 hrs)} = \frac{\text{\#infected cells (24 hrs)}}{\text{200 cells}} \times 100$$

C. $$\text{Infection Index (II)} = \frac{\text{\%infected (B)}}{\text{\%surviving cells (A)}}$$

growth. Strains TML and B had a high index compared to the laboratory strain LT-2 (Figure 1). We arbitrarily chose strain LT-2 as the presumed non-pathogenic, prototroph, since many laboratories have utilized this strain in physiological, biochemical and genetic studies. Universally, this strain has been accepted as a non-pathogen, but there are reports of laboratory-associated infections, hence our conclusion that strain LT-2 has maintained some low level of pathogenicity (see reference 2 as an example). We exposed strains LT-2 and TML to ultraviolet irradiation and gentamicin to gently kill the cultures with minimal surface structure alteration and utilized these killed cultures in the invasion model (Figure 2). Uptake of these killed bacteria was exceedingly low and led to the conclusion that the salmonellae were, in some manner, actively involved in the mediation of penetrating the HeLa cells. The recent studies from Dr. Falkow's laboratory substantiate that conclusion (3, 4, 5). The inclusion of the lectin ConA (concanovalin A) or mannose in the attachment media significantly decreased attachment/penetration by all isolates tested (eight different *S. typhimurium* isolates tested). These were expected results and contributed little new information. At this point in our studies, we were interested in determining the impact of lysogeny on pathogenicity. Phage were isolated from strains TML and B and were utilized to establish lysogeny in strain LT-2. These strains were tested in the invasion assay along with non-lysogenic variants of strains LT-2 and TML (Figure 3). The assay was repeated several times and the results remained consistent, implying that some factor or factors relative to invasion are encoded in the phage genome. Restriction endonuclease-generated maps of these phage genomes are different, as is the antibody inactivation kinetics utilizing antibodies directed against phage P22 (data not shown).

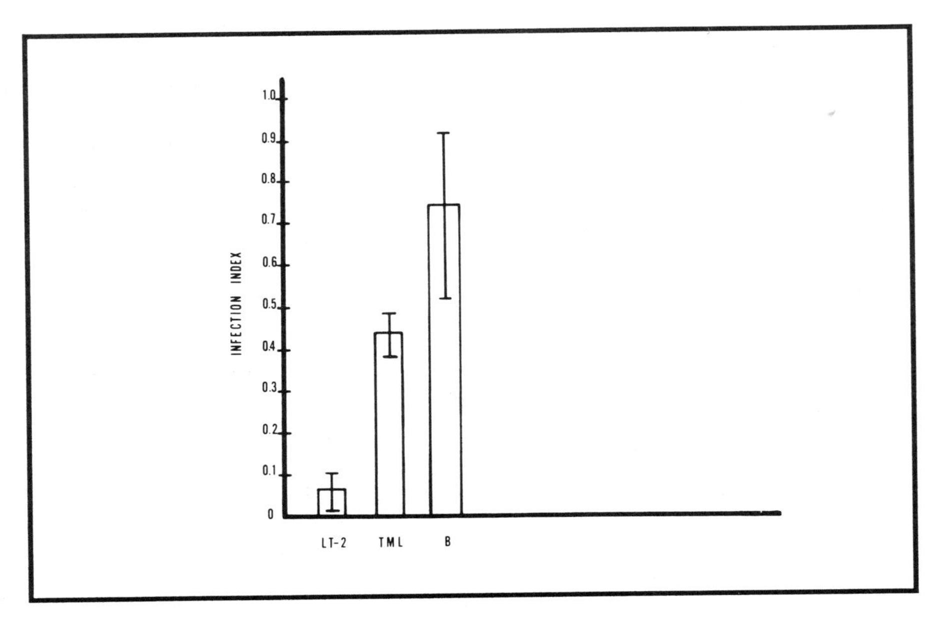

Fig. 1. Infection of HeLa cells by *S. typhimurium*.

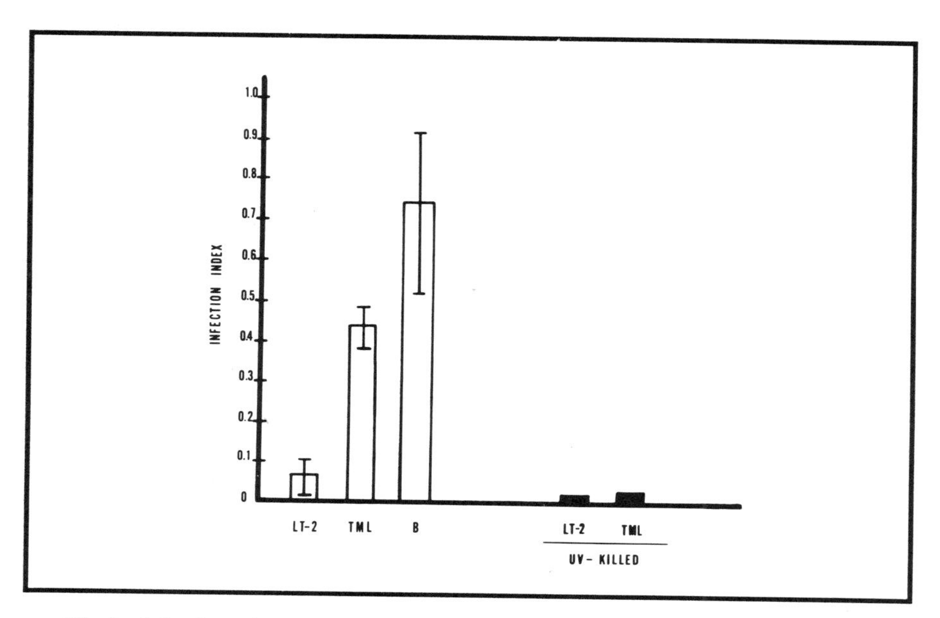

Fig. 2. Infection of HeLa cells by *S. typhimurium* (live and UV-irradiated cells).

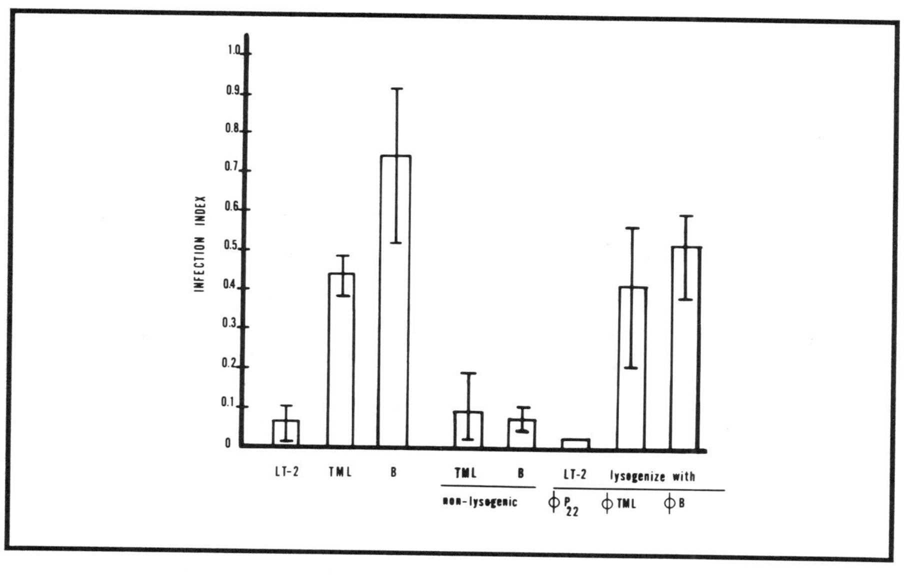

Fig. 3. Infection of HeLa cells by *S. typhimurium* (lysogeny and infectivity).

Table 3. Infection Index: Summary.

Strains	# Of Trials	Infection Index With Endogenous Phage	Infection Index Nonlysogenic Isolates
LT-2	18	(P22) = 0.13 (0.05 - 0.2) TML = 0.35 B = 0.36 Wi18 = 0.12	0.079 (0 - 0.177)
TML	18	0.85 (0.3 - 3.0)	0.26
Wi18	2	1.285 (0.5 - 2.07)	ND
B	12	0.625 (0.26 - 2.14)	0.8
PH*	1	1.6 (0.16 - 5.7)	ND
NBC**	1	2.2 (0.22 - 15.4)	ND

*12 different isolates
**14 different isolates
ND = Not done

A summary of the *S. typhimurium* data is presented in Table 3. At the time these data were collected, we were fairly comfortable with our interpretation and proceeded to screen 12 different isolates obtained from humans (PH strain) and 14 different isolates from horses and cows (NBC isolates). The vast range of values were difficult to explain and this line of study discontinued. Now, with the experience of *S. enteritidis* to learn from, we believe that the variance among the PH and NBC isolates is an expression of different virulence factors. Thus, this data needs to be resurrected and reevaluated.

When problems with the *S. enteritidis* became evident, the experiences described above provided insights to the development of protocols to assess attachment, invasion and intracellular growth. Table 4 lists the modifications of the original approach. Our first alteration was to use round coverslips in 24-well plates, and more recently we have begun to use 8-well chamber slides. One chamber serves as an uninoculated control. The inoculum can be standardized (approximately 10^7 cfu/well) or any other permutation can be performed. In the data presented, a single slide was removed for each time period, the seven chambers were inoculated with seven different strains. Our routine mandates that one of those strains is a repeat assay of a previous run and that the strain is inoculated into two different wells.

Table 4. Protocol for attachment: Invasion studies of *Salmonella enteritids*.

TIME (HRS)	STEP	MANIPULATION
-24	1.	Inoculate chamberslip with 10^4 HeLa Cells in antibiotic-free Dulbeccos minimal Eagles medium (ABF-DMEM).
	2.	Prepare 5 ml BHI cultures of bacterial isolates. Incubate 37 ° C/aerator.
-1.5	3.	Harvest and wash bacteria. Resuspend to standard number (10^9 cfu/ml).
0	4.	Inoculate 0.01 ml of desired suspension into each chamber well. One well is kept as uninoculated control.
1	5.	Remove media from wells. Add fresh, prewarmed DMEM containing gentamicin (100 μg/ml). (N.B. 1 slide culture removed, fixed and stained.)
24	6.	Remove media from slides, cell layer washed with PBS, fixed and stained.

ATTACHMENT

Modifying the above procedure slightly permitted collection of data on attachment mechanisms (6, 7). Overnight bacterial cultures were inoculated into the 24-well plates (containing coverslips with HeLa cells) and coverslips were removed at one hr and three hrs, rinsed in PBS, fixed and stained as previously described. The pattern of attachment for each strain was determined for 100 cells. In Table 5, Y8-P2 (also known as the Benson Strain) was the only isolate which appeared to have a loss of association in the three hr evaluation. Other investigators (Baker *et al.*, unpublished observations) have reported at a variety of conferences that Y8-P2 is less virulent than other strains and our results would seem to support such observations. This particular isolate has a claim to fame as Y8-P2 was the first *S. enteritidis* isolated from yolks during our investigation of eggs implicated in the 1987 New York outbreak. The other four strains were derived from a variety of sources: ovary, oviduct, intestine, and a fecal swab of a cat. The pattern of attachment was both diffuse (random attachment of one or more salmonellae around the surface of the cell membrane) and localized (five or more bacteria at a single site). The presence of mannose in the incubation medium did not alter the pattern of attachment, nor did the number of cells with bacteria attached.

Table 5. Diffuse vs. localized attachment.

Strain	#Cells With Attached Bacteria (100 Cells Examined)		Type of Attachment
	1 HR	3 HR	
Y8-P2	100	71	D + L*
575	100	96	D + L
52-1	88.5	98	D + L
31	99.5	93	D + L
32	98.5	87.5	D + L

*Diffuse
L = Localized

INVASION

We utilized the infection protocol to further assess attachment and penetration of HeLa cells on the coverslips or slides after one hour attachment and after one hour incubation in gentamicin. The second sample would be reflective of those bacteria which had become irreversibly associated with the HeLa cell. Finally, a 24 hr specimen would reflect those strains able to penetrate and survive within the eukaryotic cell. Attachment would appear to be an activity separate from invasion. In these assessments, the invasion of strain 52-1 may be unique and could be indicative of constitutive gene expression. The calculated infection index (Table 6) reflects a higher value: although interpretation is hampered by the lack of a control non-invasive strain. At the moment a naturally-occurring non-invasive variant of *S. enteritidis* has not been found. The use of fluorescent dye-labeled antibody directed to the common cell wall antigen to stain the bacteria showed a perinuclear predisposition of the intracellular microcolonies.

The issue of intracellular survival and pathogenicity was approached through the modification of an agar-overlay procedure(8). HeLa cells were grown in ABF-DMEM in 24-well plates and inoculated with varying dilutions of the bacteria. After one hour to permit attachment, the plates were washed with sterile PBS, and DMEM containing gentamicin (100 μg/ml) added to each well. Following 24 hrs of incubation this medium was removed; the wells washed twice with PBS and BHI agar containing 0.012% SPS (polyanthole-sulfonic acid, sodium salt) and tetrazolium chloride (0.5%) was added to each well. The agar was allowed to solidify, the plates are inverted and incubated at

Table 6. Exotoxin evaluation.

Cell Lines Utilized:	Toxin Production Tested For:
vero, 4-a, CHO	a) Cell-free CAYE broth
P388, L929, HeLa	b) Cleared sonicates

Toxin Sources:

35 different *S. enteritidis* = Negative in all cell lines and source

Commercial cholera toxin-positive in Y-1; CHO

37 ° C for 18 to 24 hrs. One lane of four wells was reserved as an uninoculated control. We hypothesize that colonies that develop are an expression of the collective attributes of a particular strain to attach, invade, and replicate within a hostile eukaryotic environment. Preliminary data indicates that the highest dilution of a culture with 25 or less colonies developing is within one log of a mouse MLD dose. There are encouraging indicators that this assay may be a good approximation of pathogenicity and data may be acquired quickly and inexpensively without employment of the traditional animal models. A rapid screening tool may be useful in assessing characteristics employing more extensive experiments with animals.

EXOTOXIN PRODUCTION

Some strains of *S. typhimurium* have been shown to produce an exotoxin similar to the LT of *Escherichia coli* and *Vibrio cholerae*. Thirty-five different isolates of *S. enteritidis* were found to be exotoxin negative for both LT-like toxin and verotoxin (Table 6). Two different approaches were utilized to harvest toxin from these strains. Crude cholera toxin were included in all assays. Verotoxin was unavailable for inclusion in the vero cell assays.

ANTIMICROBIAL SUSCEPTIBILITY

Thirty-four *S. enteritidis* isolates were evaluated in our routine screen for antimicrobial susceptibility. Thirty-five *Salmonella* Group B isolates and one Group D isolate, originally isolated from horses and cows admitted with a diagnosis of salmonellosis to the George D. Widener Hospital for Large Animals at New Bolton Center were included for comparison (Table 7). There is a remarkably uniform susceptibility among the poultry and egg isolates while most of the clinical samples were highly resistant to most antibiotics in the routine drug panel.

Table 7. Antimicrobiol susceptibility of *Salmonella*.

	% SUSCEPTIBILITY TO:*								
# OF ISOLATES	AN	AM	CM	CB	GM	K	SSS	SXT	TG
34-Poultry/egg isolates	100	97	100	88	100	94	0	100	94
35-NBC isolates (group B)	100	6	6	6	6	6	3	3	6
1-NBC isolate (group D)	100	100	100	100	0	100	0	100	100

*Determined by standard Kirby-Bauer method.

PLASMID ANALYSIS

The same 35 isolates utilized in the previous experiments evaluated by plasmid fingerprinting using the method of Platt (9). A species-specific plasmid (SSP) has been identified in 98% of these isolates. Fingerprints of the plasmid DNA with restriction endonuclease Pstl, Sma and AvaII are homogeneous.

Dr. Platt and I have found some strains of *S. enteritidis* which appear to have plasmids with similarities to the SSP of *S. typhimurium* and may be transitionary forms. Most strains carry only one plasmid, but recently we have begun to find isolates carrying a low molecular plasmid, in addition to the SSP.

GENOME FINGERPRINTS

Genomic DNA of the 35 isolates mentioned above have been digested by 16 different restriction endonucleases. Particularly interesting are digests by Hind III and Bgl. We are still evaluating the degree of homogeneity between these isolates; the initial impressions are that all the isolates confirmed by conventional serotyping are not homogeneous relative to the organization of genetic material. This conclusion was tested by probing southern blots of the DNA digests for the location of the ribosomal genes. A commercial preparation of rRNA was radiolabelled with p^{32} by terminal translocation. Differences in the location of rRNA genes have been found among some Bgl digests. We are still working through the 16 different digests and blots of the 35 strains and may find other dissimilarities as the study continues.

One should not be so presumptuous to believe that within a few short years all the mysteries of *S. enteritidis* pathogenesis will be laid open for inspection. The necessary systems for evaluation described here will facilitate the delineation of the virulence factors. There are differences among *S. enteritidis* relative to attachment, cell penetration and intracellular survival. No exotoxin has thus far been detected in cell-free culture fluids or sonicates. A species specific plasmid is present in most strains. Fingerprints and rRNA blots of genomic DNA are incomplete but current data indicates some differences in the orientation of the DNA.

ACKNOWLEDGEMENTS

This research was supported by grants from the Southeastern Poultry and Egg Association and the Commonwealth of Pennsylvania.

The following individuals have contributed their talents in these studies: Drs. Paul Strzemienski, Charles Lewis, Sherrill Davison, Mariano Salem, Linda Keller, and technicians Sue Erney, Lisa Barker, Kim Sprout, Caroline Yarbrough, Nancy Fitzkee, Gary Lotz, Gail Benson, Sheri Turner, and Sue Haske. Collaborators included Drs. David Kradel, Dennis Fernando, Wilson Miller, Pierre Brunett, and Michael Opitz.

This presentation is dedicated to the memory of my mother who died shortly before the symposium.

REFERENCES

1. **Giannella, R.A., O. Washington, P. Gemski, and S.B. Formal.** Invasion of HeLa cells by *Salmonella typhimurium*: A model for study of invasiveness of *Salmonella.* J. Infect. Dis. **128:**69-75.
2. **Baumberg S., and R. Freeman.** 1971. *Salmonella typhimurium* strain LT-2 is still pathogenic for man. J. Gen. Microbiol. **65:**99-100.
3. **Finlay, B.B., and S. Falkow.** 1988. Comparison of the invasion strategies used by *Salmonella choleraesuis, Shigelle flexneri* and *Yersinia enterocolitica* to enter cultured animal cells: endosome acidification is not required for bacterial invasion or intracellular replication. Biochimie. **70:**1089-1099.
4. **Small, P.L.C., R.R. Isberg, and S. Falkow.** 1987. Comparison of the ability of enteroinvasive *Escherichia coli*, *Salmonella typhimurium*, *Yersinia pseudotuberculosis* and *Yersinia enterocolitica* to enter and replicate within HEp-2 cells. Infect. Immun. **55:**1674-1679.
5. **Finlay, B.B., F. Heffron, and S. Falkow.** 1989. Epithelial cell surfaces induce *Salmonella* proteins required for bacterial adherence and invasion. Science **243:**940-943.
6. **Kihlstrom, E.** 1977. Infection of HeLa cells with *Salmonella typhimurium* 395 MS and MR 10 bacteria. Infect. Immun. **17:**290-295.

7. **Scaletsky, I.C.A., M.L.M. Silva, and L.R. Trabulsi.** 1984. Distinctive Patterns of adherence of enteropathogenic *Escherichia coli* to HeLa cells. Infect. Immun. **45:**534-536.
8. **Niesel, D.W., C.E. Chambers, and S.L. Stockman.** 1985. Quantitation of HeLa cell monolayer invasion by *Shigella* and *Salmonella* species. J. Clin. Microbiol. **22:**897-902.
9. **Platt, D.J., J.S. Chesham, D.J. Brown, C.A. Kraft, and J. Taggart.** 1986. Restriction enzyme fingerprinting of enterobacterial plasmids: A simple strategy with wide application. J. Hygiene **97:**205-210.

DISCUSSION

R. CURTISS: My first question has to do with the invasiveness of *Salmonella* strains in HeLa cells, as there is inconsistency in the literature. What was the osmolarity of the growth medium used for your *S. enteritidis* isolates?

C. BENSON: We use brain heart infusion broth. This medium does contain 0.5% sodium chloride. One thing that we have looked at is whether you do this as an aeroebic culture or a stationary culture and it makes no difference. Does it make a difference at what temperature I grow this thing at? *S. enteritidis* is an indiscriminate pathogen. No matter how you grow your innoculum, whether you grow it at 20, 37, or 42, you get the same results for invasivity. When I coat the slides and I have different people reading them, we have to break the code and figure out that the invasive properties are all the same.

R. CURTISS: Your conclusion about the constitutive expression of invasion is probably right with *S. typhimurium*. You certainly can show a temperature effect. That is, if you grow the organism at 30 ° C. It takes more time to become invasive as compared with when the organism is grown at 37 ° C or higher. This is similar to *Shigella*.

EXPERIMENTAL *SALMONELLA ENTERITIDIS* INFECTIONS IN CHICKENS

C. W. Beard
R. K. Gast

Southeast Poultry Research Laboratory
USDA, Agricultural Research Service
Athens, Georgia 30605

ABSTRACT

The epidemiologic association of human illnesses and deaths from *Salmonella enteritidis* with the consumption of Grade A table eggs was an unexpected development for public health officials and for the table-egg industry. Eggs had not been significantly involved as the source of human salmonellosis since the implementation of regulations on the pasteurization of egg products and the prohibition of the sale of cracked and/or dirty eggs nearly twenty years ago. In attempts to understand these developments, research was initiated to determine if strains of *Salmonella enteritidis* were different from other paratyphoids in the way they gained access to the internal contents of eggs. This paper presents data on the responses of laying hens experimentally exposed to *S. enteritidis* related to recovery of the bacterium from eggs and internal organs, effects on egg production, and serology.

The association of eggs and egg products with human salmonellosis was not an unusual event before 1970. Because eggs were so frequently incriminated as the origin of human salmonellosis in the 1960's, the public health officials of the Communicable Disease Center (now known as the Centers for Disease Control) utilized the egg-salmonellosis connection as one of the models in their short courses on the epidemiology of foodborne illness.

The passage of the Egg Products Inspection Act in late 1970 and its implementation in 1971 removed cracked and dirty eggs from the retail markets and required the pasteurization of egg breaking plant products. Because the contamination of the internal contents of the eggs was by penetration of the salmonellae through cracks and checks in the shell from debris on the external surface, the prohibition of the sale of such eggs and the requirement for egg product pasteurization resulted in a marked decrease in the problem.

It was a report published in 1988 by state and federal public health workers that brought eggs back into the forefront as a source of foodborne salmonellosis in humans. St. Louis *et al.* (8) reported a sixfold increase in foodborne illness due to *Salmonella enteritidis* in the Northeast United States. Epidemiologic investigations of an outbreak in a New York hospital implicated Grade A shell eggs used in uncooked mayonnaise prepared in the hospital kitchen as the source of the *S. enteritidis* outbreak. Numerous outbreaks of *S. enteritidis* in humans, mostly in institutional settings, were traced to grade A eggs in a wide variety of foods containing raw or partially cooked eggs. The incidence of the human illnesses was essentially confined to the warm weather months. Improper handling and/or storage of the egg-containing foods was a frequent finding in epidemiologic investigations (6). The number of outbreaks and states involved increased each year and extended into the Mid-Atlantic states.

Dr. Lisa Lee of the Centers for Disease Control (CDC) reported at the 1989 meeting of the *Salmonella* Committee of the U. S. Animal Health Association that in the U. S. during the period 1985-88, there were 140 outbreaks of *S. enteritidis* in humans in 12 states, producing 4,976 known illnesses, 896 hospitalizations, and 30 deaths. Sixty-five of 89 outbreaks where the food source of the *S. enteritidis* was determined included the consumption of Grade A shell eggs. Five of the 12 states that were the source of the eggs were outside the Northeast United States. Dr. Lee reported 56 outbreaks through July in 1989 with 1,270 illnesses and 12 deaths (11/12 in nursing homes). Where the food was identified, 20 of 29 diets included eggs.

In November, 1989, Dr. Sean Altercruse reported in a meeting of the General Conference Committee of the National Poultry Improvement Plan that there had been 68 human outbreaks in the U. S. in 1989 involving 33 restaurants, 15 nursing homes, six home settings, and four banquets.

The tracing of a significant number of human outbreaks of gastroenteritis caused by any paratyphoid salmonellae such as *S. enteritidis* to the consumption of food containing grade A shell eggs was a marked departure from earlier experiences. Because the history of egg-associated salmonellosis in humans had previously been connected to egg-breaking plant products which had not been pasteurized or to cracked and or dirty eggs, the initial reports of the association of human illness with the consumption of Grade A shell eggs was met with considerable skepticism in many professional and egg industry circles. The occurrence of the outbreaks in warm weather months and the high proportion of institutional kitchens involved led many to believe that the *S. enteritidis* problem was simply a repeat of events of the 1960's: cracked/dirty eggs becoming internally contaminated and improper handling and storage in warm kitchens and inadequate cooking amplifying the problem to result in the human outbreaks.

Although convincing reports describe the important role of product abuse in some outbreaks (6), there were enough occurrences of the problem without histories of product abuse and instances of laboratory isolation of *S. enteritidis* from eggs with apparently intact shells to indicate that this particular paratyphoid was gaining entry into the egg by means other than penetration through defective shells. The most logical hypothesis was that *S. enteritidis* was incorporated into the internal contents of the egg during its formation within the reproductive tract of the hen.

Salmonella pullorum and *Salmonella gallinarium* are two non-motile species of the genus that are well-known for their ability to cause severe disease in chickens and to be transmitted vertically through the eggs to progeny. Both of these species along with *S. enteritidis* belong to the *Salmonella* serogroup D. Over 80% of the salmonellae isolated from turkeys and chickens are members of the antigenic serogroups B, C, D, or E. Most species belong to serogroup C, followed by E, B, and lastly, D. Most poultry isolates belong to serogroup B. Because eggs have not been considered an important source of salmonellae for humans since the early 1970's, there has been very little recent research on the egg-*Salmonella* connection accounting for the paucity of data in this presentation. This lack of research may have led to the surprise associated with the unexpected developments whereby clean, Grade A eggs with intact shells were identified as the source of *S. enteritidis* in humans. Numerous unanswered questions concerning the contamination of eggs by infected hens complicated the development of effective and practical programs of prevention and control for *S. enteritidis*.

We initiated research at the Southeast Poultry Research Laboratory to address some of the more urgent questions. The first efforts were to determine if we could experimentally expose hens to isolates of *S. enteritidis* and recover the microorganism from the internal contents of their eggs (4). Initial attempts were unsuccessful. Hens were given large oral doses of broth cultures of several different *S. enteritidis* isolates. They exhibited no marked signs of disease while shedding the *S. enteritidis* in their feces, and no recoveries were made from eggs. It was a more recent isolate which was obtained from Dr. Charles Benson, University of Pennsylvania, that finally yielded positive results. Dr. Benson had been evaluating isolates in an *in vitro* cell culture assay that is based on cell invasiveness. He felt that his isolate 19299-52-1 was the most likely candidate with which we could simulate the field experience in the laboratory. Following oral inoculation, isolate 52-1 did result in some diarrhea, a transient decline in egg production and contamination of the internal contents of eggs, particularly those laid within the first two weeks. The *S. enteritidis* was recovered from both albumen and whole yolks. When the yolks were seared with a red hot spatula so that yolk could be aseptically removed using a syringe and needle, the yolk material was free of *S. enteritidis*. All eggs were collected and stored for four days at room temperature (25-27 C) before they were opened and the contents cultured. Several ages of hens have yielded similar results.

Control hens in an adjacent room which had been sham-inoculated with sterile growth medium were included in all experiments to compare egg productivity and to supply negative control eggs for isolation attempts. Sham-inoculated hens were also placed in layer cages adjacent to inoculated hens to determine if the *S. enteritidis* would spread horizontally. Contact hens shared waterers and feed troughs with inoculated hens. The contact hens became infected and also produced *S. enteritidis*-contaminated eggs.

Several serologic procedures were evaluated to determine their sensitivity and specificity in detecting infected hens. The microagglutination test using four antigens prepared with the homologous *S. enteritidis* strain, another avian *S. enteritidis* isolate, a human *S. enteritidis* isolate obtained from the CDC and *S. pullorum* were evaluated. The homologous antigen and the other avian *S. enteritidis* antigen yielded the highest titers, but all produced positive results with titers peaking at one to two weeks and declining slowly thereafter. Titers were still greater than 1:20 after ten weeks.

The whole blood plate test using two commercially available *S. pullorum* antigens were evaluated and found to yield results which were compatible with the microagglutination tests. The plate tests, because *S. enteritidis* and *S. pullorum* are in the same serogroup D, were an effective and rapid method of detecting antibodies against group D salmonellae.

When groups of inoculated hens were serially sacrificed and their internal organs cultured, *S. enteritidis* was recovered from 25% to 67% of ovaries, liver, spleen, and oviduct during the first two weeks. In two trials approximately 7% to 10% of hens yielded positive cloacal swab recoveries through 18 weeks following inoculation when they were terminated. Attempts to recover *S. enteritidis* from the eggs of inoculated hens after three weeks post inoculation were negative even though the hens remained colonized and continued to shed the organisms in the feces.

These results, however, do not absolutely preclude the possibility that *S.enteritidis* gained entry into the egg contents through the shell. The narrow window of positive recoveries from egg contents and the long period of fecal shedding could, however, be interpreted as an indirect indication that the egg contamination is occurring during egg formation. Additional support for a unique means of egg contamination during formation within the hen is the emergence of only S. enteritidis as a Grade A egg-associated problem while the other paratyphoids have not been a frequent problem since the early 1970's when the Egg Inspection Act was implemented.

Even before the *S. enteritidis* problem became known in the U. S., a similar problem was emerging in Europe and the U. K. (1, 2, 3, 5). An interesting distinction between the U. S. and European experiences is that although the problem in both areas is due to *S. enteritidis*, they are of different phage types. The *S. enteritidis* European isolates are predominately phage type 4 while there is no phage type 4 in the U. S. The U. S. outbreaks have been caused by several phage types of *S. enteritidis* including 8, 13, 13a, and 14. Another distinction is the apparently greater virulence of the phage type 4 isolates which has resulted in up to 20% mortality in young chickens (7).

The public knowledge of eggs as the source of human outbreaks of *S. enteritidis* in England followed with statements by a government health official concerning the probable widespread nature of the problem in layer flocks apparently resulted in a marked decline in egg sales in that country. Some estimates of the declines in egg consumption were up to 65%.

The concerns of this symposium appear to be mainly directed toward intestinal colonization as it relates to the problem of contamination of broiler meat. It is generally held that the public health hazard from contaminated broiler meat comes from cross-contamination in the kitchen or from improper cooking, in that custom and recipes for broiler meat preparation usually specify that it be well cooked and not served raw or even rare. Eggs, however, are frequently consumed raw or only partially cooked. Many widely accepted food dishes include completely raw eggs. These food products are often the foods that *S. enteritidis* epidemiologic tracebacks determine to be the source of human outbreaks.

It is highly unlikely that raw and/or partially cooked eggs will cease to be consumed. Personal tastes and custom will probably negate any efforts to change these eating habits. That leaves the burden of this public health problem on the poultry industry to produce eggs free of internal *Salmonella* contaminants and upon others in the marketing chain to assure the proper storage and handling of eggs should low levels of contamination exist. The final and probably a very significant role in preventing egg-source public health problems will rest with the user. Improper preparation and storage of egg dishes in the kitchen and on serving lines has been frequently associated with outbreaks of *S. enteritidis* (6). Had proper and accepted practices of refrigeration been utilized in the storage of egg products following break-out, many cases of illness would likely have been avoided.

The solution to the egg-associated *S. enteritidis* problem is not going to be accomplished with any simple or superficial measures. It will require a sustained effort within the layer breeder industry, the table egg industry, the egg distribution and marketing systems, and finally, the egg user. To avoid some of the potential problems associated with large institutional users, consideration should be given to the use of pasteurized egg products from egg breaking plants instead of the practice of large-scale breaking of eggs in institutional kitchens. It may eventually be a feasible and an accepted marketing concept to make pasteurized egg products readily available to the home user in containers such as one or two liter waxed paper cartons. With the use of these products, traditional recipes and customs using raw eggs could be preserved without any risk of eggborne salmonellosis.

REFERENCES

1. **Cowden, J.M., D. Chisholm, M. O'Mahony, D. Lynch, S.L. Mawer, G.E. Spain, L. Ward, and B. Rowe.** 1989. Two outbreaks of *Salmonella enteritidis* phagetype 4 infection associated with the consumption of fresh shell-egg products. Epidemiol. Infect. **103**:47-52.
2. **Cowden, J.M., D. Lynch, C.A. Joseph, M. O'Mahony, S.L. Mawer, B. Rowe, and C.L.R. Bartlett.** 1989. Case control study of infections with *Salmonella enteritidis* phage type 4 in England. Brit. Med. J. **299**:771-773.
3. **Coyle, E.F., C.D. Ribeiro, A.J. Howard, S.R. Palmer, H.I. Jones, L. Ward, and B. Rowe.** 1988. *Salmonella enteritidis* phage type 4 infection: Association with hens' eggs. Lancet **2**:1295-1297.
4. **Gast, Richard K., and C.W. Beard.** 1990. Production of *Salmonella enteritidis*-contaminated eggs by experimentally infected hens. Avian Dis. **34(2)**: 438-446.
5. **Hopper, S.A., and S. Mawer.** 1988. *Salmonella enteritidis* in commercial layer flock. Vet. Rec. **123**:351.
6. **Lin, F.C., J.G. Morris, D. Trump, D. Tilghman, P.K. Wood, N. Jackman, E. Israel, and J. P. Libonati.** 1988. Investigation of an outbreak of *Salmonella enteritidis* gastroenteritis associated with consumption of eggs in a restaurant chain in Maryland. Amer. J. Epidem. **128**:839-844.
7. **Lister, S.A.** 1988. *Salmonella enteritidis* infection in broilers and broiler breeders. Vet. Rec. **123**:350.

DISCUSSION

R. CURTISS: It appears that with this strain, SE6, the transmission is a consequence of the acute infection, since after three to four weeks you essentially get no *enteritidis* in any of the eggs sampled. If you were to take that strain and immunize birds prior to commencement of egg laying, although they would all get colonized, would any of them ever lay eggs with *enteritidis*. A related question is: are all *enteritidis* strains going to behave in this way, because if some can chronically infect and cause infections in ovaries and oviduct and transmission, it could be different than the strain that you have studied here?

C. BEARD: We do have to characterize the various strains in order to determine whether it is only in the acute phase. There are some people in this audience who have been out in the field fighting this problem on a day-to-day basis that can tell you stories that are unbelievable about one floor of a house being infected and another floor not infected, hens removed from the infected floor and ovaries sampled, and 23 out of 25 are positive. These are traced back to flocks where there have been human health problems. This is a small laboratory experiment that I am describing.

G. MEAD: We in the U. K. have been very interested to compare our experiences with this particular organism with your own. Dr. Humphrey with our PHLS has shown that there is significantly greater heat resistance with phage type 4 strains, which is certainly important in relation to the cooking of eggs. Is there any evidence of heat resistance among your isolates?

C. BEARD: We have not examined that. Talking about cooking eggs and the influence that the information could have, it also could be very significant if we do not find it in the internal liquid yolk contents because even when a sunny-side-up egg is cooked, the top of the egg is coagulated. So it may have real practical significance to know whether it is in the membrane or the liquid yolk, or with time and improper storage can get into the yolk content. No, we have not looked at the heat sensitivity.

SESSION IV: IMMUNIZATION

Convener: *Richard J. Meinersmann*

USDA, ARS , Russell Research Center
Athens, Georgia

The classic definition of immunity is based on the onanistic response: individuals with prior immunological knowledge, that is to say, exposure to a foreign substance or microorganism, will have an immune memory and respond more readily than an individual without memory. Responses to enteropathogens are evidently onanistic, with one important restriction. The restriction is that the duration of the memory is often very short, perhaps about six months.

The definitions of immunity evolved and came to be based on the demonstration of antibody responses, i.e. antibody titers, and cell mediated immune response. Antibody in the form of secretory IgA clearly is important in gastrointestinal immunity. Secretory IgA apparently has effect by inhibiting adherence, forming aggregates in the mucin layer which are swept away, or by inhibiting motility. The effector arm of the cell mediated immune response is not easily measured in mucosal immunity, but afferent processing is critical. T helper, suppressor, contrasuppressor, and switch cells are all involved and their manipulations may be critical in successful immunization.

The most modern definition of immunity is based on genetic linkage to the major histocompatibility complex (MHC). Linkage to the MHC is not *a priori* evidence of an immune function, but it strongly indicates that study of the immune response is appropriate. Dr. Hyun Lillehoj has been prominent in defining linkages of the MHC to responses by chickens against intestinal pathogens.

The gastrointestinal tract has special problems for the immune system. The lumen of the gut is not 'in' the body. Also, the immune system has to be selective so that the beneficial organisms need to be maintained. Enteric organisms do not make willing antigens, waiting for the immune response to clear them. The mechanisms that microorganisms use to establish and maintain colonization were the focus of the last session. In this session we will focus on methods of specifically enhancing the immune clearance mechanisms to eliminate enteropathogens (which may not actually be pathogenic to the chicken) from the gastrointestinal tract of the chickens.

NONRECOMBINANT AND RECOMBINANT AVIRULENT *SALMONELLA* LIVE VACCINES FOR POULTRY

R. Curtiss III
S. B. Porter
M. Munson
S. A. Tinge
J. O. Hassan
C. Gentry-Weeks
S. M. Kelly

Department of Biology
Washington University
St. Louis, Missouri 63130

ABSTRACT

Deletion (Δ) mutations in the genes for adenylate cyclase (*cya*) and cyclic AMP receptor protein (*crp*) render several *Salmonella* species completely avirulent and highly immunogenic for several animal hosts. Δ*cya*, Δ*crp*, or Δ*cya* and Δ*crp* mutations were introduced into four *S. typhimurium* and two *S. enteritidis* strains which are highly virulent with oral LD_{50}s of 10^4 to 10^7 CFU for one-day-old chicks. These mutants were completely avirulent such that one-day-old chicks survived oral challenge with 1×10^9 CFU. The Δ*cya* Δ*crp S. typhimurium* strains effectively colonize the intestinal tract but have limited ability to colonize deeper tissues such as the liver and spleen. Since avirulent Δ*cya* Δ*crp Salmonella* strains effectively colonize the gut-humoral lymphoid tissue (GALT) and are capable of stimulating secretory, humoral, and cellular immune responses, they can serve as effective antigen delivery vehicles for stimulating immune responses against foreign antigens specified by cloned genes. In order to achieve stable high-level expression of cloned genes expressing colonization and virulence antigens from other pathogens, we have devised a balance lethal host-vector system. The host chromosome contains a deletion for the gene encolding aspartate-β-semialdehyde dehydrogenase (*asd*) which imposes a requirement for the essential cell wall constituent diaminopimelic acid (DAP) and the cloning vector has the wild-type *asd*+ allele such that its loss leads to DAPless death of the bacterial. As a consequence, 100% of the recombinant avirulent *Salmonella* recovered from immunized animals several weeks after oral immunization continue to express the cloned gene product. We have been using this system to express colonization antigens of *Bordetella avium*, the causative

Colonization Control of Human Bacterial Enteropathogens in Poultry

agent of rhinotracheitis, to investigate induction of protective immunity to colonization by this pathogen in chickens and turkeys. Additional information on the safety, efficacy, and practicality of these oral immunization strategies will be presented.

INTRODUCTION

There are more than 1,800 serotypes of *Salmonella* combined into five major and many minor antigenic groups as defined by their O and H antigens (108). Nevertheless, many consider that only three species of *Salmonella* exist: *S. typhi, S. choleraesuis*, and *S. enteritidis*, the last of which contains the vast majority of serotypes (92). Most of the *S. enteritidis* serotypes (which we designate as species) have a relatively low host specificity and thus can infect a diversity of animal species including humans (92). *Salmonella* infection in humans affects predominantly the very young, the elderly, and immune compromised individuals (24, 28, 92) and in most cases is caused by contaminated food products. Recent studies reveal that a principle source of *Salmonella* infection in humans is contaminated poultry products (35, 173). The seven most prevalent *Salmonella* species isolated from poultry and associated with human disease are *S. enteritidis, S. typhimurium, S. heidelberg, S. infantis, S. agona, S. st. paul*, and *S. montivideo* (20, 24, 26, 78). Most of these *Salmonella* species cause gastroenteritis in humans with possible persistence and continued shedding. It is estimated (73), however, that in the U.S. less than 1% of these *Salmonella* infections are reported and accurately diagnosed. Based on numerous considerations, Chalker and Blaser (28) have estimated that there are between 800,000 and 3,700,000 *Salmonella* infections per year in the United States. Most *Salmonella* are transmitted through the food chain by fecal contamination of carcasses during the dressing operation (7). A more recent concern is associated with the transmission of *S. enteritidis* through the egg directly to the consumer (25, 26, 151), presumably because some strains of *S. enteritidis* can persistently infect the ovaries of laying hens (2, 110, 139). It is, therefore, evident that transmission of *Salmonella* to humans through persistent asymptomatic infection of farm animals and contamination of meat and eggs constitutes an important public health problem (25, 26, 78, 151, 168). An additional complication is the increasing isolation of multiple drug resistant *Salmonella* which account for 20 - 25 % of the human cases (88). It is believed that subtherapeutic amounts of antibiotics in animal feed select for resistant bacteria which eventually infect humans, thus exacerbating the health problem (88, 140, 160).

In mice and humans (and presumably in other mammals) a common mucosal immune network exists (49, 123, 125) such that presentation of antigens to the gut-associated lymphoid tissue (GALT or Peyer's patches) or to the bronchial association lymphoid tissue (BALT) triggers proliferation and dissemination of committed B cells to all secretory tissues and glands in the body with the ultimate production of secretory IgA (sIgA) (14, 15, 27, 107, 121, 172). sIgA directed against specific surface antigens of pathogens that colonize and pass through a mucosal surface serves to block their colonization and invasion (163, 174). There is also evidence to suggest that antigen-specific sIgA might facilitate antibody-dependent cytotoxicity mediated by lymphocytes in the gut epithelium and lamina propria (163, 164). Although a secretory immune response is inadequate to block infection by invasive pathogens completely, it does increase the dose of microorganisms necessary to cause disease (41, 125). Consequently, its induction should decrease the likelihood of infection and contagious spread of pathogens. *Salmonella* are invasive, facultative intracellular parasites and thus total protection against *Salmonella* infection and persistence requires the induction of humoral (149) and cellular (113) immunities to augment the secretory immune response which would lessen the likelihood for infection.

Upon oral ingestion, invasive *S. typhimurium* strains initially attach to, invade and colonize GALT prior to colonizing deeper tissues such as liver and spleen (23). Avirulent mutants of various *Salmonella* species have been isolated by numerous investigators (45, 46) and have often been shown to be capable of inducing protective immunity to a mammalian host against subsequent challenge with virulent *Salmonella* (45, 46). With the exception of avirulent mutants of *S. gallinarum* initially isolated by Smith (157, 158), all other work has been with mammalian hosts. More recently, it has been demonstrated that some avirulent *Salmonella* mutants are immunogenic and retain the ability to attach to, invade and persist in the GALT (41, 43, 51, 55, 70, 119) to thereby stimulate secretory, humoral and cellular immune responses.

Recombinant DNA techniques can be used to clone genes for colonization and virulence antigens from various microbial pathogens (77, 114, 170) and avirulent *S. typhimurium* strains can be used to synthesize these foreign proteins (31, 39, 53, 45). The recombinant avirulent *Salmonella* strains colonize the GALT and thus induce production of sIgA against the particular expressed colonization or virulence antigen in secretions which bathe mucosal surfaces (45). Humoral and cellular immune responses against the expressed cloned gene products are also detectable.

Based on the foregoing, it can be asked whether various means of attenuating *Salmonella* for mammals effectively attenuate *Salmonella* for avian species such as chickens. If so, it would then be appropriate to ask whether such attenuated *Salmonella* induce secretory, humoral and cellular immune responses against *Salmonella* antigens and if these responses in turn would diminish the ability of *Salmonella* to colonize, persist and be shed by such immunized chickens. Lastly, by use of recombinant avirulent *Salmonella* vaccine strains expressing colonization antigens for an avian respiratory pathogen, it would be possible to investigate whether chickens do have a generalized secretory immune system and whether its stimulation would block colonization of respiratory pathogens in the respiratory tract.

We review here prior studies on the genetic control over colonization and virulence of *Salmonella*, on means to render *Salmonella* avirulent yet immunogenic and on the use of recombinant avirulent *Salmonella* to elicit immune responses against expressed cloned gene products. We also report preliminary work to evaluate the applicability of information largely derived from studies of *Salmonella* in mammalian hosts to see whether these findings are applicable in chickens and to determine if it is possible to develop effective strategies for diminishing *Salmonella* infection, persistence and shedding by poultry.

GENETIC MODIFICATION OF *SALMONELLA*

S. typhimurium is a favorite microorganism for genetic studies. Because of its close genetic relationship with *Escherichia coli* (134), the most well studied organism from a molecular genetic point of view, a diversity of techniques and genetic modifications can be used to investigate the molecular genetic basis of *Salmonella* pathogenicity. Until recently, there was little concerted effort to apply all of these methodological approaches to study *Salmonella* virulence. Also, most early efforts to select attenuated *Salmonella* strains for evaluation as vaccines employed classical means of mutant induction and enrichment (5, 6, 61, 69, 157-159). Thus it was not until 1981 that Hoiseth and Stocker (87) employed transposon mutagenesis to generate avirulent immunogenic *Salmonella* strains for vaccines and not until 1987 that Hone *et al.* (89) used recombinant techniques to generate specific deletions in genes thought to engender avirulence and immunogenicity of vaccine strains. Only in the last several years have these approaches been used to analyze *Salmonella* pathogenicity (59, 62, 63, 65-67, 71, 80, 112, 128).

It is our belief that the most effective vaccine to immunize against *Salmonella* infection would be an attenuated derivative of a highly invasive *Salmonella* strain. In other words, it would be important for a vaccine strain to be fully capable of colonizing the intestinal tract and especially the GALT. In addition, although not substantiated by any rigorous experimental evidence, it could be surmised that persistence in deep tissues of the reticuloendothelial system might also be advantageous or even necessary to elicit high-level longterm protective immunity. Invasive strains of *Salmonella* with these properties are not always easy to work with or to manipulate genetically. It is therefore useful to isolate mutations in *S. typhimurium* LT-2 which was originally isolated and described by Zinder and Lederberg (177) and is very well characterized genetically with many mutant strains available (152). This strain is also reasonably avirulent, at least for healthy adults. One can then move mutations isolated and characterized in *S. typhimurium* LT-2 into the virulent invasive *Salmonella* strain to be used as a vaccine candidate. We therefore use *S. typhimurium* LT-2 strains that are restriction-deficient, modification proficient, sensitive to bacteriophage lambda because of the presence of the *E. coli* LamB receptor (84) and sensitive to the generalized transducing bacteriophages PlL4 (36, 109) and P22HT*int* (155) due to the presence of a *galE* mutation that permits reversible synthesis of core and O antigen components of the lipopolysaccharide (LPS) (137). Early studies with *Salmonella* to study virulence made use of chemical mutagens and radiation to isolate mutants that might have caused more than one mutational lesion, often in unknown genes. Such an approach is no longer necessary or desirable given the existence of very detailed genetic maps for *S. typhimurium* (152) and *E. coli* (4) and the diversity of refined methods for introducing single well-defined mutational gene defects (116).

Because of the concern for the intentional release of microorganisms modified by recombinant DNA techniques (38), it is prudent to consider developing attenuated live vaccine strain derivatives of *Salmonella* using conventional, yet precise, genetic methodologies. This can be accomplished by using the well characterized transposon Tn*10* (103) (Figure lA) which encodes resistance to tetracycline and has the possibility for transposing into essentially any gene within the *Salmonella* genome thereby interrupting its function. Random libraries with Tn*10* inserted into numerous sites within the *S. typhimurium* LT-2 genome (Figure lB) can easily be obtained upon infecting a rough (*galE*), bacteriophage lambda-sensitive *S. typhimurium* LT-2 strain with the bacteriophage lambda Tn*10* transposon vector NK561 (103). Bacteriophage lambda is unable to replicate in *Salmonella* (9, 10) and therefore all tetracycline-resistant isolates following infection with this lambda transposon vector must have the Tn*10* transposed to the chromosome or to the large virulence plasmid which *S. typhimurium* LT-2 possesses (96). One can then identify mutants with simultaneous resistance to tetracycline and a selected biochemical alteration.

Growth of the mutant with Tn*10* inserted into the desired gene in the presence of galactose allows the *galE* mutant to synthesize LPS core and sidechains to confer sensitivity to phage P22HT*int* and permits production of a transducing phage lysate. This P22HT*int* lysate can then be used to transduce the gene inactivated by Tn*10*, designated *genX*::Tn*10*, into any of a diversity of other *S. typhimurium* strains. Most *S. typhimurium* strains isolated in nature are either sensitive to P22 or are immune to it due to the presence of a temperate phage in the prophage state with the same immunity properties as P22 (176). Nevertheless, P22 can universally inject its DNA into all of these strains. P22 can thus be used to transduce any gene inactivated by insertion of Tn*10* into any *S. typhimurium* strain followed by selection for tetracycline resistance. Furthermore, since P22 attaches to and injects its DNA into strains of *Salmonella* with O antigen 12 (176), it can be used to move Tn*10*-induced mutations into species in serogroups A, B, D_1, and D_3 which include *S. typhimurium, S. typhi, S. dublin, S. enteritidis, S. gallinarum,* and *S. pullorum*.

Tn*10*-insertion mutagenesis has an additional beneficial attribute in that cells resistant to tetracycline become sensitive to the drug fusaric acid (17, 115). If one thus takes a strain with a *genX*::Tn*10* mutation and selects for growth on fusaric acid-containing media, one invariably selects for deletion mutations wherein the Tn*10* and adjacent sequences, including *genX*, are deleted from the chromosome so that the cell becomes sensitive to tetracycline and resistant to fusaric acid. Thus one can generate spontaneous deletion (Δ) mutations for *genX* which are unable to revert back to the wild-type state.

These strategies can be modified and expanded. Thus random genes in *Salmonella* can be mutated by use of the transposon Tn*5* which encodes determinants of kanamycin resistance (13) or by use of transposon fusion vectors in which the product of the gene into which the transposon inserts will be fused to the protein specified by the coding sequence on the transposon such as alkaline phosphatase encoded by the *phoA* gene (117), β-galactosidase encoded by the *lacZ* gene (104) or chloramphenicol acetyltransferase encoded by the *cat* gene (34). The Tn*phoA* (117) transposon fusion vector is very useful in the study of bacterial virulence since display of enzyme activity by alkaline phosphatase requires its transport across the cytoplasmic membrane. Thus, the only Tn*phoA* fusions that lead to positive display of alkaline phosphatase activity are those in which the Tn*phoA* is inserted into a gene whose product is normally trans- ported across the cytoplasmic membrane and is often localized to the outer membrane of the *Salmonella* cell. This is an important feature of this system since many colonization and virulence antigens are localized on the cell surface. Again, these Tn*phoA* insertions can be transduced from the strain in which they are isolated into a diversity of highly virulent strains to investigate their behavior upon infection of cells in culture (67, 71, 128) or upon infection of a suitable animal host.

These strategies have been used to isolate and characterize transposon-induced mutations in *Salmonella* genes for the synthesis of flagella (22, 82, 112, M. Carsiotis, B. A. D. Stocker, I. A. Holder, D. Weinstein, and A. D. O'Brien, Abstr. Annu. Meet. Am. Soc. Microbiol. 1987, B-169, p. 53), pili (97), and invasin (71, 93), that alter survival in macrophages (63, 128) or globally regulate metabolic as well as virulence attributes (43, 51, 62, 70). In addition, this strategy can be used to identify mutants with various other defects affecting some virulence attribute. Further study can then be conducted to delineate the biochemical nature of the defect and establish the specific process in the display of virulence that is defective.

The ability to make fusions in virulence genes such that the gene now expresses alkaline phosphatase (because of fusion to the *phoA* gene), β-galactosidase (because of fusion to the *lacZ* gene), or chloramphenicol acetyltransferase (because of fusion to the *cat* gene) enables one to investigate how the virulence gene is regulated in response to the environment (51, 120, 127) or during infection of cells in culture (66) or even to the animal host.

Recombinant techniques (116) can be used to accurately delineate the nature of deletion mutations generated by excision of Tn*10* from the chromosome following selection for fusaric acid resistance. In this case, chromosomal DNA from the mutant is isolated, restricted with several appropriate restriction enzymes, separated by electrophoresis on an agarose gel and then used for Southern blot analysis (161) with radioactively-labelled probes of cloned DNA sequences for the deleted gene or of DNA sequences flanking the deleted gene. A greater precision in defining the extent of deletions can be achieved, if desired, by using DNA sequencing strategies (153) and the polymerase chain reaction (130) to amplify sequences flanking the site of the deletion.

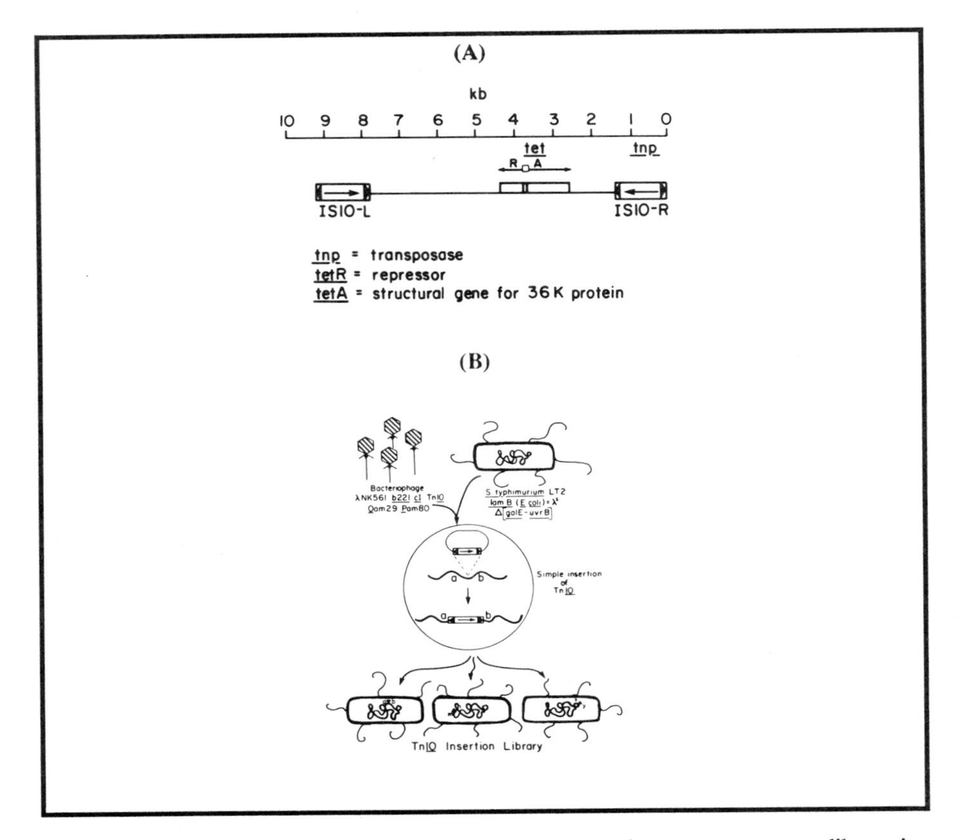

Fig. 1. The transposon Tn*10* (A) and the generation of a Tn*10* transposon library in *S. typhimurium* (B).

MECHANISMS OF *SALMONELLA* PATHOGENICITY

Studies in mice have revealed that LPS O antigen repeats on the surface of *S. typhimurium* are important not only to withstand nonspecific host defense mechanisms (148) but also for effective invasion through the mucin and glycocalyx covering the intestinal epithelium (122, 135). As a consequence, rough mutants lacking LPS O antigens, when given orally, tend to exhibit poor invasive abilities and pass right through the intestinal tract (135). The degree to which this is true in poultry has not been rigorously investigated although it is clear from early studies by Smith (157, 158) and individuals using his *S. gallinarum* and *S. typhimurium* rough strains for oral challenge of birds (81, 83, 156) that there is an impairment of intestinal colonization in the absence of wild-type smooth LPS (122, 138, 156).

The specific adhesins that enable *S. typhimurium*, after passing through the mucin and glycocalyx, to attach to enterocytes lining the intestine or to the M cells (111) overlying the GALT, are as yet unknown. Although antibody responses to flagella can be protective (82) and the presence of flagella and display of motility enhance the rate with which *Salmonella* attach to and invade cells in culture, the absence of flagella or motility

Table 1. Virulence of *inv*$^+$ and *invA*::Tn*phoA* *S. typhimurium* SL1344 strains for orally inoculated one-day-old White Leghorn chicks.

Strain	Genotype	Inoculating Dose (CFU)	Survival Live/Total	Mean Day of Death
χ3339	*inv*$^+$	2×10^4	1/4	6.0
χ3643	*invA*::Tn*phoA*	8×10^4	3/4	8.0
χ3643	*invA*::Tn*phoA*	8×10^5	4/4	-
χ3643	*invA*::Tn*phoA*	8×10^6	3/4	7.0

[a]Bacteria for inoculation were grown as overnight standing cultures at 37 ° C in L broth (109). These cultures were diluted one to twenty into prewarmed L broth and aerated at 37 ° C for two to three hours until an optical density at 600 nm of about 0.8 to 1.0 was reached. The cells were concentrated 20-fold by centrifugation at 8000 x g for 10 min at room temperature followed by suspension in buffered saline plus gelatin (BSG). Fertile White Leghorns eggs (SPAFAS, Roanoke, IL) were incubated and hatched in Humidaire incubator-hatchers. Newly hatched chicks were given 100 μl of the appropriate dilution of *Salmonella* via micropipette tip before being given food and water. Food and water were given to inoculated birds 30 minutes after infection. Birds were monitored daily for signs of disease (i.e., diarrhea, drooping, loss of appetite, weight loss, unresponsiveness, and death). Infected birds were housed in modified guinea pig cages with filter bonnet tops, wire floors, and thermostatically regulated temperatures in an animal room affording P2 level of containment. All materials leaving this room were autoclaved prior to further processing or dishwashing.

is without effect on the ability of *Salmonella* to colonize the intestinal tract and has no influence on the LD_{50} for *Salmonella* administered orally to mice (112). Similarly, type 1 pili which are present on *Salmonella* (56) and which can be shown to facilitate mannose-sensitive adherence to cells in culture (97) are also without effect on intestinal colonization (Lockman and Curtiss, unpublished). This is not to say that flagella and pili might not serve as adhesins to facilitate intestinal colonization, only that neither are essential for colonization. Ultimately, by successively inactivating genes for many potential adhesins, it should be possible to recover *Salmonella* mutants that are defective in intestinal colonization such that adding back each wild-type gene would enable one to quantitate the relative contribution of any and all putative adhesins.

Galán and Curtiss (71) have identified four genes which govern the ability of *S. typhimurium* to invade cells in culture as well as to invade cells lining the intestinal tract in mice. The invasion mechanism is specified by three proteins encoded by the *invA*, *B*, and *C* genes, which constitute an operon, and the *invD* gene which is closely linked to the *invABC* operon (71). The *invA* gene encodes a 54 kDa protein located in the outer membrane of *S. typhimurium* and which has been insertionally inactivated by fusion to the transposon Tn*phoA*. Since *S. typhimurium invA* mutants are defective in colonizing the intestinal tract and in reaching deeper tissues they are not particularly effective in inducing protective immunity. We have begun to investigate whether mutations in colonization and virulence determinants of importance for *S. typhimurium* virulence to mice have the same influence on colonization and virulence in one-day-old chicks. We have therefore introduced these mutations into *S. typhimurium* strains that are highly virulent and invasive for one-day-old chicks. Table 1 demonstrates that an *invA* mutant of the highly virulent *S. typhimurium* SL1344 strain has very much reduced virulence by the oral route of inoculation in one-day-old chicks. The results obtained are thus quite parallel to those seen in an infected mammalian host (71).

Table 2. Virulence of *phoP*+ and *phoP*::Tn*10* *S. typhimurium* strains for orally inoculated one-day-old chicks[a].

Strain	Genotype	Inoculation Dose (CFU)	Survival Live/Total
χ3761[b]	*phoP*$^+$	2×10^3	0/3
χ4126	*phoP*::Tn*10*	9×10^4	3/4
χ4126	*phoP*::Tn*10*	9×10^5	3/4
χ4126	*phoP*::Tn*10*	9×10^6	2/4
χ4126	*phoP*::Tn*10*	9×10^7	3/4

[a]See Table 1 footnotes.

[b]χ3761 was isolated from the spleen of a chick orally inoculated three days earlier with χ3663, a highly virulent *S. typhimurium* strain isolated from an infected horse. χ3761 is the parent of χ4126.

The *phoP* gene is unique to *Salmonella* and serves as a positive regulator necessary for the expression of genes including those for acid phosphatases (101, 102) and others necessary for the ability of *Salmonella* to survive in macrophages. Thus Fields *et al.* (63) originally isolated a number of mutants with transposon-induced mutations impairing their ability to survive in macrophages and the most impaired had *phoP*::Tn*10* mutations (62). Independently, we had found that *phoP* mutants were totally avirulent for mice and yet induced high-level protective immunity to challenge with virulent *Salmonella* (70). We thus isolated a Tn*10* insertion in the *phoP* gene and introduced it into the highly virulent *S. typhimurium* strain χ3761 (which had been recovered from the spleen of a chicken orally infected at one-day of age with the highly virulent *S. typhimurium* strain (χ3663). Based on numerous experiments, the oral LD_{50} of χ3761 for one-day-old chicks is 3×10^3 colony forming units (CFU). As is revealed by the data in Table 2, the *phoP*::Tn*10* mutation in χ3761 (yielding strain χ4126) was substantially avirulent but nevertheless led to the death of some chicks at all inoculating doses. It therefore appears that *S. typhimurium* with a *phoP*::Tn*10* mutation retains virulence for some chicks which in turn implies that the nature of interaction between *S. typhimurium* and avian macrophages may be different than the interaction between *S. typhimurium* and murine macrophages. Of course, the avian population challenged was outbred whereas the mice used in previous studies were inbred.

Table 3. Virulence of wild-type and plasmid-cured *Salmonella* strains for orally inoculated one-day-old chicks[a].

Strain		Origin	Vir Plasmid	LD_{50} (CFU)	Mean Day of Death
S. typhimurium	χ3761	χ3663	+	3×10^3	5.5
	χ3870	χ3761	-	5×10^5	5.5
S. enteritidis	χ3700	4937[b]	+	1×10^7	6.5
	χ3866	χ3700	-	2×10^9	7.6
S. enteritidis	χ3895	B6996[c]	+	3×10^6	6.0
	χ3996	χ3895	-	2×10^7	6.0

[a]See Table 1 footnotes.
[b]From J. Glenn Morris, Jr., University of Maryland School of Medicine.
[c]From Charles E. Benson, University of Pennsylvania.

ROLE OF THE *S. TYPHIMURIUM* AND *S. ENTERITIDIS* VIRULENCE PLASMID IN INVASIVE DISEASE

Invasive strains of *S. typhimurium, S. enteritidis, S. dublin, S. choleraesuis, S. gallinarum*, and *S. pullorum* possess large plasmids that contribute to their virulence (11, 29, 79, 86, 96, 99, 131, 132, 143, 166). In the case of *S. typhimurium* (79, 143), *S. enteritidis* (132), and *S. dublin* (29, 166) infection of mice, the virulence plasmid contributes to the ability of the strains to effectively colonize deep tissues such as mesenteric lymph nodes, liver and spleen whereas its loss in cured strains is without effect on ability to effectively colonize the intestinal tract including the GALT. The *virA* gene, which encodes a 28 kDa protein, is largely responsible for the ability of plasmid-containing strains to effectively colonize deep tissues (80). It has also been shown that the *S. typhimurium virA* gene hybridizes to the large plasmids in all isolates of the six invasive *Salmonella* species listed above (145). The virulence plasmid is not necessary, however, for *S. typhimurium* to attach to, invade or multiply in cells in culture (79). It is clear from previous studies that the virulence plasmid in *S. gallinarum* (12) and *S. pullorum* (11) contribute significantly to the potential of these avian pathogens to cause invasive infectious disease after oral challenge to chickens. Fewer studies, however, have been conducted to investigate the role of the virulence plasmids of *S. typhimurium* and *S. enteritidis* in effective colonization of avian tissues after oral administration and causation of disease. We have therefore generated isogenic plasmid-free strains using a site-specific insertion of a kanamycin-resistance (Km^r) determinant within the *parA* gene of the virulence plasmid (167). Since the *parA* gene product is responsible for precise partitioning of progeny virulence plasmid replicas to progeny cells prior to cell division, its inactivation by P22HT*int*-mediated transduction of the *parA*::Km^r insert destabilizes the virulence plasmid which results in generation of plasmid-free kanamycin-sensitive cells (167). One can also demonstrate by reintroduction of the virulence plasmid that the sole genetic change is the loss of the virulence plasmid with no other mutational lesions associated with changes in virulence. This is a vast improvement over previous curing regimens which used harsh treatments such as long-term cultivation of cultures at 42 - 43 ° C (143), incubation in the presence of low concentrations of novobiocin (79, 131) and/or treatment of cultures with mutagens (131). In these instances, secondary mutations were often selected which may have had secondary influences on virulence. The data in Table 3 reveal that plasmid-cured derivatives of both *S. typhimurium* and *S. enteritidis* have an increased LD_{50} following oral inoculation of one-day-old chicks. Therefore, the virulence plasmids of *S. typhimurium* and *S. enteritidis* play a role in contributing to invasive disease in chickens when the *Salmonella* cells are administered by the oral route.

ISOLATION AND CHARACTERIZATION OF AVIRULENT *SALMONELLA* MUTANTS

Bacon and collaborators (5, 6) were first to isolate auxotrophic mutants of *S. typhi* to evaluate their virulence in mice. Subsequent to their studies, many investigators have isolated and characterized mutants of various *Salmonella* species for virulence and immunogenicity, usually in mice and more rarely in other animal species such as chickens and humans. Table 4 enumerates the many types of avirulent mutants discovered which have mutations conferring auxotrophy, drug dependence, alterations in the utilization and/or synthesis of carbohydrates, temperature sensitivity with regard to growth, and defects in global control of gene expression. In addition, several investigators have conducted studies with virulence plasmid-cured derivatives of *S. enteritidis* (132) and *S. dublin* (64) as potential oral vaccines. The facts that plasmid-cured strains might not elicit as high a level of protective immunity as plasmid-containing strains and are quite virulent when administered by routes other than peroral (79) diminishes enthusiasm for their use, other than in conjunction with other means of attenuation.

The ideal attenuated vaccine strain should have a number of attributes. First, it should be completely avirulent and highly immunogenic. This combination of traits is often difficult to achieve since hyperattenuation often reduces immunogenicity and high-level immunogenicity may be associated with an unacceptable number of adverse symptoms in vaccinated individuals. For example, *S. typhimurium* purine-requiring mutants have been found to be totally avirulent (124) but, although they persist in tissues for a considerable period of time, induce a very poor immune response (141). On the other hand, precise excision of a *galE* gene by recombinant techniques in *S. typhi* was found to be insufficient to adequately attenuate so that some vaccinated individuals contracted typhoid fever (91). Second, the vaccine strain should retain its tissue tropism without causing disease or impairment of normal host physiology and growth. Data on the ability of the mutant *Salmonella* listed in Table 4 to effectively colonize the intestinal tract including the GALT are generally unavailable with the exception of data on mutants with Δ*aro* (141), Δ*pur* (19, 141), Δ*cya* (43), Δ*χrp* (43), and Δ*phoP* (70) mutations which all seem to retain tropism for cells in the intestinal tract. Third, for safety, the strains should have two or more attenuating deletion mutations. For this reason, vaccines with mutations to streptomycin dependency or conferring temperature-sensitivity, which are due to revertable point mutations, are not satisfactory since some vaccinated individuals could contract disease from revertants in the immunizing dose. This has in fact occurred (171). Fourth, the vaccine strains should have the attenuating phenotype unaffected by the diet or the host. This is most easily achieved by using mutations that do not confer auxotrophy but directly affect a virulence attribute or eliminate a gene function necessary for regulation of other genes. Lastly, the vaccine strain should be easy to grow and store in a manner to maintain the avirulence and immunogenic phenotypes.

Several years ago, during a review search for *Salmonella* genes that might contribute to the ability of that pathogen to infect and persist in animal tissues, it became apparent that many putatively required genes were subject to catabolite repression such that their transcription would be dependent upon the presence of cyclic AMP (cAMP) and its interaction with the cAMP receptor protein (CRP) to facilitate gene transcription (1). Subsequent work revealed that Tn*10*-induced mutations in the *cya* gene for adenylate cyclase and the *crp* gene for the CRP protein rendered *Salmonella* avirulent and immunogenic for mice (43). Subsequent studies demonstrated that oral inoculation of mice with as few as 10^7 Δ*cya* Δ*crp S. typhimurium* cells engendered immunity to challenge 30 days later with as many as $1x10^9$ virulent wild-type *Salmonella* cells which is 10,000 times the LD_{50} (44). Furthermore, it was found that protective immunity could be detected eight days after oral immunization and lasted for at least four months (47). Δ*cya* Δ*crp S. typhimurium* strains grow more slowly than wild-type cells *in vitro* as well as in cells in culture. This is undoubtedly due to the fact that the interaction between cAMP and the CRP is needed for the expression of numerous genes for carbohydrate transport and utilization, for glycogen synthesis, for amino acid transport and breakdown and for a number of other processes that would facilitate rapid growth in a diversity of environments (45).

Based on the foregoing, it was appropriate to investigate whether Δ*cya* Δ*crp* derivatives of *S. typhimurium* and *S. enteritidis* strains which were highly virulent for one-day-old chicks would be avirulent and engender some type of immune response in chickens. Highly virulent strains were identified by screening a large number of *S. typhimurium* and *S. enteritidis* strains for ability to cause lethal disease by oral inoculation of one-day-old chicks. The strains were collected from colleagues all over the world and were recovered from chickens, turkeys, pigs, calves, dogs, cats, horses and humans with clinical disease, with subclinical infections and with persistent infections. Other strains were obtained from contaminated eggs or foodstuffs and still others have an unknown etiology.

Initially, all these strains were screened for auxotrophy or prototrophy, ability to produce colicins, levels of drug resistance, metabolic activities, presence of prophages, and presence and sizes of plasmids. Since we are interested in genetically modifying strains, all strains were also evaluated for ease of genetic manipulation such as by transduction with phages P22HT*int* and PlL4, by conjugational transfer or by transformation including electroporation of plasmid DNA. In general, all strains displaying plasmid-mediated antibiotic resistance were excluded from further study since the Food and Drug Administration has mandated that live vaccine strains be free from antibiotic resistance mutations.

Based on these analyses, we selected five *S. typhimurium* and two *S. enteritidis* strains to use for study on mechanisms of *Salmonella* virulence for poultry and to attenuate for evaluation as potential vaccines to immunize poultry against *Salmonella* infection, persistance and shedding.

We next introduced *cya*::Tn*10* and Tn*10* mutations into one of the avian virulent *S. typhimurium* strains SL1344 (χ3339) and selected a diversity of fusaric acid-resistant isolates each having different amounts of the *cya* and/or *crp* gene deleted. Thorough tests of these independently isolated mutants led to the selection of several for further study. These deletion mutations were then co-transduced with a closely linked Tn*10* insertion into the selected *S. typhimurium* and *S. enteritidis* strains virulent for one-day-old chicks. It is apparent from the results presented in Table 5, as well as data from additional experiments, that Δ*cya* or Δ*crp* or Δ*cya* and Δ*crp* mutations render all *S. typhimurium* and *S. enteritidis* strains tested highly avirulent when administered by the oral route. It is striking to note that for the *S. typhimurium* strain χ3761. which has a peroral LD_{50} of 3×10^3, one-day-old chicks can be orally inoculated with 1,000,000 times this LD_{50} dose of the Δ*cya* Δ*crp* strain χ3985 (which was derived from χ3761) without any detectable signs of disease.

During the first week of life chicks become increasingly tolerant to *Salmonella* infection such that one-week-old chicks are totally resistant to oral challenge with 1×10^9 of any of the *S. typhimurium* or *S. enteritidis* strains that are capable of lethal infection when one-day-old chicks are inoculated. Because of this, it is possible to examine colonization of three- or seven-day-old chicks by oral inoculation with high titers of these virulent invasive *S. typhimurium* and *S. enteritidis* strains. The result is very efficient colonization with high-level excretion over a prolonged period of time. In contrast, birds infected with 1×10^9 of the Δ*cya* Δ*crp* strain χ3985. Although initially colonized at a substantial level, ceased to be colonized and/or to excrete the *Salmonella* vaccine strain after four to five weeks.

Table 4. Mutations rendering *Salmonella* avirulent.

Gene	Mutant Phenotype	Reference
pab	requirement of *p*ABA	Bacon *et al.* (5, 6); Brown and Stocker (19)
*asp**	requirement for aspartic acid	Bacon and *et al.* (5, 6); Kelly and Curtiss, unpublished
*his**	requirement for histidine	Bacon *et al.* (5, 6); Fields *et al.* (63)
*cys**	requirement for cystine	Bacon *et al.* (5, 6)
pur	requirement for purines	Bacon *et al* (5, 6); McFarland and Stocker (124); Fields *et al.* (63)
?	rough; defective in LPS synthesis	Smith (157, 158)
aroA	requirement for aromatic amino acids, *p*ABA and dihydroxybenzoic acid	Hoiseth and Stocker (87); Dougan *et al.* (54); Mukkur *et al.* (129); Lascelles *et al.* (106)
aroC	requirement for aromatic amino acids, *p*ABA and dihydroxybenzoic acid	Dougan *et al.* (52)
asd	requirement fro threonine, methionine, and diaminopimelic acid	Curtiss (37)
dap	requirement for diaminopimelic acid	Clarke and Gyles (30)
purA	requirement for adenine	Brown and Stocker (58)
purHD	requirement for hypoxanthine and thiamine	Edwards and Stocker (58)
nadA	requirement for quinolinic acid	A. K. Wilson and B. A. D. Stocker, Abstr. Annu. Meet. Soc. Microbiol. 1988, D-73, p. 83.
pncB	requirement for nicotinic acid or nicotinamide mononucleotide	A. K. Wilson and B .A. D. Stocker, Abstr. Annu. Meet. Soc. Microbiol. 1988.
rpsL	streptomycin-dependent	Reitman (146); Dupont *et al.* (57); Cvjetznovic (48)
galE	renders cells reversibly rough	Germanier and Furer (75, 76); Hone *et al.* (89)
Ts	decrease cell proliferation at 37 ° C	Fahey and Cooper (61); Ohta *et al.* (142)
cya	inefficient transport and use of carbohydrates and amino acids and inability to synthesize cell surface structures	Curtiss and Kelly (43)
crp	inefficient transport and use of carbohydrates and amino acids and inability to synthesize cell surface structures	Curtiss and Kelly (43)
phoP	regulates genes for acid phosphatases and virulence	Galán and Curtiss (70)
ompR	regulates synthesis of some outer membrane proteins	Dorman *et al.* (51)

*Only some mutants of these types are avirulent and the avirulent mutants have not been investigated for immunogenicity.

These results suggest that Δ*cya* Δ*crp S. typhimurium* strains might be capable of inducing an immune response in young chicks that would inhibit or diminish colonization by wild-type virulent *S. typhimurium* strains and reduce the period and titers of *Salmonella* shed.

Table 5. Virulence of *Salmonella* wild-type and attenuated mutants for orally inoculated one-day-old chicks[a].

Strain	Genotype	Origin	LD_{50} (CFU)
A. *S. typhimurium*			
χ3306	*gyrA1816*	SR-11	$>1 \times 10^9$
χ4064	Δ*cya-1* Δ*crp-2 gyrA1816*	χ3306	$>1 \times 10^9$
χ3663	wild type	30875[c]	2×10^4
χ3779	Δ*crp-10*	χ3663	$>2 \times 10^9$
χ3761[b]	wild type	χ3663	3×10^3
χ3784	Δ*crp-10*	χ3761	$>5 \times 10^8$
χ3954	Δ*crp-11*	χ3761	$>2 \times 10^8$
χ3962	Δ*cya-12*	χ3761	$>3 \times 10^8$
χ3985	Δ*crp-11* Δ*cya-12*	χ3954	$>4 \times 10^9$
χ3739	wild type	3860C[d]	2×10^5
χ3780	Δ*crp-10*	χ3739	$>1 \times 10^9$
B. *S. enteritidis*			
χ3700	wild type	4937	1×10^7
χ3779	Δ*crp-10*	χ3739	$>1 \times 10^9$

[a]See Table 1 foonotes.
[b]See Table 2 footnote b.
[c]From Patrick McDonough, Cornell University.
[d]From Robert C. Clarke, University of Guelph.

RECOMBINANT AVIRULENT *SALMONELLA* HOST-VECTOR SYSTEMS

The discovery that delivery of antigens to the GALT elicited a generalized secretory immune response (14, 15, 27, 107, 121, 172) and the subsequent findings that this route of antigen delivery also elicited humoral and cellular immune responses (16, 18, 41, 53-55, 60, 118, 119, 162) stimulated research to develop systems for efficient delivery of antigens to the GALT to stimulate all three branches of the immune system. Since *Salmonella*, including some of their avirulent derivatives attach to, invade and colonize the GALT as part of the normal infective process (23, 43) and since genes for colonization and virulence antigens from other pathogens can be cloned into and expressed by avirulent *Salmonella*, there has been significant effort to develop recombinant avirulent *Salmonella* for oral vaccination. Table 6 lists various recombinant avirulent *Salmonella* vaccines that have been constructed and enumerates the types of immune response detected. In most instances, however, protective immunity against infection by the pathogen whose cloned genes were expressed by the recombinant avirulent *Salmonella* were not reported. In some cases, this was undoubtedly due to the instability of the constructs which could not be stably maintained in the immunized animal.

We have found that Δ*cya* Δ*crp* *S. typhimurium* strains efficiently express colonization and virulence antigens specified by cloned genes from other pathogens to a higher level than observed for other avirulent *Salmonella* strains and for many of the standard *E. coli* hosts (40). Another advantage was the discovery that some cloning vectors were maintained more stably in Δ*cya* Δ*crp* strains than in other *E. coli* and *S. typhimurium* strains (40). Both of these features might be due to the slower growth rate of the Δ*cya* Δ*crp* *Salmonella*. Nevertheless, achieving stable high-level expression when the recombinant avirulent *Salmonella* is in the animal host is essential in order to induce an early immune response. Constructs made by numerous individuals (Table 6) including ourselves (41) were not particularly stable such that the majority of *Salmonella* recovered from animals several days after oral immunization no longer produced the cloned gene product. A further problem encountered has been the decision of governmental regulatory agencies to preclude use of live vaccine strains displaying resistance to antimicrobial agents. We therefore designed and constructed a balanced-lethal host-vector system (133) with: (1) a deletion mutation in the *Salmonella* chromosome blocking synthesis of an essential metabolite that is not readily available in nature and not synhesized by animals and, (2) a plasmid cloning vector without any drug resistance gene which contains a gene to complement the chromosome deletion mutation and permit the recombinant cells to synthesize the essential metabolite. In this system, loss of the plasmid in any cell leads to cell death (133). For this purpose, we used mutants containing a deletion mutation of the *asd* gene encoding aspartic-β-semialdehyde dehydrogenase, an enzyme required for the synthesis of diaminopimelic acid (DAP), an essential constituent of the rigid layer of the cell wall in all gram-negative and some gram-positive bacteria (154). Strains with Δ*asd* mutations when grown under conditions where protein synthesis is possible but DAP is absent continue to synthesize protein, but not cell wall, and undergo DAPless death with cell lysis and liberation of cell contents (126). The wild-type *asd* gene has been cloned from a number of microorganisms including Streptococcus mutans (21) and *S. typhimurium* (J.E. Galán, K. Nakayama, and R. Curtiss, III, Abstr. Annu. Meet. Am. Soc. Microbiol. 1989, D-46, p. 90) and plasmid cloning vectors have been constructed with the *asd*+ gene as the selectable marker (133, J.E. Galán, *et al.*, Abstr. Annu. Meet. Am. Soc. Microbiol. 1989). These plasmid vectors also contain a strong constitutive promoter, a multiple cloning site, a transcription terminator and a cassette specifying a plasmid replicon of chosen plasmid copy number per chromosome DNA equivalent (133, J.E. Galán, *et al.*, Abstr. Annu. Meet. Am. Soc. Microbiol. 1989). One such vector pYA248 is depicted in Figure 2.

Table 6. Recombinant avirulent *Salmonella* to induce immune responses.

Salmonella Mutant	Expressed Foreign Antigen	Type of Immune Response	References
S. typhi Ty21a *galE*	*Shigella sonnei* O-antigen	secretory & humoral	Formal *et al.* (68); Tramont *et al.*; Black *et al.*
S. typhi Ty21a *galE*	B subunit of *Escherichia coli* enterotoxin	humoral	Clements & El-Morshidy (32)
S. typhimurium *galE*	*E. coli* K88 fimbriae	secretory & humoral	Stevenson & Manning (162)
S. dublin Δ*aroA*	B subunit of *E. coli* enterotoxin	secretory & humoral	Clements *et al.* (33)
S. typhi Ty21a *galE*	*Streptococcus mutans* surface protein antigen (SpaA) and glucosyl-transferase (GtfA)	N.I.[a]	Curtiss *et al.* (42)
S. typhimurium Δ*asd*, Δ*aroA*	*Streptococcus sobrinus* spaA	secretory & humoral	Curtiss (37); Curtiss *et al.* (41, 42)
S. typhimurium Δ*thyA*, Δ*asd* and Δ*aroA*	*S. mutans* surface protein antigen (SpaA) and glucosyl-transferase (GtfA)	secretory & humoral	Katz *et al.* (98)
S. typhimurium Δ*aroA*	*E. coli* K88 fimbriae enterotoxin	secretory & humoral	Dougan *et al.* (55)
Δ*aroA*	B subunit of *E. coli*	secretory & humoral	Dougan *et al.*. (53)
Δ*aroA*	*β*-galactosidase	humoral & cellular	Dougan *et al.*. (53)

(Continued)

Table 6. (Continued)

Salmonella Mutant	Expressed Foreign Antigen	Type of Immune Response	References
S. typhimurium Δ*aroA*	B subunit of *E. coli* enterotoxin	secretory & humoral	Maskell *et al.* (119)
Δ*aroA*	*E. coli* K88 fimbriae	humoral	Maskell *et al.* (118)
S. typhimurium Δ*aroA*	*Shistosoma mansoni*	N.I.[a]	Taylor *et al.* (165)
S. typhi Ty21a *galE*	*Vibrio cholerae* LPS	secretory & humoral	Labrooy *et al.* (105)
S. typhimurium Δ*aroA*	*E. coli* β-galactosidase	humoral & cellular	Brown *et al.* (18)
S. typhimurium Δ*aroA*	*Streptococcus pyogenes* M protein	secretory & humoral	Poirier *et al.* (144)
S. typhi Ty21a *galE*	*Shigella flexneri* 2a O-antigen	humoral	Baron *et al.* (8)
S. typhimurium Δcya Δcrp	*S. sobrinus* SpaA	U.I.[b]	Curtiss *et al.* (40, 44, 47)
S. typhimurium Δ*cya* Δ*crp* Δ*asd*	*S. sobrinus* SpaA	secretory & cellular	Nakayama, Lue, & Curtiss, unpublished
S. typhimurium Δ*cya* Δ*crp*	*Streptococcus equi* M protein	U.I.[b]	Galán *et al.* (72)
S. typhimurium Δ*cya* Δ*crp* Δ*asd*	*Mycobacterium leprae* protein antigens	cellular	Mundayoor, Clark-Curtiss, & Curtiss, unpublished
S. typhimurium	*Plasmodium berghei* circumsporozoite protein	humoral & cellular	Sadoff *et al.* (15)
S. typhimurium *galE*	*E. coli* K-88 fimbriae	humoral	Hone *et al.* (90)
S. dublin Δ*aroA*	*V. cholerae* toxin epitope	humoral	Newton *et al.* (136)

[a] Not investigated
[b] Under investigation

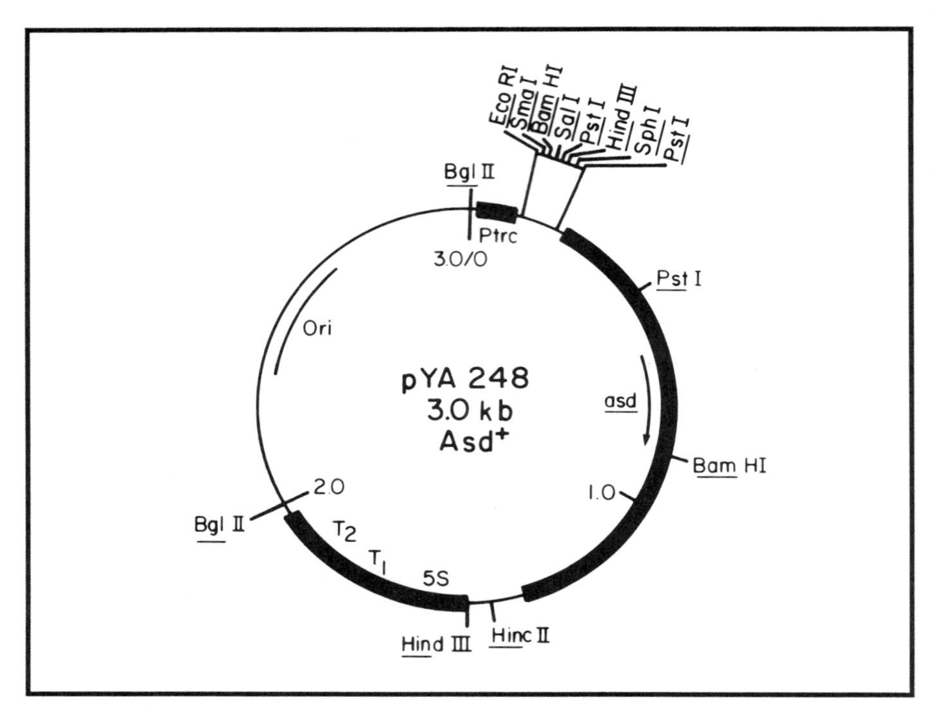

Fig. 2. The pYA248 *asd*+ cloning vector for cloning genes for colonization and virulence antigens and having them expressed in Δ*cya* Δ*crp* Δ*asd S. typhimurium* vaccine strains (133).

S. typhimurium Δ*asd* mutants are totally avirulent and after oral administration to mice briefly colonize the GALT and elicit a secretory immune response (41). The introduction of the pYA248 *asd*+ plasmid into a Δ*asd S. typhimurium* mutant restores full virulence in mice as demonstrated by obtaining the same LD_{50} as for the *asd*+ wild-type parental *S. typhimurium* strain (133).

We have been cloning genes for putative colonization antigens on *asd*+ vectors using Δ*cya* Δ*crp* Δ*asd S. typhimurium* recipient strains followed by immunization of animals with these constructs to see whether the immune response induced is protective against subsequent infection by the pathogen whose genes are expressed by the recombinant avirulent vaccine strain. We have been particularly interested in analyzing *Bordetella avium*, the causative agent of rhinotracheitis in poultry (100). This pathogen colonizes the ciliated tracheal epithelial cells of young birds (3) and produces a tracheal cytotoxin (74) and a dermonecrotic toxin (147) which undoubtedly account for the observed tissue pathology. The ability of *B. avium* to colonize on ciliated tracheal epithelial cells in the bird is undoubtedly due to the presence of pili (94) and/or to outer membrane proteins (85). As a means to identify important colonization antigens, we have shotgun cloned *B. avium* genes into various bacteriophage and cosmid vectors such as λgtll (175) and pCP13 (50), respectively, and used antibodies raised in rabbits against *B. avium* surface structures as well as convalescent sera from turkeys surviving *B. avium* infection to screen the libraries to detect expression of important antigens. We have thus identified clones expressing a number of surface proteins. The genes for some of these have been inactivated by transposon mutagenesis and returned to *B. avium* to investigate whether such mutants are still capable of colonizing the trachea of young birds (J.- L. Boucaud and C. Gentry-Weeks, Abstr. Annu. Meet. Am. Soc. Microbiol. 1989, B-135, p. 53). In other cases the cloned genetic determinants have been placed into an *asd*+ vector and introduced into a Δ*cya* Δcrp Δ*asd S. typhimurium* vaccine strain. These constructs which stably express the *B. avium* surface protein at high level and in a stable manner can now be used to investigate induction of a protective immune response in young chickens. These studies will also reveal whether antigen delivery to the GALT, as accomplished by oral administration of a recombinant avirulent *Salmonella* capable of colonizing the GALT, will induce a generalized secretory immune response with production of sIgA in the respiratory tract and whether such a response is capable of blocking colonization by *B. avium*. A particular problem is whether an immune response could be elicited to confer protection at a time when birds are susceptible to *B. avium* infection. There are *B. avium* strains, however, capable of colonization and infection in older birds (95). It is also possible to investigate, using the recombinant avirulent *S. typhimurium* strains expressing *B. avium* antigens for immunization of hens, egg transmission of immunity to the newly hatched chick.

SUMMARY

Studies have commenced to delineate the genetic control over attributes in *S. typhimurium* and *S. enteritidis* necessary for these invasive strains of *Salmonella* to infect, colonize, and cause death in one-day-old chicks. Thus, the large virulence plasmid present in invasive strains of *S. typhimurium* and *S. enteritidis* contributes to virulence when these strains are administered orally to one-day-old chicks. It is also evident that the *inv* genes necessary for *S. typhimurium* to invade cells in culture and to effectively colonize the intestinal tract of mice are also required to cause invasive disease in one-day-old chicks. The *phoP* gene which governs traits necessary for survival of *S. typhimurium* in murine macrophages is also necessary for complete virulence of *Salmonella* in chicks. There is some indication, however, that there might be heterogeneity in the ability of chicks to fend off an infection from what should be an avirulent mutant.

Deletion(Δ) mutations in the genes for adenylate cyclase (*cya*) and cAMP receptor protein (*crp*) render several *Salmonella* species completely avirulent and highly immunogenic for several animal hosts. These mutations were introduced into four *S. typhimurium* and two *S. enteritidis* strains which are highly virulent with oral LD_{50} of 10^3 to 10^7 CFU for one-day-old chicks. These mutants were completely avirulent such that one-day-old chicks survived oral challenge with 1×10^9 CFU. The Δ*cya* Δ*crp S. typhimurium* strains effectively colonized the intestinal tract but have limited ability to colonize deeper tissues such as the liver and spleen. They cease to be detectable four to five weeks after oral inoculation.

It is now possible to investigate whether using Δ*cya* Δ*crp*, *phoP* and/or virulence plasmid-cured strains to orally vaccinate newly hatched chicks can serve to diminish colonization, persistence and shedding of various *Salmonella* species by these immunized birds. It also appears that the recombinant avirulent constructions designed will also enable a fuller analysis of the avian immune system and will likely lead to strategies for immunization to prevent infection of birds with pathogens that are a problem to the poultry industry as well as to prevent infection and colonization with pathogens such as *Salmonella* that are transmitted through contaminated poultry carcasses and eggs to humans.

ACKNOWLEDGEMENTS

Research was supported by grants ROIDE06669 and ROIAI24533 from the U. S. Public Health Service, National Institutes of Health and 88-34116-3792 from the United States Department of Agriculture.

REFERENCES

1. **Alper, M. D., and B. N. Ames.** 1978. Transport of antibiotics and metabolite analogs by systems under cyclic AMP control: positive selection of *Salmonella typhimurium cya* and *crp* mutants. J. Bacteriol. **133:**149-157.
2. **Anon.** 1988. *Salmonella enteritidis* phage type 4: chicken and egg. The Lancet. **ii:**720-722.
3. **Arp, L. H., and N. F. Cheville.** 1984. Tracheal lesions in young turkeys infected with *Bordetella avium*. Am. J. Vet. Res. **45:**2196-2200.
4. **Bachmann, B. J.** 1987. Linkage map of *Escherichia coli* K-12, Edition 7, pp. 807-876. *In* F. C. Neidhardt (ed.), *Escherichia coli* and *Salmonella typhimurium* cellular and molecular biology, Vol. 2. American Society for Microbiology, Washington, DC.
5. **Bacon, G. A., T. W. Burrows, and M. Yates.** 1950. The effects of biochemical mutation on the virulence of *Bacterium typhosum*: the virulence of mutants. Br. J. Exp. Pathol. **32:**714-724.
6. **Bacon, G. A., T. W. Burrows, and M. Yates.** 1951. The effects of biochemical mutation on the virulence of *Bacterium typhosum*: the loss of virulence of certain mutants. Br. J. Exp. Pathol. **32:**85-96.
7. **Bailey, J. S., J. E. Thomson, and N. A. Cox.** 1987. Contamination of poultry during processing, pp. 193-211. *In* F. E. Cunningham and N. A. Cox (eds.), The microbiology of poultry meat products. Academic Press, Orlando, FL.
8. **Baron, L. S., D. J. Kopecko, S. B. Formal, R. Seid, P. Guerry, and C. Powell.** 1987. Introduction of *Shigella flexneri* 2a type and group antigen genes into oral typhoid vaccine strain *Salmonella typhimurium* Ty21a. Infect. Immun. **55:**2797-2801.
9. **Baron, L. S., E. Penido, I. R. Ryman, and S. Falkow.** 1970. Behavior of coliphage lambda in hybrids between *Escherichia coli* and *Salmonella*. J. Bacteriol. **102:**221-233.
10. **Baron, L. S., I. R. Ryman, E. M. Johnson, and P. Gemski, Jr.** 1972. Lytic replication of coliphage lambda in *Salmonella typhosa* hybrids. J. Bacteriol. **110:**1022-1031.
11. **Barrow, P. A., and M. A. Lovell.** 1988. The association between a large molecular mass plasmid and virulence in a strain of *Salmonella pullorum*. J. Gen. Micro. **134:**2307-2316.

12. **Barrow, P. A., J. M. Simpson, M. A. Lovell, and M. M. Binns.** 1987. Contribution of *Salmonella gallinarum* large plasmid toward virulence in fowl typhoid. Infect. Immun. **55:**388-392.
13. **Berg, D. E., and C. M. Berg.** 1983. The prokaryotic transposable element Tn*5*. Bio/Tech. **1:**417-435.
14. **Bergmann, K.-C., and R. H. Waldman.** 1988. Stimulation of secretory antibody following oral administration of antigen. Rev. Infect. Dis. **10:**939-950.
15. **Bienenstock, J., M. McDermott, D. Befus, and M. O'Neill.** 1978. A common mucosal immunologic system involving the bronchus, breast and bowel. Adv. Exp. Med. Biol. **107:**53-59.
16. **Black, R. E., M. M. Levine, M. L. Clements, G. Losonsky, D. Herrington, S. Berman, and S. B. Formal.** 1987. Prevention of shigellosis by a *Salmonella typhi-Shigella sonnei* bivalent vaccine. J. Infect. Dis. **155:**1260-1265.
17. **Bochner, B. R., H.-C. Huang, G. L. Schieven, and B. N. Ames.** 1980. Positive selection for loss of tetracycline resistance. J. Bacteriol. **143:**926-933.
18. **Brown, A., C. E. Hormaeche, R. D. de Hormaeche, M. Winther, G. Dougan, D. J. Maskell, and B. A. D. Stocker.** 1987. An attenuated *aro*A *Salmonella typhimurium* vaccine elicits humoral and cellular immunity to cloned β-galactosidase in mice. J. Infect. Dis. **155:**86-92.
19. **Brown, R. F., and B. A. D. Stocker.** 1987. *Salmonella typhi* 205aTy, a strain with two attenuating auxotrophic characters, for use in laboratory teaching. Infect. Immun. **55:**892-898.
20. **Bryan, F .L.** 1981. Current trends in foodborne salmonellosis in the United States and Canada. J. Food Prot. **44:**394-401.
21. **Cardineau, G. A., and R. Curtiss, III.** 1987. Nucleotide sequence of the *asd* gene of *Streptococcus mutans*: identification of the promoter region and evidence for attenuator like sequences preceding the structural gene. J. Biol. Chem. **262:**3344-3353.
22. **Carsiotis, M., D. L. Weinstein, H. Karch, I. A. Holder, and A. D. O'Brien.** 1984. Flagella of *Salmonella typhimurium* are a virulence factor in infected C57BL/6J mice. Infect. Immun. **46:**814-818.
23. **Carter, P. B., and F. M. Collins.** 1974. The route of enteric infection in normal mice. J. Exp. Med. **139:**1189-1203.
24. **Centers for Disease Control.** 1984. Human *Salmonella* isolates-United States, 1983. Morbid. Mortal. Weekly Rep. **33:**693-695.
25. **Centers for Disease Control.** 1987. Increasing rate of *Salmonella enteritidis* infections in the northeastern United States. J. Am. Med. Assoc. **257:**600-601.
26. **Centers for Disease Control.** 1988. Update: *Salmonella enteritidis* infections and grade A shell eggs-United States. Morbid. Mortal. Weekly Rep. **37:**490-496.
27. **Cebra, J. J., P. J. Gearhart, R. Kamat, S. M. Robertson, and J. Tseng.** 1976. Origin and differentiation of lymphocytes involved in the secretory IgA response. Cold Spring Harbor Symp. Quant. Biol. **41:**201-215.
28. **Chalker, R. B., and M. J. Blaser.** 1988. A review of human salmonellosis III. Magnitude of *Salmonella* infection in the United States. Rev. Infect. Dis. **10:**111-124.
29. **Chikami, G. K., J. Fierer, and D. G. Guiney.** 1985. Plasmid-mediated virulence in *Salmonella dublin* demonstrated by use of a Tn*5*-*oriT* construct. Infect. Immun. **50:**420-424.

30. **Clarke, R. C., and C. L. Gyles.** 1987. Vaccination of calves with a diaminopimelic acid mutant of *Salmonella typhimurium*. Can. J. Vet. Res. **51:**32-38.
31. **Clements, J. D.** 1987. Use of attenuated mutants of *Salmonella* as carriers for delivery of heterologous antigens to the secretory immune system. Pathol. Immunopathol. Res. **6:**137-146.
32. **Clements, J. D., and S. El-Morshidy.** 1984. Construction of a potential live oral bivalent vaccine for typhoid fever and cholera-*Escherichia-coli*-related diarrheas. Infect. Immun. **46:**564-569.
33. **Clements, J. D., F. L. Lyons, K. L. Lowe, A. L. Farrand, and S. El-Morshidy.** 1986. Oral immunization of mice with attenuated *Salmonella enteritidis* containing a recombinant plasmid which codes for production of the B subunit of heat-labile *Escherichia coli* enterotoxin. Infect. Immun. **53:**685-692.
34. **Close, T. J., and R. L. Rodriguez.** 1982. Construction and characterization of the chloramphenicol-resistance gene cartridge: a new approach to the transcriptional mapping of extrachromosomal elements. Gene. **20:**305-316.
35. **Cunningham, F. E.** 1987. Types of microorganisms associated with poultry carcasses, pp. 29-42. *In* F. E. Cunningham and N. A. Cox (eds.), The microbiology of poultry and meat products. Academic Press, Inc., Orlando, FL.
36. **Curtiss, R., III.** 1981. Gene transfer, pp. 243-265. *In* P. Gerhardt, R. G. E. Murray, R. Costilow, E. W. Nester, W. A. Wood, N. R. Kreig, and G. B. Phillips (eds.), Manual of methods for general bacteriology. American Society for Microbiology, Washington, DC.
37. **Curtiss, R., III.** 1986. Genetic analysis of *Streptococcus mutans* virulence and prospects for an anticaries vaccine. J. Dent. Res. **65:**1034-1045.
38. **Curtiss, R., III.** 1988. Engineering organisms for safety: What is necessary?, pp. 7-20. *In* M. Sussman, C. H. Collins, F. A. Skinner, and D. E. Stewart-Tull (eds.), The release of genetically-engineered microorganisms. Academic Press, London.
39. **Curtiss, R., III.** 1989. Live attenuated *Salmonella* strains as vectors for antigen deliver. *In* G. C. Woodrow and M. M. Levine (eds.), New generation vaccines: the molecular approach. Marcel Dekker, Inc., New York. (In Press).
40. **Curtiss, R., III, R. M. Goldschmidt, N. B. Fletchall, and S. M. Kelly.** 1988. Avirulent *Salmonella typhimurium* Δ*cya* Δ*crp* oral vaccine strains expressing a streptococcal colonization and virulence antigen. Vaccine **6:**155-160
41. **Curtiss, R., III, R. Goldschmidt, S. M. Kelly, M. Lyons, S. Michalek, R. Pastian, and S. Stein.** 1987. Recombinant avirulent *Salmonella* for oral immunization to induce mucosal immunity to bacterial pathogens, pp. 261-271. *In* H. Kohler and P. T. LoVerde (eds.), Vaccines: new concepts and developments. Proceedings of the Tenth International Convocation on Immunology. Longman Scientific Technical, Harlow, Essex, Great Britain.
42. **Curtiss, R., III, R. Goldschmidt, R. Pastian, M. Lyons, S. M. Michalek, and J. Mestecky.** 1986. Cloning virulence determinants from *Streptococcus mutans* and the use of recombinant clones to construct bivalent oral vaccine strains to confer protective immunity against *S. mutans*-induced dental caries, pp. 173-180. *In* S. Hamada S. M. Michalek, H. Kiyono, L. Menaker, and J. R. McGhee (eds.), Molecular Microbiology and Immunology of *Streptococcus mutans*. Elsevier Science Publishers B.V. (Biomedical Division), Amsterdam.
43. **Curtiss, R., III, and S. M. Kelly.** 1987. *Salmonella typhimurium* deletion mutations lacking adenylate cyclase and cyclic AMP receptor protein are avirulent and immunogenic. Infect. Immun. **55:**3035-3043.

44. **Curtiss, R., III, S. M. Kelly, P. A. Gulig, C. R. Gentry-Weeks, and J. Galán.** 1988. Avirulent salmonellae expressing virulence antigens from other pathogens for use as orally administered vaccines, pp. 311-328. *In* J. A. Roth (ed.), Virulence mechanisms of bacterial pathogens. American Society for Microbiology, Washington, DC.
45. **Curtiss, R., III, S. M. Kelly, P. A. Gulig, and K. Nakayama.** 1989. Selective delivery of antigens by recombinant bacteria. Curr. Top. Microbiol. Immunol. **146:**35-49.
46. **Curtiss, R., III, S. M. Kelly, P. A. Gulig, and K. Nakayama.** 1989. Stable recombinant avirulent *Salmonella* vaccine strains, pp. 33-47. *In* M. Z. Attasi (ed.), Immunobiology of Proteins and Peptides. V. Vaccines. Proceedings of the Fifth International Symposium on the Immunobiology of Proteins and Peptides. Plenum Pub. Corp., NY.
47. **Curtiss, R., III, R. Nakayama, and S. M. Kelly.** 1989. Recombinant avirulent *Salmonella* vaccine strains with stable maintenance and high level expression of cloned genes *in vivo*. *In* B. Albini, R. J Genco, D. L. Ogra, and M. M. Weiser (eds.), Immunology and Immunopathology of the Alimentary Canal. Proceedings of the Eleventh International Convocation on Immunology. Marcel Dekker, New York and Basel. Immunol. Invest. **18:**583-596.
48. **Cvjetanovic, B., D. M. Mel, and O. Felsenfeld.** 1970. Study of live typhoid vaccine in chimpanzees. Bull. W.H.O. **42:**499-507.
49. **Czerkinsky, C., S. J. Prince, S. M. Michalek, S. Jackson, M. W. Russell, Z. Moldoveanu, J. R. McGhee, and J. Mestecky.** 1987. IgA antibody-producing cells in peripheral blood after antigen ingestion: evidence for a common mucosal immune system in humans. Proc. Natl. Acad. Sci. USA **84:**2449-2453.
50. **Darzins, A., and A. M. Chakrabarty.** 1984. Cloning of genes controlling alginate biosynthesis from a mucoid cystic fibrosis isolate of *Pseudomonas aeruginosa*. J. Bacteriol. **159:**9-18.
51. **Dorman, C. J., S. Chatfield, C. F. Higgins, C. Hayward, and G. Dougan.** 1989. Characterization of porin and *ompR* mutants of a virulent strain of *Salmonella typhimurium*: *ompR* mutants are attenuated *in vivo*. Infect. Immun. **57:**2136-2140.
52. **Dougan, G., S. Chatfield, D. Pickard, J. Bester, D. O'Callaghan, and D. Maskell.** 1988. Construction and characterization of vaccine strains of *Salmonella* harboring mutations in two different *aro* genes. J. Infect. Dis. **158:**1329-1335.
53. **Dougan, G., C. E. Hormaeche, and D. J. Maskell.** 1987. Live oral *Salmonella* vaccines: potential use of attenuated strains as carriers of heterologous antigens to the immune system. Parasite Immunol. **9:**151-160.
54. **Dougan, G., D. Maskell, D. Pickard, and C. Hormaeche.** 1987. Isolation of stable *aroA* mutants of *Salmonella typhi* Ty2: properties and preliminary characterisation in mice. Mol. Gen. Genet. **207:**402-405.
55. **Dougan, G., R. Sellwood, D. Maskell, K. Sweeney, F. Y. Liew, J. Beesley, and C. Hormaeche.** 1986. *In vivo* properties of a cloned K88 adherence antigen determinant. Infect. Immun. **52:**344-347.
56. **Duguid, J. P., E. S. Anderson, and I. Campbell.** 1966. Fimbriae and adhesive properties in salmonellae. J. Pathol. Bacteriol. **92:**107-137.
57. **DuPont, H. L., R. B. Hornick, M. J. Snyder, J. P. Libonati, and T. E. Woodard.** 1971. Immunity in typhoid fever: evaluation of live streptomycin-dependent vaccine, pp. 236-239. Antimicrob. Agents Chemother. 1970.

58. **Edwards, M. F., and B. A. D. Stocker.** 1988. Construction of Δ*aroA his* Δ*pur* strains of *Salmonella typhi*. J. Bacteriol. **170:**3991-3995.
59. **Elsinghorst, E. A., L. S. Baron, and D. J. Kopecko.** 1989. Penetration of human intestinal epithelial cells by *Salmonella*: molecular cloning and expression of *S. typhi* invasion determinants in *Escherichia coli*. Proc. Natl. Acad. Sci. USA **86:**5173-5177.
60. **Elson, C. O., and W. Ealding.** 1984. Generalized systemic and mucosal immunity in mice after mucosal stimulation with cholera toxin. J. Immunol. **132:**2736-2741.
61. **Fahey, K. J., and G. N. Cooper.** 1970. Oral immunization against experimental salmonellosis. I. Development of temperature sensitive mutant vaccines. Infect. Immun. **1:**263-270.
62. **Fields, P. I., E. A. Groisman, and F. Heffron.** 1989. A *Salmonella* locus that controls resistance to microbicidal proteins from phagocytic cells. Science **243:**1059-1062.
63. **Fields, P. I., R. V. Swanson, C. G. Haidaris, and F. Heffron.** 1986. Mutants of *Salmonella typhimurium* that cannot survive within the macrophage are avirulent. Proc. Natl. Acad. Sci. USA **83:**5189-5193.
64. **Fierer, J., G. Chikami, L. Hatlen, E. J. Heffernan, and D. Guiney.** 1988. Active immunization with LD842, a plasmid-cured strain of *Salmonella dublin*, protects mice against group D and group B *Salmonella* infection. J. Infect. Dis. **158:**460-463.
65. **Finlay, B. B., B. Gumbiner, and S. Falkow.** 1988. Penetration of *Salmonella* through a polarized Madin-Darby canine kidney epithelial cell monolayer. J. Cell Biol. **107:**221-230.
66. **Finlay, B. B., F. Heffron, and S. Falkow.** 1989. Epithelial cell surfaces induce *Salmonella* proteins required for bacterial adherence and invasion. Science. **243:**940-943.
67. **Finlay, B. B., M. N. Starnbach, C. L. Francis, B. A. D. Stocker, S. Chatfield, G. Dougan, and S. Falkow.** 1988. Identification and characterization of Tn*phoA* mutants of *Salmonella* that are unable to pass through a polarized MDCK epithelial cell monolayer. Mol. Microbiol. **2:**757-766.
68. **Formal, S. B., L. S. Baron, D. J. Kopecko, O. Washington, C. Powell, and C. A. Life.** 1981. Construction of a potential bivalent vaccine strain: introduction of *Shigella sonnei* form I antigen genes into the *galE Salmonella typhi* Ty21a typhoid vaccine strain. Infect. Immun. **34:**746-750.
69. **Furness, G., and D. Rowley.** 1956. Transduction of virulence within the species *Salmonella typhimurium*. J. Gen. Microbiol. **15:**140-145.
70. **Galán, J. E., and R. Curtiss, III.** 1989. Virulence and vaccine potential of *phoP* mutants of *Salmonella typhimurium*. Microb. Pathogen. **6:**433-443.
71. **Galán, J. E., and R. Curtiss, III.** 1989. Cloning and molecular characterization of genes whose products allow *Salmonella typhimurium* to penetrate tissue culture cells. Proc. Natl. Acad. Sci. USA **86:**6383-6387.
72. **Galán, J. E., J. F. Timoney, and R. Curtiss, III.** 1988. Expression and localization of the *Streptococcus equi* M protein in *Escherichia coli* and *Salmonella typhimurium*, pp. 34-41. *In* D. G. Powell (ed.), Equine Infectious Diseases V. Proceedings of the Fifth International Conference. The University of Kentucky Press, Lexington, KY.
73. **Gangarosa, E. J.** 1978. What have we learned from 15 years of *Salmonella* surveillance? *In* "National Salmonellosis Seminar." Washington, DC.

74. **Gentry-Weeks, C. R., B. T. Cookson, W. E. Goldman, R. B. Rimler, S . B. Porter, and R. Curtiss, III.** 1988. Dermonecrotic toxin and tracheal cytotoxin, putative virulence factors of *Bordetella avium*. Infect. Immun. **56:**1698-1707.
75. **Germanier, R., and E. Furer.** 1971. Immunity in experimental salmonellosis. II. Basis for the avirulence and protective capacity of *galE* mutants of *Salmonella typhimurium*. Infect. Immun. **4:**663-673.
76. **Germanier, R., and E. Furer.** 1975. Isolation and characterization of *galE* mutant Ty 21a of *Salmonella typhi*: a candidate strain for a live, oral typhoid vaccine. J. Infect. Dis. **131:**553-558.
77. **Goebel, W.** 1985. Genetic approaches to microbial pathogenicity. Curr. Top. Microbiol. Immunol. Volume **118.**
78. **Green, S. S., A. B. Moran, R. W. Johnston, P. Uhler, and J. Chiu.** 1982. The incidence of *Salmonella* species and serotypes in young whole chicken carcasses in 1979 as compared with 1967. Poult. Sci. **61:**288-293.
79. **Gulig, P. A., and R. Curtiss, III.** 1987. Plasmid-associated virulence of *Salmonella typhimurium*. Infect. Immun. **55:**2891-2901.
80. **Gulig, P. A., and R. Curtiss, III.** 1988. Cloning and transposon insertion mutagenesis of virulence genes of the 100-kilobase plasmid of *Salmonella typhimurium*. Infect. Immun. **56:**3262-3271.
81. **Gupta, B. R., and B. B. Mallick.** 1976. Immunization against fowl typhoid. 1. Live oral vaccine. Indian J. Anim. Sci. **46:**502-505.
82. **Hackett, J., S. Attridge, and D. Rowley.** 1988. Oral immunization with live, avirulent *fla* + strains of *Salmonella* protects mice against subsequent oral challenge with *Salmonella typhimurium*. J. Infect. Dis. **157:**78-84.
83. **Harbourne, J. F.** 1957. The control of fowl typhoid in the field by use of live vaccines. Vet. Rec. **69:**1102-1107.
84. **Harkki, A., and E. T. Palva.** 1985. A *lamB* expression plasmid for extending the host range of phage λ to other enterobacteria. FEMS Microbiol. Lett. **27:**183-187.
85. **Hellig, D. H., L. H. Arp, and J. A. Fagerland.** 1988. A comparison of outer membrane protein and surface characteristics of adhesive and non-adhesive phenotypes of *Bordetella avium*. Avian Dis. **32:**787-792.
86. **Helmuth, R., R. Stephan, C. Bunge, B. Hoog, A. Steinbeck, and E. Bulling.** 1985. Epidemiology of virulence-associated plasmids and outer membrane protein patterns within seven common *Salmonella* serotypes. Infect. Immun. **48:**175-182.
87. **Hoiseth, S. K., and B. A. D. Stocker.** 1981. Aromatic-dependent *Salmonella typhimurium* are non-virulent and effective as live vaccines. Nature **291:**238-239.
88. **Holmberg, S. D., M. T. Osterholm, K. A. Senger, and M. L. Cohen.** 1984. Drug-resistant *Salmonella* from animals fed antimicrobials. N. Engl. J. Med. **311:**617-622.
89. **Hone, D., R. Morona, S. Attridge, and J. Hackett.** 1987. Construction of defined *galE* mutants of *Salmonella* for use as vaccines. J. Infect. Dis. **156:**167-174.
90. **Hone, D., S. Attridge, L. van den Bosch, and J. Hackett.** 1988. A chromosomal integration system for stabilization of heterologous genes in *Salmonella* based vaccine strains. Microb. Pathogen. **5:**407-418.
91. **Hone, D. M., S. R. Attridge, B. Forrest, R. Morona, D. Daniels, J. T. LaBrooy, R. C. A. Bartholomeusz, D. J. C. Shearman, and J. Hackett.** 1988. A *galE via* (vi antigen-negative) mutant of *Salmonella typhi* Ty2 retains virulence in humans. Infect. Immun. **56:**1326-1333.

92. **Hook, E. W.** 1985. *Salmonella* species (including typhoid fever), pp. 1256-1268. *In* G. L Mandell, R. G. Douglas, and J.E. Bennett (eds.), Principles and practice of infectious diseases. John Wiley & Sons, NY.
93. **Isberg, R. R., D. L. Voorhis, and S. Falkow.** 1987. Identification of invasin: a protein that allows enteric bacteria to penetrate cultured mammalian cells. Cell **50:**769-778.
94. **Jackwood, M. W., and Y.M. Saif.** 1987. Pili of *Bordetella avium*: expression, characterization, and role in *in vitro* adherence. Avian Dis. **31:**277-286.
95. **Jensen, M. M., and M. S. Marshall.** 1982. Case report - control of turkey *Alcaligenes* rhinotracheitis in Utah with a live vaccine. Avian Dis. **25:**1053-1057.
96. **Jones, G. W., D. K. Rabert, D. M. Svinarich, and H. J. Whitfield.** 1982. Association of adhesive, invasive, and virulent phenotypes of *Salmonella typhimurium* with autonomous 60-megadalton plasmids. Infect. Immun. **38:**476-486.
97. **Jones, G. W., and L. A. Richardson.** 1981. The attachment to, and invasion of HeLa cells by *Salmonella typhimurium*: the contribution of manno-sensitive and mannose-resistant haemagglutinating activities. J. Gen. Microbiol. **127:**361-370.
98. **Katz, J., S. M. Michalek, R. Curtiss, III, C. Harmon, G. Richardson, and J. Mestecky.** 1987. Novel oral vaccines: The effectiveness of cloned gene products on inducing secretory immune responses. Adv. Exp. Med. Biol. **216:**1741-1747.
99. **Kawahara, K., Y. Haraguchi, M. Tsuchimoto, N. Terakado, and H. Danbara.** 1988. Evidence of correlation between 50-kilobase plasmid of *Salmonella choleraesuis* and its virulence. Microb. Pathogen. **4:**155-163.
100. **Kersters, K., K.-H. Hinz, A. Hertle, P. Segers, A. Lievens, O. Siegmann, and J. De Ley.** 1984. *Bordetella avium* sp. nov., isolated from the respiratory tracts of turkeys and other birds. Int. J. Syst. Bacteriol. **34**56-70.
101. **Kier, L. D., R. Weppelman, and B. N. Ames.** 1977. Resolution and purification of two phosphatases and a cyclic phosphodiesterase of *S. typhimurium*. J. Bacteriol. **130:**429-436.
102. **Kier, L. D., R. M. Weppleman, and B. N. Ames.** 1979. Regulation of non-specific acid phosphatase in *Salmonella*: *phoN* and *phoP* genes. J.Bacteriol. **138:**155-161.
103. **Kleckner, N., J. Roth, and D. Botstein.** 1977. Genetic engineering *in vivo* using translocatable drug-resistance elements. J. Mol. Biol. **116:**125-159.
104. **Kroos, L., and D. Kaiser.** 1984. Construction of Tn*5 lac*, a transposon that fuses *lacZ* expression to exogenous promoters, and its introduction into *Myxococcus xanthus*. Proc. Natl. Acad. Sci. USA **81:**5816-5820.
105. **LaBrooy, J. T., B. Forrest, and C. Bartholomeusz.** 1986. Immunisation against cholera with a Ty21a-cholera hybrid, pp. 143-147. *In* Bacterial vaccines and local immunity. Proceedings of Sclavo International Conference. Siena, Itay.
106. **Lascelles, A. K., K. J. Beh, T. K. S. Mukkur, and G. Willis.** 1988. Immune response of sheep to oral and subcutaneous administration of live aromatic-dependent mutant of *Salmonella typhimurium* (SL1479). Vet. Immunol. Immunopath. **18:** 259-267.
107. **LeFever, M. E., and D. D. Joel.** 1984. Peyer's patch epithelium: an imperfect barrier, pp. 45-46. *In* C. M. Schiller (ed.), Intestinal Toxicology. Raven Press, New York.
108. **LeMinor, L.** 1984. Genus III. *Salmonella* Lignieres 1900, 389^{AL}, pp. 427-458. *In* N. R. Krieg and J. G. Holt (ed.), Bergey's manual of systemic bacteriology. The Williams & Wilkins Co., Baltimore.

109. **Lennox, E. S.** 1955. Transduction of linked genetic characters of the host by bacteriophage Pl. Virology. **1**:190-206.
110. **Lin, F.-Y. C., J. G. Morris, Jr., D. Trump, D. Tilghman, P. K. Wood, N. Jackman, E. Israel, and J. P. Libonati.** 1988. Investigation of an outbreak of *Salmonella enteritidis* gastroenteritis associated with consumption of eggs in a restaurant chain in Maryland. Am. J. Epidemiol. **128**:839-844.
111. **Lindquist, B. L., E. Lebanthal, P.-G Lee, M. W. Stinson, and J. M. Merrick.** 1987. Adherence of *Salmonella typhimurium* to small intestinal enterocytes of the rat. Infect. Immun. **55**: 3044-3050.
112. **Lockman, H. A., and R. Curtiss, III.** 1990. *Salmonella typhimurium* lacking flagella or motility remain virulent in BALB/c mice. Infect. Immun. **58**:137-143.
113. **Mackaness, M. B.** 1964. The immunological basis of acquired cellular resistance. J. Exp. Med. **120**:105-120.
114. **Macrina, F. L.** 1984. Molecular cloning of bacterial antigens and virulence determinants. Annu. Rev. Microbiol. **38**:193-219.
115. **Maloy, S. R., and W. D. Nunn.** 1981. Selection for loss of tetracycline resistance by *Escherichia coli*. J. Bacteriol. **145**:1110-1112.
116. **Maniatis, T., E. F. Pritsch, and J. Sambrook.** 1982. Molecular cloning: a laboratory manual. Cold Spring Harbor Laboratory. Cold Spring Harbor, NY.
117. **Manoil, C., and J. Beckwith.** 1985. Tn*phoA*: a transposon probe for protein export signals. Proc. Natl. Acad. Sci. USA **82**:8129-8133.
118. **Maskell, D., F. Y. Liew, K. Sweeney, G. Dougan, and C. Hormaeche.** 1986. Attenuated *Salmonella typhimurium* as live oral vaccines and carriers for delivering antigens to the secretory immune system, pp. 213-217. *In* F. Brown, R. M. Chanok, and R. A. Lerner (eds.), Vaccines 86: New approaches to immunization developing vaccines against parasitic, bacterial, and viral disease, Cold Spring Harbor Laboratory, Cold Spring Harbor, NY.
119. **Maskell, D. J., K. J. Sweeney, D. O'Callaghan, C. E. Hormaeche, F. Y. Liew, and G. Dougan.** 1987. *Salmonella typhimurium aroA* mutants as carriers of the *Escherichia coli* heat-labile enterotoxin B subunit to the murine secretory and systemic immune systems. Microb. Pathogen. **2**:211-221.
120. **Maurelli, A. T., B. Blackmon, and R. Curtiss.** 1984. Temperature dependent expression of virulence genes in *Shigella* species. Infect. Immun. **43**:195-201.
121. **McCaughan, G. and A. Basten.** 1983. Immune system of the gastrointestinal tract. Internal. Rev. Physiol. **28**:131-157.
122. **McCormick, B. A., B. A. D. Stocker, D. C. Laux, and P. S. Cohen.** 1988. Roles of motility, chemotaxis, and penetration through and growth in intestinal mucus in the ability of an avirulent strain of *Salmonella typhimurium* to colonize the large intestine of streptomycin-treated mice. Infect. Immun. **56**:2209-2217.
123. **McDermott, M. R., and J. Bienenstock.** 1979. Evidence for a common mucosal immunologic system I. Migration of B immunoblasts into intestinal, respiratory, and genital tissues. J. Immunol. **122**:1892-1989.
124. **McFarland, W. C., and B.A.D. Stocker.** 1987. Effect of different purine auxotrophic mutations on mouse-virulence of a Vi-positive strain of *Salmonella dublin* and of two strains of *Salmonella typhimurium*. Microb. Pathogen. **3**:129-141.
125. **McNabb, P. C., and T. B. Tomasi.** 1981. Host defense mechanisms at mucosal surfaces. Annu. Rev. Microbiol. **35**:477-496.
126. **Meadow, P., and E. Work.** 1956. Interrelationships between diaminopimelic acid, lysine and their analogues in mutants of *Escherichia coli*. Biochem. J. **64**:llp.

127. **Miller, J. F., J. J. Mekalanos, and S. Falkow.** 1989. Coordinate regulation and sensory transduction in the control of bacterial virulence. Science. **243:**916-922.
128. **Miller, S.I., A. M. Kukrak, and J. J. Mekalanos.** 1989. A two-component regulatory system (*phoP phoQ*) controls *Salmonella typhimurium* virulence. Proc. Natl. Acad. Sci. USA **86:**5054-5058.
129. **Mukkur, T. K. S., G. H. McDowell, B. A. D. Stocker, and A. K. Lascelles.** 1987. Protection against experimental salmonellosis in mice and sheep by immunisation with aromatic-dependent *Salmonella typhimurium*. J. Med. Microbiol. **24:11-19.**
130. **Mullis, K. B., and F. A. Faloona.** 1987. Specific synthesis of DNA *in vitro* via a polymerase-catalyzed chain reaction. Methods Enzymol. **155:**335-350.
131. **Nakamura, M., S. Sato, T. Ohya, S. Suzuki, and S. Ikeda.** 1985. Possible relationship of a 36-megadalton *Salmonella enteritidis* plasmid to virulence in mice. Infect. Immun. **47:**831-833.
132. **Nakamura, M., S. Sato, S. Suzuki, Y. Tamura, O. Itoh, T. Koeda, and S. Ikeda.** 1988. Virulence and immunogenicity of plasmid-cured *Salmonella* serovar Enteritidis AL1192 against cattle. Jpn. J. Vet. Sci. **50:**706-713.
133. **Nakayama, K., S. M. Kelly, and R. Curtiss, III.** 1988. Construction of an *asd* + expression-cloning vector: stable maintenance and high level expression of cloned genes in a *Salmonella* vaccine strain. Bio/Tech. **6:**693-697.
134. **Neidhardt, F. C.** 1987. *Escherichia coli* and *Salmonella typhimurium* cellular and molecular biology. Vol. 1 and 2. American Society for Microbiology, Washington, DC.
135. **Nevola, J. J., B. A. D. Stocker, D. C. Laux, and P. S. Cohen.** 1985. Colonization of the mouse intestine by an avirulent *Salmonella typhimurium* strain and its lipopolysaccharide-defective mutants. Infect. Immun. **50:**152-159.
136. **Newton, S. M. C., C. O. Jacob, and B. A. D. Stocker.** 1989. Immune response to cholera toxin epitope inserted in *Salmonella* flagellin. Science **244:**70-72.
137. **Nikaido, H.** 1961. Galactose-sensitive mutants of *Salmonella*. I. Metabolism of galactose. Biochim. Biophys. **Acta. 48:**460-469.
138. **Nnalue, N. A., and B. A. D. Stocker.** 1987. The effects of O-antigen character and enterobacterial common antigen content on the *in vivo* persistence of aromatic-dependent *Salmonella* sp. live-vaccine strains. Microb. Pathogen. **3:**31-44.
139. **O'Brien, J. D. P.** 1988. *Salmonella enteritidis* infection in broiler chickens. Vet. Rec. **122:**214.
140. **O'Brien, T. F., J. D. Hopkins, E. S. Gilleece, A. A. Medeiros, R. L. Kent, B. O. Blackburn, M. B. Holmes, J. P. Reardon, J. M. Vergeront, W. L. Schell, E. Christenson, M. L. Bissett, and E. V. Morse.** 1982. Molecular epidemiology of antibiotic resistance in *Salmonella* from animals and human beings in the United States. N. Engl. J. Med. **307:**1-6.
141. **O'Callaghan, D., D. Maskell, F. Y. Liew, C. S. F. Easmon, and G. Dougan.** 1988. Characterization of aromatic-and purine-dependent *Salmonella typhimurium*: attenuation, persistence, and ability to induce protective immunity in BALB/c mice. Infect. Immun. **56:**419-423.
142. **Ohta, M., N. Kido, Y. Fujii, Y. Arakawa, T. Komatsu, and N. Kato.** 1987. Temperature-sensitive growth mutants as live vaccines against experimental murine salmonellosis. Microbiol. Immunol. **31:**1259-1265.

143. **Pardon, P., M. Y. Popoff, C. Coynault, J. Marly, and I. Miras.** 1986. Virulence-associated plasmids of *Salmonella* serotype typhimurium in experimental murine infection. Ann. Inst./Pasteur Microbiol. **137B**:47-60.
144. **Poirier, T. P., M. A. Kehoe, and E. H. Beachey.** 1988. Protective immunity evoked by oral administration of attenuated *aroA Salmonella typhimurium* expressing cloned streptococcal M protein. J. Exp. Med. **168**:25-32.
145. **Poppe, C., R. Curtiss, III, P. A. Gulig, and C. L. Gyles.** 1989. Hybridization studies with a DNA probe derived from the virulence region of the 60 Mdal plasmid of *Salmonella typhimurium*. Can. J. Vet. Res. **53**:378-384.
146. **Reitman, M.** 1967. Infectivity and antigenicity of streptomycin-dependent *Salmonella typhosa*. J. Infect. Dis. **117**:101-106.
147. **Rimler, R. B.** 1985. Turkey coryza: toxin production by *Bordetella avium*. Avian Dis. **29**:1043-1046.
148. **Roantree, R. J.** 1971. The relationship of lipopolysaccharide structure to bacterial virulence. pp. 1-37. *In* G. Weinbaum, S. Kadis, and S. J. Ajl (eds.), Microbial Toxins, Vol. V. Academic Press, New York.
149. **Robbins, J. B., R. Schneerson, I. L. Archaya, C. U. Lowe, S. S. Szu, E. Daniels, Y. H. Yang, and B. Trollfars.** 1988. Protective roles of mucosal and serum immunity against typhoid fever. Monogr. Allergy. **34**:315-320.
150. **Sadoff, J. C., W. R. Ballou, L. S. Baron, W. R. Majarian, R. N. Brey, W. T. Hockmeyer, J. F. Young, S. J. Cryz, J. Ou, G. H. Lowell, and J. D. Chulay.** 1988. Oral *Salmonella typhimurium* vaccine expressing circumsporozoite protein protects against malaria. Science. **240**:336-338.
151. **St. Louis, M. E., D. L. Morse, M. E. Potter, T. M. DeMelfi, J. J. Guzewich, R. V. Tauxe, and P. A. Blake.** 1988. The emergence of grade A eggs as a major source of *Salmonella enteritidis* infections. New implications for the control of Salmonellosis. J. Am. Med. Assoc. **259**:2103-2107.
152. **Sanderson, K. E., and J. R. Roth.** 1988. Linkage map of *Salmonella typhimurium*, Edition VII. Microbiol. Rev. **52**:485-532.
153. **Sanger, F., S. Nicklen, and A. R. Coulson.** 1977. DNA sequencing with chain-terminating inhibitors. Proc. Natl. Acad. Sci. USA. **74**:5463-5467.
154. **Schleifer, K. H., and O. Kandler.** 1972. Peptidoglycan types of bacterial cell walls and their taxonomic implications. Bacteriol. Rev. **36**:407-477.
155. **Schmeiger, H.** 1972. Phage P22-mutants with increased or decreased transduction abilities. Mol. Gen. Genet. **119**:75-88.
156. **Silva, E. N., G. H. Snoeyenbos, O. M. Weinack, and C. F. Smyser.** 1981. Studies on the use of 9R strain of *Salmonella gallinarum* as a vaccine in chickens. Avian Dis. **25**:38-52.
157. **Smith, H. W.** 1956. The use of live vaccines in experimental *Salmonella gallinarum* infection in chickens with observations on their interference effect. J. Hyg. **54**:419-432.
158. **Smith, H. W.** 1956. The immunity to *Salmonella gallinarum* infection in chickens produced by live cultures of members of the *Salmonella* genus. J. Hyg. **54**:433-439.
159. **Smith, H. W.** 1965. The immunization of mice, calves and pigs against *Salmonella dublin* and *Salmonella cholerae-suis* infections. J. Hyg. **63**:117-135.
160. **Snydman, D. R., and S. L. Gorbach.** 1982. Salmonellosis: Nontyphoidal, pp. 463-485. *In* A. S. Evans and H. A. Feldman (eds.), Bacterial Infections of Humans: Epidemiology and Control. Plenum Publishing, New York.

161. **Southern, E. M.** 1975. Detection of specific sequences amoung DNA fragments separated by gel electrophoresis. J. Mol. Biol. **98:**503-517.
162. **Stevenson, G., and P. A. Manning.** 1985. Galactose epimeraseless (*galE*) mutant G30 of *Salmonella typhimurium* is a good potential live oral vaccine carrier for fimbrial antigens. FEMS. Microbiol. Lett. **28:**317-321.
163. **Tagliabue, A., D. Boraschi, L. Villa, D. F. Keren, G. H. Lowell, R. Rappuoli, and L. Nencioni.** 1984. IgA-dependent cell-mediated activity against enteropathogenic bacteria: distribution, specificity, and characterization of the effector cells. J. Immunol. **133:**988-992.
164. **Tagliabue, A., L. Nencioni, L. Villa, D. F. Keren, G. H. Lowell, and D. Boraschi.** 1983. Antibody-dependent cell-mediated antibacterial activity of intestinal lymphocytes with secretory IgA. Nature. **306:**184-186.
165. **Taylor, D. W., J. S. Cordingley, D. W. Dunne, K. S. Johnson, W. J. Haddow, C. E. Hormaeche, V. Nene, and A. E. Butterworth.** 1986. Molecular cloning of shistosome genes. Parasitology **91:**S73-S81.
166. **Terakado, N., T. Sekizaki, K. Hashimoto, and S. Naitoh.** 1983. Correlation between the presence of a fifty-megadalton plasmid in *Salmonella dublin* and virulence for mice. Infect. Immun. **41:**443-444.
167. **Tinge, S. A., and R. Curtiss, III.** 1990. Conservation of *Salmonella typhimurium* virulence plasmid maintenance regions among *Salmonella* serovars as a basis for plasmid-curing. Infect. Immun. (In Press).
168. **Todd, E. C. D.** 1980. Poultry-associated foodborne disease - its occurrence, cost, sources and prevention. J. Food Prot. **43:**129-139.
169. **Tramont, E. C., R. Chung, S. Berman, D. Keren, C. Kapfer, and S. B. Formal.** 1984. Safety and antigenicity of typhoid-*Shigella sonnei* vaccine (Strain 5076-1C). J. Infect. Dis. **149:**133-136.
170. **Urbaschek, B.** 1987. Perspectives on bacterial pathogenesis and host defense. *In* Reviews of Infectious Diseases, Volume 9, Supplement 5. Infectious Diseases Society of America, The University of Chicago Press, Chicago.
171. **Vladoianu, I. R., F. Dubini, and A. Bolloli.** 1975. Contribution to the study of live streptomycin-dependent *Salmonella* vaccines: the problem of reversion to a virulent form. J. Hyg. **75:**203-213.
172. **Weisz-Carrington, P., M. Roux, M. McWilliams, J. M. Phillips-Quagliata, and M. E. Lamm.** 1979. Organ and isotype distribution of plasma cells producing specific antibody after oral immunization, evidence for a generalized secretory immune system. J. Immunol. **123:**1705-1708.
173. **Williams, J. E.** 1984. Paratyphoid infections, pp. 91-129. *In* M. S. Hofstad, J. H. Barnes, B. W. Calnek, W. M. Reid, and R. W. Yoder, Jr. (eds.), Diseases of poultry. Iowa State University Press, IA.
174. **Williams, R. C., and R. J. Gibbons.** 1972. Inhibitions of bacterial adherence by secretory immunoglobulin A: A mechanism of antigen disposal. Science. **177:**697-699.
175. **Young, R. A., and R. W. Davis.** 1983. Efficient isolation of genes by using antibody probes. Proc. Natl. Acad. Sci. USA **80:**1194-1198.
176. **Zinder, N. D.** 1958. Lysogenization and superinfection immunity in *Salmonella*. Virology. **5:**291-326.
177. **Zinder, N. D., and J. Lederberg.** 1952. Genetic exchange in *Salmonella*. J. Bacteriol. **64:**679-699.

IMMUNOLOGICAL CONTROL OF *SALMONELLA* IN POULTRY

P. A. Barrow

Department of Microbiology
Institute for Animal Health, Houghton Laboratory
Houghton, Huntingdon
Cambridgeshire, PE17 2DA
United Kingdom

ABSTRACT

For a number of reasons the empirical approach to developing vaccines to control intestinal salmonellosis in poultry has produced variable or indifferent results, and it is still unclear whether vaccination is a feasible means of control. An attempt will be made to explain this by comparing "food-poisoning", host non-specific serotypes with the host-specific serotypes where a similar approach has been successful. The requirements for a successful vaccine for poultry will be discussed. Whereas killed bacterins are relatively ineffective, some live attenuated vaccines including *galE*, *aroA*, and streptomycin-requiring mutants have produced some reduction in faecal excretion of challenge *S. typhimurium* strains. Other mutants are becoming available. The advantages of using transposon-engineered mutants will be emphasized. Work at Houghton has included studying the protection induced by an "ideal" strain of *S. typhimurium*, highly invasive for chickens, but non-virulent for mammals. Vaccination studies with the parent and avirulent mutants will be presented. In these studies humoral and cell-mediated immune responses were produced against several *Salmonella* antigens. Recent work on preventing ovarian localization of *S. enteritidis* by vaccination will be presented. Future work will include the microbial and molecular basis of colonization and its relationship to immunity.

Colonization Control of Human Bacterial Enteropathogens in Poultry

INTRODUCTION

An attempt is made to explain the relatively poor success in immunizing poultry against host non-specific *Salmonella* serotypes that usually produce food-poisoning compared with the success obtained with the small number of serotypes that more typically produce systemic "typhoid-like" diseases in a restricted range of host-species. Most of our understanding of immunity to salmonellosis arises from experimental work with the latter group of organisms, usually from *S. typhimurium* infections in mice. Such work may not be entirely relevant to the largely disease-free intestinal colonization by most *Salmonella* serotypes. Humoral and cellular immune responses to *S. typhimurium* infection can be detected in the chicken by either agglutination or ELISA reactions (humoral) or by delayed-type hypersensitivity reactions (cellular). The role of these in intestinal clearance of *Salmonella* organisms remains to be determined. Whereas live, attenuated vaccines against host-specific serotypes are highly protective, similarly developed vaccinal strains have been shown to be less effective in protecting chickens against intestinal colonization by host non-specific serotypes. The reasons for this may include the inappropriate choice of strain from which to construct the mutant. Killed vaccines produce a poor or inconsistent response. Recent work in our laboratory shows that chickens which have cleared themselves of an invasive avian *S. typhimurium* infection are highly resistant to reinfection and intestinal colonization. Vaccines have been developed from this strain. The criteria for an ideal vaccine are listed. Newer methods of attenuation are now becoming available which may be exploited by poultry bacteriologists. The strong humoral immune response to infection opens the way to rapid, large scale screening of flocks for *Salmonella* infection by using ELISA tests. Most work on the development of vaccines for poultry has been empirical and little progress can be made without basic information on the microbial and molecular basis of colonization, the relationship between colonization and immunity and the identification of protective antigens.

HOST-SPECIFIC (HOST-ADAPTED) AND NON-SPECIFIC (NON-ADAPTED) SEROTYPES

From the point of view of pathogenesis the *Salmonella* genus can be divided into two groups. One group typically produces systemic disease and is rarely involved in human food-poisoning while the other typically produces food-poisoning and only produces systemic disease under certain circumstances. A comparison of the biological aspects of the two groups might explain our success in immunological control of the former group and our failure with the latter.

The first group consists of a small number of serotypes that characteristically produce severe disease, initially involving the reticuloendothelial system, in a restricted number of host species. They include *Salmonella typhi* and *S. paratyphi* producing disease in man, *S. gallinarum* and *S. pullorum* in poultry, *S. dublin* in cattle, *S. cholerae-suis* in pigs, in addition to a few others. Infection is normally by the oral route. The organisms do not multiply in the alimentary tract to any extent but are highly invasive for the appropriate host species and enter the sub-mucosa of the intestine where some bacteria are killed and others are ingested by macrophages or pass into the blood stream. They are removed rapidly by macrophage-like cells of the liver, spleen, and bone marrow. Multiplication of the facultatively intracellular bacteria proceeds with cell death and, in the non-immune, animal ultimately host death ensues. Although during disease bacteria enter the blood and other organs including the gut, they are not excreted in the faeces in large numbers and thus rarely enter the food chain. They are infrequently associated with food-poisoning.

The second group comprises the remaining 2,000 or so serotypes. They are not restricted to particular host species and their epidemiology can therefore be complex. They do not typically produce systemic disease in adult animals. Newly hatched chickens and young calves can be particularly susceptible to infection with some strains of *S. typhimurium* and *S. enteritidis* and high mortality can ensue from the resulting systemic disease. Survivors will excrete large numbers of organisms in the faeces for several weeks since organisms in this group colonize the gut well. Extensive carcass contamination can result in organisms entering the food chain in large numbers. Ingestion by man of these strains can result in diarrhoea although the pathogenesis of this disease has not, as yet, been fully elucidated. Economically, the occurrence of mortality and morbidity in chickens as a result of *Salmonella* infection is not a major problem. The main reason for our interest in this group is the increasing number of cases of human food poisoning. Many Western countries have some measure of control over the host-specific group of serotypes whereas the host non-specific "food-poisoning" serotypes remain a major source of morbidity for man.

CURRENT UNDERSTANDING OF IMMUNITY TO SALMONELLOSIS

Most of our understanding of the basis of immunity to salmonellosis has arisen from work on systemic disease, much of it using *Salmonella typhimurium* infections in mice (mouse typhoid) as a model. Caution must therefore be exercised in trying to extrapolate findings from such work to infections of poultry with host non-specific serotypes.

Following infection of mice, poultry or other animal species with the appropriate serotype, a strong immunity to reinfection develops. Work with mice indicates that post-infection carriers are particularly resistant. Not only are large numbers of activated macrophages present in the tissues, resulting in rapid clearance from the blood and intracellular bacterial destruction, but, in the case of carriers, increased resistance to other pathogens is also observed. Attempts are made to reproduce this level of resistance by vaccination. However, depending on whether killed vaccines or live, attenuated vaccines are used, not only is the degree of immunity affected but the relative contributions of cell mediated and humoral immunity also differ.

Because of several advantages including safety and quality control, many attempts have been made to develop non-living vaccines for use with host-specific serotypes. Because of our microbiological fixation with O-antigens many preparations have consisted of heat killed bacterins in which many protein antigens are obviously denatured. Such vaccines do not usually produce good protection against infection by the natural, oral route. This has been shown with *S. typhimurium, S. cholerae-suis, S. enteritidis*, and *S. dublin* in mice (32, 47, 56) and *S. gallinarum* in chickens (46). In some cases, however, as with *S. dublin* infection in cattle (1) and *S. typhi* in man (18), killed vaccines have been demonstrated to produce some protection. Killed vaccines stimulate the production of high levels of antibody and in some of the experimental work bacterial challenge has been by the intra-peritoneal route where antibodies are particularly effective (30). However, this route is highly artificial and unnatural. The general ineffectiveness of killed vaccines is variously attributed to rapid destruction and elimination of the organisms, destruction of the relevant antigens during preparation, and inappropriate presentation. However, none of these reasons militate against the use of a non-living vaccine should a suitable one be developed. Although the high levels of antibody produced are opsonizing, killed vaccines are less effective at controlling intracellular multiplication in the liver and spleen (13, 14, 61).

Recently, there has been increased interest in the use of outer membrane proteins (OMPs) as vaccines. This has arisen from the realization that earlier good results on protectiveness by using ribosomal vaccine preparations were probably the result of contamination of these preparations by membrane proteins (24, 50). Such proteins are not only protective in mice (60) but also stimulate a delayed type hypersensitivity in addition to a humoral immune response.

In contrast, animals that have recovered from infection or have been vaccinated with a live attenuated organism, can be protected against up to 1000 50% lethal doses and are thus highly resistant (21, 22). Protection is good against a parenteral or oral route of challenge.

Live vaccines stimulate a strong, cell mediated immunity (CMI) which some groups consider to be essential for protection (14). Again, this supposition arises mainly from work with parenteral *S. typhimurium* infection in mice. The evidence for the predominant role of CMI in protection includes the poor protection afforded by vaccinating T cell depleted mice and successful transfer of immunity with adoptive transfer of T cells from immune mice (40). Transfer of serum from immune animals is not sufficient to transfer full immunity (12). Typically, CMI has been indicated by the strength of a delayed type hypersensitivity (DTH) reaction. These two manifestations of T cell activity are, however, not completely correlated. This has been demonstrated by immunizing mice with an *aroA* mutant of *S. typhimurium* (26). It may well be that *in vitro* cellular bactericidal assays are better indicators of cellular immunity (53). However, several groups have indicated that antibodies play a major role in the expression of immunity even after parenteral challenge. This has been done again by adoptive transfer experiments, this time of B and T cells (22) and by using animals in which either B or T cells have been depleted (33). Also, some experiments indicate that transfer of immune serum can increase resistance (47). Undoubtedly, CMI plays a major part in immune control of systemic salmonellosis but until studies are carried out using natural routes of infection and in which the expression of immunity at the level of the mucosa and its secretions is included, a major contribution by humoral immunity cannot be dismissed. It is quite likely that in many cases both CMI and humoral immunity act in concert (36).

The situation with host non-specific serotypes is more confusing again since we are not only interested in preventing bacteria from entering and multiplying in the tissues but in eliminating large numbers of them from the alimentary tract.

Infection of poultry with strains which are obviously invasive, such as *S. typhimurium* or *S. enteritidis*, induces both DTH and a humoral immune response (29, 34, 55, 59). *Salmonella*-specific antibodies can be detected in serum, bile and in the intestine. Serum antibodies are agglutinating and complement-fixing but all the antigens responsible for their induction are not known. Identification of minor antigens would be difficult with the insensitive agglutination reactions upon which most immunoglobulin studies of salmonellosis in poultry have hitherto been based. The use of ELISAs and immunoblots should improve this situation. Since most serotypes are largely confined to the alimentary tract it is tempting to suggest, in the absence of hard evidence, that secretory IgA plays a major role in intestinal clearance but whether it is locally produced or is serum IgA secreted in the bile is unknown. Perhaps following infection with an invasive strain both sources of IgA contribute to producing maximum immunity. The contributions of IgG and IgM to intestinal clearance are unknown, if they are involved at all. Similarly, the role of CMI in gut clearance can only be guessed at since information on the behaviour of macrophages and natural killer cells, or their equivalent, in the intestinal mucosa and lumen is almost non-existent. It must also be borne in mind that results obtained with *S. typhimurium* or other invasive strains cannot necessarily be extrapolated to other, less-invasive strains and serotypes.

Work at Houghton on the immune response to *Salmonella* infection of poultry (20) was carried out using infection of four-day-old chickens with *S. typhimurium* strain F98 (5, 48). When chickens are infected at this age they do not become ill but excrete the organism in the faeces for several weeks (49). Because this strain is highly invasive and stimulates a good protective immunity (see section entitled "Vaccination of Poultry Against Host-Non-Specific [Non-Adapted] Serotypes"), we expected a strong systemic and local humoral response in addition to a DTH reaction. Humoral responses were measured by indirect ELISA, using a number of antigens. Titres of serum IgG, IgM, and IgA rose within a few days of infection and peaked at four weeks. IgM rose slightly quicker than IgG or IgA. IgM and IgA titres decreased after a few weeks whereas IgG titres persisted. A similar picture was observed by examining the intestinal washings of these birds, except that here specific IgA titres were higher than those of the other two classes. Much higher titres of specific-IgA were found in bile.

DTH was measured as a crude indicator of CMI by inoculating antigens into the foot-pad of chickens at various times after oral infection. Inoculation of whole bacterial cell soluble protein and OMPs produced a strong reaction whereas flagella and lipopolysaccharide produced limited responses suggesting that it might be in such outer membrane preparations that we should look for useful immunogens.

VACCINATION AGAINST HOST-SPECIFIC (HOST ADAPTED) SEROTYPES

Infection of poultry, man or other animal species with one of the host species-specific serotypes induces a strong protective immunity against reinfection. Live, attenuated vaccines have been developed for economically important animals, which can be administered parenterally, since the disease is primarily systemic, and which are protective (Table 1).

These vaccines are used extensively. They are rough and therefore do not stimulate the production of anti-somatic-antigen antibodies. This is an advantage since the use of the vaccines does not interfere with serological tests based on the production of such antibodies to detect natural infection.

In contrast, immunological control of the host-non-specific serotypes in poultry has been far less successful. The reasons for this are manifold. They include antigenic heterogeneity and the fact that they are less host-specific and therefore have a complex epidemiology. Variable characteristics such as virulence, colonization ability, and invasiveness also contribute to a poorly understood immune response. Historically, therefore, the value of vaccination for these serotypes has been questionable.

Table 1. Protective ability of attenuated vaccines for host specific serotypes.

		mortality caused by *S. dublin* (calves), *S. cholerae-suis* (pigs), or *S. gallinarum* (fowl) when animals are	
Live Vaccine	Animal Species	Vaccinated and Challenged	Challenged Only
S. gallinarum 9R	fowl	0/40	25/40
S. dublin 51	calves	1/15	11/15
S. cholerae-suis 6	pigs	0/12	7/12

Six-week-old chickens and calves and 11-week-old pigs vaccinated subcutaneously and challenged orally three weeks later.
(After Smith, 46, 47).

VACCINATION OF POULTRY AGAINST HOST-NON-SPECIFIC (NON-ADAPTED) SEROTYPES

Because of the paucity of information on colonization and immunity relating to the *Salmonella* serotypes that are usually associated with food poisoning, the development of vaccines for use with poultry has been almost exclusively empirical. Various types of non-living vaccines have been used experimentally and in the field. They certainly generate an immune response but produce a poor or inconsistent protective effect. Some of this work is described below.

In several experiments Truscott immunized chickens with heated sonicates, prepared from different combinations of serotypes and incorporated in the feed (57). This was followed some weeks later by oral challenge. The extent of protection, measured by isolation of the challenge strains from cloacal swabs, was variable (Table 2). In some of these experiments protection is very good while in others it is poor. For a vaccine to be of potential use in the field experimental results must show consistent, excellent protection.

Other groups (11, 54) have found that heated, whole cell bacterins have little effect on faecal shedding either in vaccinated birds or in their progeny (Table 3). Following such vaccination of the dams, newly hatched chickens have been challenged and mortality observed. Some reductions have been produced (31, 58), but the biological significance of these findings in relation to faecal shedding and subsequent carcass contamination is unknown.

Table 2. Immunization against host-non-specific serotypes with heated sonicates incorporated in the feed.

Expt.	Antigens in Vaccine	Challenge	Isolation of challenge strains (%) from cloacal swabs from: Vaccinated and Challenged	Challenged Only
1	3 serotypes*	3 serotypes: oral challenge	14	83
2	3 serotypes	3 serotypes: in-feed challenge	1.7	71
3	4 serotypes+	4 serotypes: contact challenge	51	71
4	9 serotypes[s]	9 serotypes: contact and oral challenge	0	13
5	*S. typhimurium*	*S. typhimurium*: contact and in-feed challenge	9	30
6	*S. typhimurium*	*S. typhimurium*: contact and in-feed challenge	12	27

* *S. typhimurium, S. infantis, S. enteritidis.*
\+ *S. typhimurium, S. infantis, S. enteritidis, S. anatum.*
s *S. typhimurium, S. infantis, S. enteritidis, S. montevideo, S. arizonae, S. sandiego, S. schwarzengrund, S. anatum, S. heidelberg.*

Vaccination in feed, challenge 2 - 7 weeks later.
(After Truscott, 57).

Some live, attenuated vaccines have been adapted for use with poultry to attempt to mimic natural infection, but whether they behave in the same way as the parent or field strains in stimulating immunity is not known. Most of this work has been done with *S. typhimurium* which may not necessarily be representative of other serotypes in *in vivo* behaviour.

A mutant of a mammalian *S. typhimurium* strain defective in the enzyme UDP-galactose epimerase (*galE*) and which is non-viable *in vivo*, was originally developed for use in mice and cattle (63), but has also been tested in poultry. Pritchard *et al.* (41) vaccinated one-day-old chickens with this strain and challenged two weeks later with an

Table 3. Faecal excretion of *S. hadar* progeny of vaccinated and unvaccinated turkey hens.

ISOLATION OF *S. HADAR* FROM CLOACAL SWABS OF POULTS THAT WERE	
Vaccinated and Challenged	Challenged Only
268/373	243/312
(72%)	(78%)

Vaccination of hens with heat inactivated *S. hadar*, challenged of poults at 0 day. (After Thain *et al.*, 54).

Table 4. Immunization of chickens with a *galE S. typhimurium*.

	Isolation of challenge *S. typhimurium* strain from samples	
Samples	Vaccinated and Challenged	Challenged Only
Cloacal Swabs*	28/95	42/97
Liver*	0/95	3/97
Spleen*	0/95	5/97
Caecal Junction*	3/95	5/97

*At slaughter.

Vaccination orally at 0 day with *galE* mutant, challenge orally at 14 days with challenge strain.

(After Pritchard *et al.*, 41).

avian *S. typhimurium* strain. Small reductions were obtained in faecal excretion and in the number of isolations from the viscera (Table 4). Similar reductions were obtained by the same laboratory when chickens wre vaccinated twice, at four and six weeks of age, followed two weeks later by challenge. Surprisingly, protection was better when chickens were inoculated intramuscularly rather than orally (52). One drawback of the use of any *galE* mutant is the question of attenuation. Some mutants are known to retain some virulence for the host (23, 37).

Others have used a murine streptomycin-dependent *S. typhimurium* strain (44). After vaccination of very young chicks and challenge eight days later, protection was poor and inconclusive. The value of such experiments in which animals are challenged when still relatively immunologically immature, so soon after vaccination and with extremely large challenge doses (in this case 10^{10} organisms), must be questioned. In order to produce an effective vaccine it may finally be essential to choose an avian strain from which to construct mutants. Failure to do so might explain the relatively poor protective ability of the *galE* and streptomycin-requiring strains described above.

Work at Houghton on reduction of faecal excretion by vaccination arose out of investigations into the basic microbiology of intestinal colonization. We, like many others, wondered whether immunity might be partly responsible for clearance of *Salmonella* organisms from the alimentary tract after experimental infection.

We have found that some strains and serotypes are excreted in the faeces for longer periods than are others (7, 49). We have found a general inverse correlation between virulence for chickens and the duration of faecal excretion. From previous work (5) on *S. typhimurium* infection in young chickens we know that invasiveness is the characteristic of overriding importance in virulence, more so than the ability to multiply in the tissues. Invasiveness can also be detected in older chickens by quantitative bacteriology (7). As might be expected, *S. typhimurium* strains are more invasive than other serotypes and they are excreted for shorter periods.

Our hypothesis was that the more invasive strains will stimulate a systemic, in addition to a local, immune response. These should combine to clear such strains from the alimentary tract quicker than might occur following colonization by less invasive strains. However, it might eventually follow, to our advantage, that after immunization with an invasive strain clearance of both invasive and less invasive challenge strains should occur to the same extent.

We thought that in looking for a strain from which to construct a vaccine we should choose a highly invasive avian strain which colonized the gut well. We chose our strain, *S. typhimurium* F98, which is highly virulent for newly hatched chicks (7, 49), but which produces no disease in mammals. We sought to establish the degree of protection we might expect by vaccinating, as it were, with the parent strain. We inoculated four-day-old chickens, orally, with a spectinomycin resistant mutant (Spc^r) of F98. When the chickens had virtually ceased to excrete this strain several weeks later, these birds and an uninfected control group were challenged orally with a nalidixic acid resistant mutant (Nal^r) of the same strain (see Table 5) (4). Good protection against faecal excretion was obtained; only one isolation of the challenge strain was made from the vaccinated birds. Similar results were obtained when the first infection was administered by the intramuscular route. In this case, the Spc^r strain was also isolated from the faeces so that it was impossible to ascertain which route was preferable.

Although in the long term we sought to study systematically the microbial characteristics and antigens involved in generating this level of immunity, we succumbed to the temptation to produce a vaccine quickly. We immunized birds with large numbers of formalin-killed organisms either intramuscularly or incorporated in the feed, in numbers that might be expected in the gut following oral inoculation. Some reduction in faecal excretion occurred by "in-feed" immunization but intramuscular administration had virtually no effect. We produced two attenuated vaccine strains from F98. One of these had been cured of its virulence plasmid (7) and was additionally made *aroA* and a second was made rough by bacteriophage. By intramuscular vaccination both strains produced an initial good reduction in faecal excretion which did not persist. By oral inoculation the *aroA* strain produced limited protection. By contrast (Table 6), the rough strain induced very good protection. Because this strain is rough it does not generate serum agglutinins to the O antigen. Neither is it virulent for man. Oral ingestion by volunteers of 10^8 organisms resulted in no ill effects, and the strains persisted in the faeces for no longer than three days.

Table 5. Faecal excretion after infection and reinfection with *S. typhimurium* F98.

	Percentage of 30 chickens excreting *S. typhimurium* F98 at (days)								
Initially Infected	11	18	25	32	39	44	53	60	67
with	period of excretion of F98 Spc^r				period of excretion of F98 Nal^r				
Nothing	0	0	0	0	53	48	15	0	0
F98 Spc^r	94	55	16	6	3	0	0	0	0

Infection orally at 4 days of age, reinfection orally at 35 days of age with F98 Nal^r. (from Barrow *et al.*, 4).

Vaccines developed for controlling host-specific serotypes are less effective in reducing the faecal excretion of host-non-specific serotypes possibly because local intestinal immunity is not stimulated. The *S. dublin* 51 vaccine has little effect in reducing excretion of *S. typhimurium* by chickens, although it reduces systemic multiplication of the challenge strain (27, 28). Such vaccines might, however, be profitably used to reduce systemic infection in birds when it occurs, such as with the present epidemic of *S. enteritidis* phage type 4 in laying hens and breeders in the United Kingdom. This serotype has now (late 1989) overtaken *S. typhimurium* as the serotype responsible for most outbreaks of human food poisoning. The strain involved is highly virulent for chickens. After experimental oral infection of newly hatched chickens we can produce 100% mortality with characteristic polyserositis (including the pericarditis seen in the field). While

Table 6. Oral immunization of chickens with a rough *S. typhimurium* strain.

Immunized With	11	18	25	32	39	46	53	60	67	74	81	88	95	102
	Percentage of chickens excreting *S. typhimurium* at (days)													
	Excretion of immunizing strain (F98 Nal^r rough)									Excretion of challenge strain (F98 Nal^r smooth)				
Nothing	0	0	0	0	0	0	0	0	0	93	64	27	7	0
Rough Strain	70	62	51	43	37	30	26	17	17	17	0	0	0	0

Vaccination orally at 4 days of age, challenge orally at 70 days.
(from Barrow *et al.*, 4).

Table 7. Immunization of laying hens against *S. enteritidis* phage type 4 with live, attentuated vaccine.

Vaccinated With	Liver	Spleen	Ovary	Oviduct	Caeca	Cloaca	Faeces	Laid Eggs
	Isolation of challenge *S. enteritidis* strain from							
Nothing	12/20	17/20	8/20	0/20	11/20	9/20	13/19	43/221
S. gallinarum 9R[1]	4/20	4/20	0/20	0/20	16/20	9/20	5/19	16/203
S. enteritidis aroA[2]	9/20	12/20	7/20	0/20	12/20	10/20	9/19	13/205

[1]Chickens immunized intramuscularly twice at two week intervals.
[2]Chickens immunized intramuscularly and orally twice at two week intervals.
All chickens challenged orally two weeks after last immunization with 10^8 fully virulent *S. enteritidis* phage type 4.
(from Barrow *et al.*, 6).

much of the food poisoning associated with this phage type is linked to the consumption of contaminated meat, a considerable proportion is connected with the consumption of infected eggs. We have developed a model infection in laying hens in which, following oral infection, the ovary becomes infected and contaminated eggs are produced. We vaccinated hens, in lay, with the *S. gallinarum* 9R vaccine (46) by the intramuscular route, twice, at two week intervals. A second group of hens was vaccinated both orally and intramuscularly, at the same times, with a rough, aroA mutant of the phage type 4 strain, which we produced in our laboratory. Two weeks after the last vaccination these and unimmunized chickens were challenged with a Nal^r mutant of the parent strain by the oral route (6).

Isolations from the ovary were reduced to zero by vaccination with 9R (Table 7). There was also some reduction in the number of isolations from the liver and spleen. The *aroA* vaccine appeared to be less effective. The effect of either of these vaccines on isolations from the caeca, cloaca, or faeces is less clear. Both vaccines produced a reduction in the number of contaminated eggs, although this picture is probably confused by egg infection resulting from both ovarian infection and from faecal contamination.

Whether such reductions would be sufficient to reduce the public health risk when eating undercooked eggs is unclear. However, it does show that some current problems can still be tackled empirically with some success.

RELEVANT NEW DEVELOPMENTS IN VACCINES AND CRITERIA FOR THE IDEAL VACCINE

Much recent work has shown that large molecular weight plasmids in host-specific serotypes are associated with virulence (2, 8, 25). Plasmid-cured strains are avirulent or of greatly reduced virulence and are also immunogenic and protective in the cases of *S. enteritidis* (35) and *S. gallinarum* (3). Since the plasmid of *S. typhimurium* is not associated with intestinal colonization in poultry (7), it is unclear whether such vaccine strains would have a role in reducing faecal excretion. Problems of reversion to virulence might arise in using such strains even though the plasmids are non-self transmissible. The problem could be overcome by introducing into the plasmid-cured vaccine strain a different plasmid not associated with virulence but which is in the same incompatibility group.

New attenuated vaccines, generated by transposon mutagenesis, are becoming available for *S. typhimurium*. By deletion of the transposon together with part of the gene into which it has inserted, thus producing a deletion mutant, the danger of reversion to virulence is virtually eliminated. By this method, mutants can be produced which are defective in aromatic amino acid synthesis and thus require p-amino benzoic acid (the most useful so far being the *aroA* mutation). *S. typhimurium aroA* mutants have been tested successfully in cattle (42). Other auxotrophic mutants, including the *pur* and *thy* mutations, are of less certain value, mainly because they have not, as yet, been tested as thoroughly (38, 39, 45). Transposon mutants have also been produced which are defective in adenylate cyclase activity (*cya*) or cyclic AMP receptor protein (15) or in specific outer membrane proteins (16). These mutants are of reduced virulence and induce immunity in laboratory mice. It will be interesting to see whether any of them can be used for poultry. It must be remembered that vaccines effective in mice may not be so against the same pathogen but in a different host (47). It should soon be possible to delete from the chromosome the gene responsible for enterotoxigenicity, thereby producing a mutant which should behave like the parent strain in the alimentary tract of poultry but which would be avirulent for man. The characteristics of an ideal vaccine for reducing faecal excretion of food poisoning host-non-specific serotypes have been summarized (41), but are also discussed below (41).

Strong protection against intestinal and systemic infection is the first requirement. An additional requirement is avirulence for man. In view of the current increased public awareness of *Salmonella* food poisoning, there may be considerable resistance to using a live vaccine that is *Salmonella*-derived, unless it can be shown that the vaccine is no more virulent than the *E.coli* strains also present on the carcass at slaughter, a more realistic criterion than complete avirulence. In the long term, it may be necessary to clone genes encoding appropriate colonization and/or other protective antigens into harmless vector organisms. However, in the absence of detailed information on these characteristics this approach is some way off. Avirulence for chickens and ease of administration are, in fact, linked. Vaccines should be administered by mass methods to minimize cost. The ideal route would be orally via the drinking water or food or by spray. However, parenteral administration may be an additional requirement for maximum protection. This is already done for a number of avian virus vaccines. Although the ideal vaccine should be avirulent for chickens, oral vaccination could require the use of an invasive strain to stimulate maximum immunity because immunogenicity may be correlated with invasiveness. Whichever route of inoculation is used, residual virulence may result in vertical transmission as occurs occasionally with the *S. gallinarum* 9R vaccine. This should present few problems providing that the vaccine produces no disease in the progeny. More importantly, to be economically viable for use in broilers the vaccine must not reduce the growth rate.

Protection should obviously last as long as possible. Protection of broilers is required for a matter of weeks, but protection of breeders and layers must last for months.

Like competitive exclusion, immunity is likely to have little protective effect against infections that are extant at the time of immunization. Following vaccination, a protective immunity takes several days to develop. The delay in development of immunity could be overcome by using a live strain that shows the colonizing-blocking effect, a form of competitive exclusion, that one strain of *Salmonella* can produce against another (9, 10). This might mean that an appropriate live, attenuated strain, administered to day-old chicks could protect by colonization-inhibition in the first week of life before full immunity develops.

The extent of cross-protection of one serotype against others is also important to determine. Where outbreaks involve one serotype this issue does not arise, but for general use, a vaccine must protect against a wide range of serotypes. If colonization or protective antigens are common to many serotypes, this does not present a problem. This again demonstrates our ignorance of the microbiological basis of colonization.

Present legislation in the United Kingdom requires that all isolations of *Salmonella* from poultry be reported (51). Thus, wild type strains must be easily differentiated from vaccine strains. This can be accomplished by the vaccine strain possessing genetic markers, such as a combination of auxotrophy and antibiotic resistance selected because they are unlikely to be encountered in the field. If monitoring is done serologically the vaccine must not express the antigens responsible for stimulating antibodies detected in the test used. In the case of agglutination reactions the use of rough strains might be sufficient. Strains can easily be made rough by transposon mutagenesis or bacteriophage activity. If ELISAs are used for serology, the vaccine should not express the antigen used for detection by the ELISA.

Vaccination against salmonellosis can produce a degree of protection over and above that possessed by unimmunized chickens even with a fully mature intestinal flora. Vaccination should thus be compatible with, and complementary to, competitive exclusion. In addition to this, no problems should occur analogous to the development of antibiotic resistance following chemotherapy, and immunization should be compatible with growth promoting antibiotics.

IMPROVED SEROLOGICAL SCREENING

A strong serum IgG response to *S. typhimurium* infection in chickens, detectable by ELISA, opens the way to rapid, large scale serological screening of flocks for *Salmonella* infection. Many serotypes are sufficiently invasive for poultry to stimulate detectable levels of specific IgG. Until recently, serological detection of salmonellosis has only extensively been used for *S. pullorum-S. gallinarum* using the crude plate agglutination test (43). It is possible to detect *S. typhimurium* serologically by using agglutination or micro-antiglobulin tests (62), but these are lengthy procedures and not suitable for large scale screening. Recent work with *S. typhi* infection in humans has indicated that the ELISA can be used to identify carriers using the Vi antigen (17). Our work (19) showed that experimental infection of chickens at different ages with different strains of *S. typhimurium* stimulated the production of high titres of IgG. These could be detected with a whole-cell soluble protein, lipopolysaccharide, flagella antigen, or outer membrane protein. With the ELISA, chickens infected with *S. typhimurium* could be differentiated from chickens infected with ten other serotypes. The test can differentiate flocks infected with *S. typhimurium* from those that are uninfected. Because of the ease with which large numbers of samples can be processed and because of the problem of intermittent faecal shedding, the ELISA should be superior to bacteriological sampling and to other serological methods for screening chickens for *S. typhimurium* infection. The system should be easily adaptable to other invasive serotypes such as *S. enteritidis*.

THE BASIS OF INTESTINAL COLONIZATION

The work described in this review does not attempt to understand the basic mechanisms of immunogenicity, colonization, and virulence that is essential for real progress in the development of vaccines.

At Houghton, we have started some work to try to identify colonization determinants in *S. typhimurium* and *S. infantis* (7) by producing mutants defective in easily recognizable characteristics and analyzing their faecal excretion patterns (Table 8). We have produced several individual mutants of *S. typhimurium* F98 which are separately rough, non-motile, and deficient in the virulence-associated plasmid, and most of these colonize the gut to the same extent as the parent. However, one rough mutant, produced by prolonged culture at room temperature was not excreted for more than two weeks. Similarly, rough, non-motile, and non-piliated mutants of *S. infantis* colonized the gut well. In contrast, one mutant of this serotype generated by nitrosoguanidine, possessed all its major antigens but was not excreted for more than four days. Further work is under way to characterize this chromosomal determinant in greater detail.

Table 8. Faecal excretion of *S. typhimurium* mutants.

Mutant of *S. typhimurium* F98	No. of Chickens	Percentage of chickens excreting *S. typhimurium* at (d)						
		7	14	21	28	35	42	49
Parent Strain	33	97	69	13	9	9	6	3
Non-Motile Strain	33	100	69	19	9	13	3	3
Rough Mutant-1	33	56	41	16	25	28	16	16
Rough Mutant-2	33	38	3	0	0	0	0	0
Virulence Plasmid Cured	33	97	91	45	12	12	9	0

All strains inoculated orally into three-week-old chickens.
(After Barrow *et al.*, 7).

In summary, much remains to be done before vaccination is a practical means of preventing *Salmonella* infections in poultry. This cannot be done on a rational basis without basic information on the microbial and molecular basis of colonization, the relationship between colonization and immunity, and the identification of protective antigens.

ACKNOWLEDGEMENTS

The author would like to acknowledge J.O. Hassan, University of Washington, and A. Berchieri, University of Sao Paulo, for their contributions to the work carried out at Houghton, and Mrs. S. Pilcher for typing the manuscript.

REFERENCES

1. **Aitken, M.M., P.W. Jones, and G.T.H. Brown.** 1982. Protection of cattle against experimentally induced salmonellosis by intradermal injection of heat-killed *Salmonella dublin*. Res. Vet. Sci. **32:**368-373.

2. **Baird, G.D., E.J. Manning, and P.W. Jones.** 1985. Evidence for related virulence sequences in plasmids of *Salmonella dublin* and *Salmonella typhimurium*. J. Gen. Microbiol. **131:**1815-1823.
3. **Barrow, P.A.** 1990. Immunity to experimental fowl typhoid in chickens induced by a virulence plasmid-cured derivative of *S. gallinarum*. Infect. Immun. **58:**2283-2288.
4. **Barrow, P.A., J.O. Hassan, and A. Berchieri.** 1990. Reduction in faecal excretion of *Salmonella typhimurium* strain F98 in chickens vaccinated with live and killed *S.typhimurium* organisms. Epidemiol. Infection. (In Press).
5. **Barrow, P.A., M.B. Huggins, M.A. Lovell, and J.M. Simpson.** 1987. Observations on the pathogenesis of experimental *Salmonella typhimurium* infection in chickens. Res. Vet. Sci. **42:** 194-199.
6. **Barrow, P.A., M.A. Lovell, and A. Berchieri.** 1990. Immunisation of laying hens against *Salmonella enteritidis* phage type 4 with live, attenuated vaccines. Vet. Rec. **126:**241-242.
7. **Barrow, P.A., J.M. Simpson, and M.A. Lovell.** 1988. Intestinal colonization in the chicken by food-poisoning *Salmonella* serotypes; microbial characteristics associated with faecal excretion. Avian Path. **17:**571-588.
8. **Barrow, P.A., J.M. Simpson. M.A. Lovell, and M.M. Binns.** 1987. Contribution of *Salmonella gallinarum* large plasmid toward virulence in fowl typhoid. Infect. Immun. **35:**388-392.
9. **Barrow, P.A., J.F. Tucker, and J.M. Simpson.** 1987. Inhibition of colonization of the chicken alimentary tract with *Salmonella typhimurium* by Gram-negative facultatively anaerobic bacteria. Epidemiol. Infect. **98:**311-322.
10. **Berchieri, A., and P.A. Barrow.** 1990. Further studies on the inhibition of colonization of the chicken alimentary tract with *Salmonella typhimurium* by pre-colonization with an avirulent mutant. Epidemiol. Infect. (In Press).
11. **Bisping, W., I. Dimitriadis, and M. Seippel.** 1971. Versuche zur oralen Immunisierung von Huhnen mit hitzeinaktivierter Salmonella-Vakzine 1. Mitteilung: Impf-und Infektionversuche an Huhnerkuken. Z. Veterinarmed. **B, 18:**337-346.
12. **Collins, F.M.** 1969. Effect of specific immune mouse serum on the growth of *Salmonella enteritidis* in non-vaccinated mice challenged by various routes. J. Bacteriol. **97:**667-675.
13. **Collins, F.M.** 1969. Effect of specific immune mouse serum on the growth of *Salmonella enteritidis* in mice preimmunized with living or ethylalcohol-killed vaccines. J. Bacteriol. **97:**676-683.
14. **Collins, F.M.** 1974. Vaccines and cell-mediated immunity. Bacteriol. Rev. **38:**371-402.
15. **Curtiss, R., and S.M. Kelly.** 1987. *Salmonella typhimurium* mutants lacking adenylate cyclase and cyclic AMP receptor protein are avirulent and immunogenic. Infect. Immun. **55:**3035-3043.
16. **Dorman, C.J., S. Chatfield, C.F. Higgins, C. Hayward, and G. Dougan.** 1989. Characterisation of porin and *ompR* mutants of a virulent strain of *Salmonella typhimurium*: *ompR* mutants are attenuated *in vivo*. Infect. Immun. **57:**2136-2140.
17. **Engleberg, N.C., T.J. Barrett, H. Fisher, B. Porter, E. Hurtado, and J.M. Hughes.** 1983. Identification of a carrier by using Vi enzyme-linked immunosorbent assay serology in an outbreak of typhoid fever on an Indian reservation. J. Clin. Microbiol. **18:**1320-1322.

18. **Germanier, R.** 1984. Typhoid fever. *In* R. Germanier, (Ed). Bacterial Vaccines. Academic Press Inc., London.
19. **Hassan, J.O., P.A. Barrow, A.P.A. Mockett, and S. McLeod.** 1990. Antibody response to experimental *Salmonella typhimurium* infection in chickens measured by ELISA. Vet. Rec. **126:**519-522.
20. **Hassan, J.O., A.P.A. Mockett, and P.A. Barrow.** 1990. Protective immunity against faecal excretion by chickens of *Salmonella typhimurium* by preinfection with *S. typhimurium*; bacteriology and humoral and cellular immune responses. (Manuscript in preparation, to be submitted to Infection and Immunity).
21. **Hobson, D.** 1957. Resistance to reinfection in experimental mouse typhoid. J. Hygiene Camb. **55:**334-343.
22. **Hochadel, J.F., and K.F. Keller.** 1977. Protective effects of passively transferred immune T- or B-lymphocytes in mice infected with *Salmonella typhimurium*. J. Infect. Dis. **135:**813-823.
23. **Hone, D.M., S.R. Attridge, B. Forrest, R. Morona, D. Daniels, J.L. LaBrooy, R.C.A. Bartholomeusz, D.J.C. Shearman, and J. Hackett.** 1988. A *galE via* (Vi antigen-negative) mutant of *Salmonella typhi* Ty2 retains virulence in humans. Infect. Immun. **56:**1326-1333.
24. **Johnson, W.** 1973. Ribosomal vaccines II. Specificity of the immune response to ribosomal ribonucleic acid and protein isolated from *Salmonella typhimurium*. Infect. Immun. **8:**395-400.
25. **Jones, G.W., D.K. Rabert, D.M. Svinarich, and H.J. Whitfield.** 1982. Association of adhesive, invasive and virulent phenotypes of *Salmonella typhimurium* with autonomous 60-megadalton plasmids. Infect. Immun. **38:**476-486.
26. **Killar, L., and T.K. Eisenstein.** 1986. Delayed-type hypersensitivity and immunity to *Salmonella typhimurium*. Infect. Immun. **52**: 503-508.
27. **Knivett, V.A., and W.K. Stevens.** 1971. The evaluation of a live *Salmonella* vaccine in mice and chickens. J. Hygiene Camb. **69:**233-245.
28. **Knivett, V.A., and J.F. Tucker.** 1972. Comparison of oral vaccination or furazolidone prophylaxis for *Salmonella typhimurium* infection in chicks. Brit. Vet. J. **128:**24-34.
29. **Lee, G.M., G.D.F. Jackson, and G.N. Cooper.** 1983. Infection and immune responses in chickens exposed to *Salmonella typhimurium*. Avian Dis. **27:**577-583.
30. **MacKaness, G.B., R.V. Blanden, and F.M. Collins.** 1966. Host-parasite relations in mouse typhoid. J. Exp. Med. **124:**573-583.
31. **McCapes, R.H., R.T. Coffland, and L.E. Christie.** 1967. Challenge of turkey poults originating from hens vaccinated with *Salmonella typhimurium* bacterin. Avian Dis. **11:**15-24.
32. **MacLeod, D.R.E.** 1954. Immunity to *Salmonella* infection in mice. J. Hygiene Camb. **52:**: 9-17.
33. **Morris, J.A., C. Wray, and W.J. Sojka.** 1976. The effect of T and B lymphocyte depletion on the protection of mice vaccinated with a *galE* mutant of *Salmonella typhimurium*. Brit. J. Exp. Pathol. **57:**354-360.
34. **Nagaraja, K.V., J.A. Newman, and B.S. Pomeroy.** 1982. Leukocyte migration inhibition in chickens inoculated with *Salmonella typhimurium*. Amer. J. Vet. Res. **43:**916-918.
35. **Nakamura, M., S.Sato, T. Ohya, S. Suzuki, S. Ikeda, and T. Koeda.** 1985. Plasmid-cured *Salmonella enteritidis* AL 1192 as a candidate for a live vaccine. Infect. Immun. **50:**586-587.

36. **Nencioni, L., L. Vila, D. Boraschi, B. Berti, and A. Tagliabue.** 1983. Natural and antibody-dependent cell-mediated activity against *Salmonella typhimurium* by peripheral and intestinal lymphoid cells in mice. J. Immunol. **130:** 903-907.
37. **Nnalue, N.A., and B.A.D. Stocker.** 1986. Some *galE* mutants of *Salmonella cholerae-suis* retain virulence. Infect. Immun. **54:**635-640.
38. **Nnalue, N.A., and B.A.D. Stocker.** 1987. Tests of the virulence and live vaccine efficacy of auxotrophic and *galE* derivatives of *Salmonella cholerae-suis*. Infect. Immun. **55:**955-962.
39. **O'Callaghan, D., D. Maskell, F.Y. Liew, C.S.F. Easmon, and G. Dougan.** 1988. Characterization of aromatic- and purine-dependent *Salmonella typhimurium*: attenuation, persistence and ability to induce protective immunity in Balb/c mice. Infect. Immun. **56:**419-423.
40. **Paul, C., K. Shalala, R. Warren, and R. Smith.** 1985. Adaptive transfer of murine host protection to salmonellosis with T-cell factor-dependent, *Salmonella*-specific T cell lines. Infect. Immun. **48:**40-43.
41. **Pritchard, D.G., S.C. Nivas, M.D. York., and B.S. Pomeroy.** 1978. Effect of Gal-E mutant of *Salmonella typhimurium* on experimental salmonellosis in chickens. Avian Dis. **22:**562-575.
42. **Robertsson, J.A., A.A. Lindberg, S. Hoiseth, and B.A.D. Stocker.** 1983. *Salmonella typhimurium* infection in calves: protection and survival of virulent challenge bacteria after immunization with live or inactivated vaccines. Infect. Immun. **41:**742-750.
43. **Schaffer, J.M., A.D. MacDonald, W.J. Hall, and H. Bunyea.** 1931. A stained antigen for the rapid whole blood test for pullorum disease. J. Amer. Vet. Med. Assoc. **79:**236-240.
44. **Schlimmel, D., K. Linde, G. Marx, and K. Ziedler.** 1974. Zum Einsatz einer Smd-*Salmonella-typhimurium* Mutante bei Kucken. Arch. Exp. Veterinarmed. **28:**551-558.
45. **Sigwart, D., B.A.D. Stocker, and J.D. Clements.** 1989. Effect of a purA mutation on efficacy of *Salmonella* live vaccine vectors. Infect. Immun. **57:**1858-1861.
46. **Smith, H.W.** 1956. The use of live vaccines in experimental *Salmonella gallinarum* infection in chickens with observations on their interference effect. J. Hygiene Camb. **54:**419-432.
47. **Smith, H.W.** 1965. The immunization of mice, calves and pigs against *Salmonella dublin* and *Salmonella cholerae-suis* infections. J. Hygiene Camb. **63:**117-135.
48. **Smith, H.W., and J.F. Tucker.** 1975. The effect of antibiotic therapy on the faecal excretion of *Salmonella typhimurium* by experimentally infected chickens. J. Hygiene Camb. **75:**275-292.
49. **Smith, H.W., and J.F. Tucker.** 1980. The virulence of *Salmonella* strains for chickens: their excretion by infected chickens. J. Hygiene Camb. **84:**479-488.
50. **Smith, R.A., and N.J. Bigley.** 1972. Ribonucleic acid-protein fractions of virulent *Salmonella typhimurium* as protective immunogens. Infect. Immun. **6:**377-383.
51. **Statutory Instruments.** 1989. No. 285. The Zoonoses Order. Her Majesty's Stationery Office. London, U.K.
52. **Subhabphant, W., M.D. York, and B.S. Pomeroy.** 1983. Use of two vaccines (Live G30D or killed RW16) in the prevention of *Salmonella typhimurium* infections in chickens. Avian Dis. **27:**602-615.

53. **Tagliabue, A., L. Villa, M.T. de Magistris, M. Romano, S. Silvestri, D. Boraschi, and L. Nencioni.** 1986. IgA-driven T cell-mediated anti-bacterial immunity in man after live oral Ty 21a vaccine. J. Immunol. **137:**1504-1510.
54. **Thain, J.A., C. Baxter-Jones, G.P. Wilding, and G.A. Cullen.** 1984. Serological response of turkey hens to vaccination with *Salmonella hadar* and its effect on their subsequently challenged embryos and poults. Res. Vet. Sci. **36:**320-325.
55. **Thain, J., and G.A. Cullen.** 1978. Detection of *Salmonella typhimurium* infection in chickens. Vet. Rec. **102:**143-145.
56. **Topley, W.W.C.** 1929. Natural acquirement of immunity. Lancet **1:**1337-1343.
57. **Truscott, R.B.** 1981. Oral *Salmonella* antigens for the control of *Salmonella* in chickens. Avian Dis. **25:**810-820.
58. **Truscott, R.B., and G.W. Friars.** 1972. The transfer of endotoxin induced immunity from hens to poults. Can. J. Comp. Med. Vet. Sci. **36:**64-68.
59. **Turnbull, P.C.B., and G.H. Snoeyenbos.** 1974. Experimental salmonellosis in the chicken. 1. Fate and host response in alimentary canal, liver and spleen. Avian Dis. **18:**153-177.
60. **Udhayakumar, V., and V.R. Muthukkaruppan.** 1987. Protective immunity induced by outer membrane proteins of *Salmonella typhimurium* in mice. Infect. Immun. **55:**816-821.
61. **Ushiba, D., K. Saito, T. Akiyama, M. Nakano, T. Sugiyama, and S. Shirono.** 1959. Studies in experimental typhoid: bacterial multiplication and host cell response after infection with *Salmonella enteritidis* in mice immunized with live and killed vaccines. Japan. J. Microbiol. **3:**231-242.
62. **Williams, J.E.** 1978. Paratyphoid infections in Hofstad, M.S., B.W. Calnek, C.F. Helmboldt, W.M. Reid, and H.W. Yoder (Eds). Diseases of Poultry. Iowa State University Press, Ames, Iowa.
63. **Wray, C., W.J. Sojka, J.A. Morris, and W.J. Brinley Morgan.** 1977. The immunization of mice and calves with *galE* mutants of *Salmonella typhimurium*. J. Hygiene Camb. **79:**17-24.

INTESTINAL IMMUNITY AND GENETIC FACTORS INFLUENCING COLONIZATION OF MICROBES IN THE GUT

H. S. Lillehoj
K. S. Chung

Protozoan Diseases Laboratory
USDA, ARS, Beltsville Agricultural Research Center
Beltsville, Maryland 20705

ABSTRACT

Recent technical advances in the molecular and cellular immunology have enhanced our understanding of the ontogeny, structure, and function of gut-associated lymphoid tissues (GALT). Isolation of single cells from the intestine and flow cytometric analysis of lymphoid populations now enable researchers to dissect various components of the GALT and study the role of subpopulations of intestinal lymphocytes in various aspects of disease processes. The molecular characterization of cytokines involved in inflammation and immunity have allowed us to devise immunotherapeutic protocols. Recent *in vivo* trials of recombinant cytokines demonstrate promising uses of the lymphokines in immunoprophylaxis and immunoprevention. The availability of molecular probes and genetically defined chicken lines make it possible to study the role of host genetic factors influencing immunity to intestinal diseases. Genes of the major histocompatibility complex (MHC) and non-MHC genes have been shown to influence disease resistance and the host immune response to infectious agents. This paper will review: (1) our current understanding of the intestinal immune system, (2) immunogenetics of the mucosal immune system, and (3) development of recombinant vaccines against intestinal infection.

Colonization Control of Human Bacterial Enteropathogens in Poultry

INTRODUCTION

The gut-associated lymphoid tissue (GALT) has evolved with specialized cytological features that reflect its role as the first line of defense in mucosal surfaces. These include antigen presenting cells, immunoregulatory cells, and effector cell types distinct from their counterparts in the systemic immune system. Due to the uniqueness of location of the GALT and its constant exposure to environmental antigenic challenge, investigation of its immune system is crucial in our understanding of food allergy, tolerance, and the immune response to intestinal infections. Recent technical advances in molecular and cellular immunology have facilitated understanding of the ontogeny, structure and function of the GALT. Isolation and flow cytometric analyses of intestinal lymphoid cells now enables researchers to dissect various components of the GALT and to investigate the role of subpopulations of intestinal lymphocytes in disease processes.

In mammals, it is well known that major histocompatibility complex (MHC) genes and non-MHC genes influence the host immune response to infectious agents and disease resistance against microbial infections. Although immunogenetic studies of the avian MHC lag behind, the recent availability of molecular probes and genetically defined chicken strains now enable the study of the role of host genetic factors influencing immunity to intestinal diseases in this species. This paper will review current understanding of the avian intestinal immune system and provide a conceptual overview of the complex cellular and molecular events involved in the intestinal immune response to enteric pathogens. Specific examples will be used to illustrate how new biotechnical developments have enhanced our understanding of the immunobiology of host-parasite interactions. For more detailed reviews on this subject, readers should refer to recent publications (11, 64).

GASTROINTESTINAL IMMUNE SYSTEM AND ANTI-MICROBIAL RESPONSES

Introduction. Host responses to intestinal microbial infections involves the complex interplay of soluble factors, leukocytes, epithelial, and endothelial cells and other physiological factors of the GALT. The GALT represents only one component of the mucosa-associated lymphoid tissues (MALT) which also includes the bronchial, salivary, naso-pharyngial and genito-urinary lymphoid tissues. Following infection, macrophage-dependant antigenic activation of T and B cells initiates a series of antigen-specific and non-specific responses involving secretory immunoglobulins, local cells and locally produced cytokines. One of the salient features of the secretory immune system is the intra-epithelial transport of immunoglobulins into the lumen of the gut. External transport of secretory immunoglobulins from blood or tissue fluids into external mucosae of the alimentary, respiratory, genito-urinary and naso-lacrimal tracts contribute a major source of immunoglobulins to the intestine. In contrast to mammals, limited information is available concerning the intestinal immune system of chickens. However as the development of vaccines against intestinal infections such as coccidiosis, colibacillosis, salmonellosis and other intestinal viral infections become an industrial priority, research of the avian intestinal immune system assumes high priority. The advent of new molecular techniques to manipulate the genomes of various pathogens and an enhanced understanding of the interactions of the GALT with peripheral lymphoid organs will soon enable new approaches to vaccination against enteric pathogens.

Gut-Associated-Lymphoid-Tissues. The GALT in chickens includes the bursa of Fabricius, cecal tonsils (CT), Peyer's patches (PP), and lymphocyte aggregates in the intraepithelium and in the lamina propria (LP) of the intestinal wall of the gastrointestinal tract. PP are lymphoid aggregates in the intestine which possess a morphologically distinct lymphoepithelium with microfold (M) cells, follicles and a B cell-dependant subepithelium zone and a T-dependant central zone (4, 68). The GALT contains unique phagocytic cells, such as M cells, possessing numerous vacuoles reflecting active pinocytosis (9). Within the gastrointestinal mucosa, lymphocytes present in the epithelium (termed intraepithelial lymphocytes, IEL) are morphologically separated by a basement membrane from the underlying lamina propria lymphocytes (LPL). The GALT, largely represented by PP, is an important site of IgA synthesis in mammals and contains a large subpopulation of B lymphocytes committed to IgA secretion (46). Recent studies showed that administration of oral antigen leads to the appearance of clonal precursors to progeny secreting mostly IgA (45). Moreover, such precursors were found to be more numerous in mucosal tissues than in the spleen and in conventional animals versus germfree animals (63). The polymeric IgA antibody is produced by plasma cells in the GALT and selectively transported through epithelial cells into external secretions. Following immunization with gut luminal antigens, antigen-sensitized, IgA-committed B cells and T cells leave the PP via efferent lymphatics, pass through mesenteric lymph nodes and enter the bloodstream through the thoracic duct. From the blood, IgA-committed B cells migrate to and selectively localize in distant mucosal tissues, e.g., mammary, salivary, and lacrimal glands and in the lamina propria regions of the gastrointestinal (GI) and upper respiratory tract where they differentiate into plasma cells. The GALT has also been shown to contain complex regulatory T cell networks with T helper and T suppressor cells that direct mucosal immune responses (46).

Tissue Distribution and Differentiation Antigens of Macrophages, NK Cells, B, and T lymphocytes in the Intraepithelium and Lamina Propria. Intestinal IEL are mainly T cells and, to a lesser extent, non-T, non-B cells (29, 36), whereas LPL are relatively enriched with immunoglobulin-producing B cells (2, 21). IEL from chickens contain 80% T cells, 10-15% B cells, approximately 5% of mononuclear phagocytes (4) and, less than 1% polymorphonuclear leukocytes and plasma cells. The cells isolated from mechanical scraping of the muscularis mucosa showed that small intestinal LPL contain 80% lymphocytes, 20% monocytes and less than 1% polymorphonuclear leukocytes. Mononuclear cells isolated from the epithelium and lamina propria were reported to contain immunoglobulin positive cells but the percentage was higher in LPL (29.5%) than in IEL (7.9%) or spleen cells (19.4%).

The PP represents the major inductive site for IgA responses to ingested antigens and pathogenic microrganisms in the gastrointestinal tract. In mammals, this tissue contains approximately 40% B-lymphocytes, a high percentage (12-16%) of which bear surface IgA and are committed to IgA synthesis (46). 40% of the PP cells are regulatory T cells which include 40-50% Lyt-1^+ inducer and helper cells, and 15-20% Lyt-2^+ cytotoxic and suppressor cells. A significant percentage of PP T cells bear Fc receptors for IgA and are important in the regulation of IgA isotype responses. The PP also contains accessory cells such as macrophages (5-9%) and functional dendritic cells. In contrast, caecal tonsil lymphocytes are mainly IgG and IgM staining B cells and relatively little IgA expressing B-cells were found in this organ (4).

The phenotypes of intestinal IEL T cells have been examined in chickens. Molecules similar to the human and murine CD3, CD4 and CD8 antigens have been identified (17, 20, 57). The predominant subset of IEL T lymphocytes express the CD3 polypeptides (gamma, delta, epsilon,and zeta) noncovalently associated with the gamma-delta chain receptor heterodimer of the antigen specific T cell receptor (TCR) referred as TCR1 (14). Another subset of T lymphocytes expresses the CD3 polypeptide chains in association with the alpha-beta chain receptor heterodimer called TCR2. The ontogeny of T cells bearing different TCR has been studied. Early in fetal development, precursor thymocytes pass through the thymus and acquire distinct functions (86). TCR1 is expressed on a small percentage of thymocytes by embryonic day 11, increases to 30% of thymocytes by day 15, and then declines to about 5% of the cells by hatching. Thymocytes bearing TCR2 and CD3 appear after 15 days of fetal development and then quickly increases in number to exceed the level of TCR1 lymphocytes. Among the 85% of adult blood T cells that express CD3, 16% are TCR1 positive (81). TCR1 cells in the blood and the thymus lack both CD4 and CD8 molecules and approximately 75% of splenic TCR1 cells express the avian CD8 (14). Most of the TRC1 cells localized in the splenic sinusoids and in the intestinal epithelium express the CD8 homologue (86). In contrast, TCR2 cells, a majority of which express the CD4 homologue, are found primarily in the splenic periarteriolar sheath (PAL) and in the LP of the intestine. The different tissue homing patterns of TCR1 and TCR2 cells suggest that they represents separate lineages of T cells with distinct physiological roles. Germinal centers seen in the CT contain scattered TCR2, CD4 cells, but no TCR1 cells. Neither subset is present in the intestinal mucosa at hatching, and only occasional TCR1 or TCR2 cells are seen in the intestine of three-day-old chickens (14). By six days post- hatching, both subsets are present and the preferential location of TCR1 cells in the epithelium and TCR2 cells in the LP are established. The number of these cells reach adult levels by one month of age.

We have recently examined TCR1- and TCR2- positive T cells in new born and adult chickens (Table 1). The level of these cells in IEL did not reach adult levels until 21 days after hatching. The genetic background influenced the numbers of IEL T cells. TK chickens in general contained a substantially higher number of TCR1 cells compared to SC chickens until 21 days of age. SC and TK chickens contained similar numbers of TCR2 cells when examined at 42 days of age. In adult SC chickens, IEL and LPL of the duodenum and jejunum contained similar percentages of TCR1 and TCR2 cells whereas those of the caeca contained a higher percentage of TCR1 cells (54).

In chickens as in mammals, T cells can also be separated into CD4 and CD8 expressing subpopulations on the basis of their function (17, 57). Chicken T cells expressing a homologue of the mammalian CD4 antigen represent 20% of spleen cells, and 40 to 45 % of blood leukocytes and 79 to 80% of thymocytes. During ontogeny, CD4 appears on thymocytes at embryonic day 13 and the frequency of CD4 bearing cells increases rapidly to approximately 90% by the end of the embryonic period (17). In the periphery, CD4 cells can be observed in substantial numbers only after hatching, and by the end of the first month adult levels have been reached. CD4 cells seem to occupy certain characteristic histological locations. They are localized mainly in the PAL tissue of the spleen and the LP in the intestine. The appearance of CD8 T cells during embryonic development of chickens is very similar to that of CD4 and by the end of embryogenesis most thymocytes express both of these molecules. In the periphery, up to 50% of spleen cells are CD8 positive whereas only 15% of blood leukocytes carry this antigen. Thymic and peripheral blood T cells expressing TCR1 are $CD4^{-}CD8^{-}$ whereas 70% of splenic and

IEL TCR1 cells in the intestine are $CD8^+$. Our recent study showed that in two-month-old SC chickens reared in a clean, but not in a germfree environment, the composition of T and B cells in the intraepithelium and LP depended upon the region from which these cells were isolated (Table 2). Intestinal cells bearing the CD8 antigen were present in both the intraepithelium and LP of the gut although IEL and LPL of the duodenum contained substantially higher numbers of CD8 cells compared to those of jejunum and caeca.

Table 1. Lymphocyte subpopulations in the jejunum IEL of two inbred strain of chickens of various ages.

		Percentage of Staining Cells				
Age (Days)	Strain	CT3	TCR1	TCR2	CD8	Macrophages
7	SC	14	5	33	11	8
	TK	47	16	25	16	8
21	SC	44	9	30	22	4
	TK	55	23	25	26	4
42	SC	58	25	20	24	5
	TK	71	24	34	45	6

A third type of cell mediating intestinal immunity is the natural killer (NK) cell. NK cells constitute a population of non-T, non-B, non-macrophage mononuclear cells with characteristic morphology and capable of spontaneous cytotoxicity against a wide variety of syngeneic, allogeneic,and xenogeneic target cells. NK-cells lack immunological memory and MHC restriction and thus their cellular lineage is debatable (27, 38). There is much confusion concerning the phenotypic characterization of human leukocytes mediating natural cytotoxicity. Although specific antigenic markers for NK-cells have been described in some mammalian species, monoclonal antibodies specific for chicken NK-cells have not been reported. Rabbit anti-asialo GM1 antibody, known to be relatively specific for NK-cells in many mammalian species (42, 43), will stain avian IEL cells (60). NK-cell activity has been reported to be present in the intestinal IEL population of mice (25, 83), rats (31), and guinea-pigs (1). In chickens, NK cell activity has been demonstrated in the spleen (47, 76) and peripheral blood (51, 30), and intestine (16). The NK cell activities of spleen, thymus, bursa, peripheral blood and gut IEL from SC and FP chickens were investigated in 4 hr and 16 hr ^{51}Cr-release assays (53). Great variability in cytotoxic potential was observed among NK cells of different lymphoid organs. Furthermore, substantial strain variation in NK-cell activity was also demonstrated. Spleen and gut IEL of SC chickens showed detectable activity in a 4 hr assay whereas NK cell cytotoxicity in gut IEL of FP chickens was not detectable until 16 hr after incubation.

Table 2. Lymphocyte subpopulations in the various intestinal subregions.

	Intra-Epithelium			Lamina Propria		
CELLS	DUO	JEJ	CAECA	DUO	JEJ	CAECA
CT3	69	69	63	71	80	72
CD8	44	26	34	40	28	21
TCR1	24	30	46	25	38	51
TCR2	29	22	14	38	27	20
SIgA	4	4	13	8	6	18
SIgM	11	22	25	10	3	26
Macrophages	5	2	10	4	1	13

Age-Dependent Maturation of Intestinal Immune System. Using a panel of monoclonal antibodies, changes in the composition of lymphocytes in the intestine were observed to occur as chickens mature presumably due to exposure to intestinal flora (Table 1). Maturation of the intestinal immune system appeared to be genetically influenced since SC chickens in general contained lower numbers of T and B cells compared to TK chickens. The percentage of mature T cells increased gradually and reached adult level at 21 to 42 days after hatching (Table 1). Similarly, CD8 and TCR2 T cells increased with age and reached adult levels at seven days. In contrast, TCR1 cells reached adult levels only after 42 days in SC chickens. Significant NK cell activity was seen in spleen and IEL of chickens less than five weeks old (53). NK cell activity then increased with age and their cytotoxic potential was not fully developed until six weeks after hatching. Significant IEL NK cell cytotoxicity was demonstrated at one week of age if the assay time was extended for 16 hrs.

The number of PP also varies depending upon age. The PP and CT of chickens were easily identified at ten days post hatching (4). As birds aged, the intestinal lymphoid aggregates underwent involution such that by 20 weeks, the lymphoid follicles became less distinct and fewer in number and there appeared to be a relative depopulation of the subepithelial zone in both the CT and PP. Not only did the morphologic characteristics of the PP vary with age, but their abundance and distribution also changed. PP were not evident at hatching, but could be identified in the intestine by day 1 or 2 and increased to a maximum of 5 at 16 weeks of age. Their number then decreased in association with morphological involution and at 52 to 58 weeks of age, only a single PP was evident. The number of IEL cells were also scarce in the newborn animal, rising in number as a consequence of exposure to environmental antigens (41). In germfree animals, few IEL were found but attained normal levels upon antigen exposure suggesting that the presence of IEL cells in the small intestine is a function of intraluminal antigenic stimulation. Postnatal development of other parameters of the intestinal immune system in chickens has also been studied (4). The proventriculus, Meckel's diverticulum, and PP contained germinal centers beginning at 12 weeks of age. At five days post hatching, CT were macroscopically visible and contained B cells with membrane bound IgM, IgG, or IgA, and some IgG and IgM plasma cells. In older chickens, the size of CT and the number of plasma cells gradually increased.

Effector Functions of IEL and LPL. The interactions of various intestinal lymphocyte effector cells during a protective immune response is very complex. During the host defense process, cells of the immune and inflammatory responses generate a microenvironment incompatible with pathogen survival through such activities as phagocytosis, antibody-dependant cellular cytotoxicity, lysosomal enzyme release, altered vascular flow and permeability, smooth muscle modulation, mucous secretion, control of myelopoiesis, and tissue repair.

1. Induction of antigen-specific responses: Two basic types of lymphocytes are involved in antigen-specific responses; B lymphocytes expressing surface immunoglobulin molecules with exquisite specificities for antigens and T lymphocytes recognizing processed antigens on APC. Upon binding of an antigen to B cells expressing sIg, cell division, and clonal expansion ensue and, immunoglobulins with the identical antigen specificity are secreted from the differentiated B cells. In contrast, T cells recognize antigens that have been processed into smaller fragments by APC and only in conjunction with gene products encoded by self-MHC genes. Recent studies have provided direct molecular evidence that antigenic peptides derived from protein antigens bind to MHC class I molecules (7). The interaction of the TCR-CD3 complex with self-MHC class I or II molecules with bound peptide antigen induces activation of these T cells, which is characterized by early changes in cytoplasmic free Ca^{2+} and inositol triphosphate, and a cascade of events resulting in T cell activation. The induction of this immune response further requires macrophage-secreted cytokines, such as interleukin 1 (IL 1), which is necessary for production of IL 2 (T cell growth factor). IL 2 initiates a cascade of interactions involving other soluble factors of T cells and macrophages to amplify and continue subsequent activation steps ultimately leading to the activation of T cells.

Immunization with viable or nonviable antigens through the gut induces the production of local antibody and cellular responses. The initial steps involve antigen uptake, processing and presentation. The mode of antigen uptake and cells involved in the gut depend upon the nature and type of antigen. Antigen presenting cells (APC) in the gut have been suggested to be dendritic cells, macrophages and epithelial cells. The dome region of the PP is covered by an epithelium containing a unique cell type, M cell (68), a follicular-associated epithelial (FAE) cell (9). M cells are actively pinocytic and phagocytic for both soluble and particulate antigens (e.g., viruses and bacteria) in the lumen of the gut. M cells can present antigen to underlying lymphoreticular cells, leading to the sensitization of lymphoid cells present in distinct T and B cell zones in the PP.

2. Antibody-mediated mechanisms: In chickens, IgA and IgM are the predominant immunoglobulins in the external intestinal secretions. Although IgG is found in the gut, it is believed to be derived from the circulation or leaked from the lymphatics following permeability changes occurring during infection. Secretory IgM which is pentameric is effective in elimination of microbes. However, several distinctive features are important for IgA to function as a secretory antibody. One is the ability of IgA monomer to polymerize. IgA and IgM heavy chains both possesses a special C-terminal extension containing an extra cysteine which cross-links monomer subunits together. Other properties of secretory IgA are its ability to associate with a 15-kDa peptide joining (J) chain and a 70-kDa protein, the secretory component (SC). The latter allows secretory IgA to cross the epithelium barrier. Polymer formation is essential for binding to the mucosal transport receptor and increases the avidity of IgA for antigen. It also directly increases the resistance of IgA to proteolytic digestion. The SC is produced by epithelial cells and remains associated with sIgA after secretion. The IgA-SC complex is internalized in endocytic vesicles, transported across the cytoplasm and exocytozed on to the external surface of the epithelium. A minor source of IgA in secretions is derived from blood via the hepatobiliary IgA transport system. In contrast to the transepithelial IgA pathway, hepatocytes express a specific receptor for blood IgA (74). Polymeric IgA injected into the blood was cleared into the bile in less than 3 hr (69).

The major functions of sIgA include prevention of environmental antigen influx into internal body compartments, neutralization of viruses and microbial toxins and prevention of adherence and colonization of mucosal surfaces by microbial pathogens. Secretory antibodies may bind to the pathogens surface and prevent binding to the epithelium by direct blocking, steric hindrance, induction of conformational changes, or reduction of motility. In this manner, microorganisms would be susceptible to the natural cleaning functions of the mucosae.

3. Cell-mediated mechanisms: T lymphocytes are comprised of functionally two distinct subpopulations distinguishable by their surface phenotypes. Cytotoxic T lymphocytes (CTL) express the CD8 antigens while helper T cells express CD4 on their surface. CTL recognize foreign antigens in the context of MHC class I molecules, whereas T helper cells recognize antigens in association with MHC class II molecules. Although CTL activity has been demonstrated in the intestine of mammals, MHC-restricted IEL CTL activity has yet to be shown in chickens. Recently, there has been an increased interest in the selective homing of TCR1 cells to the intestinal epithelium. The majority of the murine intestinal IEL bear TCR1 and CD8 antigens. However, the intestinal IEL of chickens contains a substantial percentage of TCR2 cells (54) similar to humans. The function of these IEL cells bearing the TCR and CD8 antigens is unknown.

The observation that chicken intestinal IEL contain NK cells that mediate spontaneous cytotoxicity suggests that NK cells may play an important role in local defense (16). Intestinal IEL NK activity depended not only upon the type of target cell but also on the incubation time and the host genetic background. Kinetic studies revealed that cytotoxicity was detectable from 2 hr after incubation and progressively increased to 16 to 18 hr. IEL of SC chickens revealed significantly higher levels of NK-cell activity than FP chickens, whereas their splenic NK-cell activity was not significantly different. Furthermore, NK-cell activity was higher in the jejunum or ileum than in the duodenum or caecum.

Information concerning the factors regulating the intestinal immune system is also limited as in the circulating immune response. The current consensus on the B cell development is that isotype-switching and B cell terminal differentiation is under T cell control (45, 46, 82). Activation of T helper cells results in the release of cytokines amplifying B cell differentiation and proliferation. Furthermore, evidence is emerging to indicate that the immunoregulatory mechanism in the mucosal system involves contrasuppressor cells that augument B cell responses by indirect effects on T cell suppression (46, 82).

Intestinal Effector Cells in Host Defense against Microbial Infections. Tremendous architectural changes in the intestine occur during infection and inflammation following exposure to microbes including increases in permeability, infiltration of cells, mucin, enzyme and immunoglobulin production. Specific and nonspecific factors influence the colonization of microbes, for example secretory immunoglobulins and mucin. Complex interactions between lymphocytes, epithelial cells, dendritic cells and local macrophages are involved in both secretory immunoglobulin and mucin production. The role of sIgs has been documented in agammaglobulinemic patients. In general, the persistence of microbes is prolonged and mortality is higher compared to normal counterparts. The role of secretory immunoglobulins is, however, less clear in some poultry infections. Despite the absence of immunoglobulins, agammaglobulinemic chickens are resistant to reinfection with the microorganisms responsible for coccidiosis (52) and leukocytozoonosis (39). Therefore, although the primary role of sIgA is to prevent invasion of microbes in the intestine, it is less certain whether sIgA limits the course of major infections once established. Despite the absence of a correlation of sIgA with resistance against poultry infections, the mucosal IgA system comprises an important defense designed and regulated to obtain maximum benefit from the required metabolic investment.

Considerable interest in gut mucosal lymphoid populations, particularly IEL NK cells has developed in recent years. NK-cells have been postulated to play an important role as a primary defense mechanisms against tumor cells, bacteria and viruses as well as the homeostasis of normal tissues (13). It has been suggested that IEL NK cells are active in the first line of host defense because of their close proximity to the gut where a variety of antigenic substances are introduced (27). Following the infection of chickens with *Eimeria*, the agent causing coccidiosis, there is an increase in asialo GM1 bearing cells and cells with NK markers (60). NK-cell activity increases following primary and secondary infections suggesting that these cells are involved in the elimination of eimerian parasites. Furthermore, FP chickens contain higher IEL NK cell activity compared to SC chickens correlating with a higher level of disease protection following secondary eimerian infection. Until NK cell-deficient strains of chickens and NK cell markers become available, definitive evidence that these cells play an essential role in mucosal infection can be made by a genetic analysis of backcross and inbred strains to determine the cause and effect relationship between innate disease resistance and putative host defense function.

Further analysis of innate immune effector mechanisms and their role in eimerian infections should consider other related effector cell types such as cytotoxic T lymphocytes and macrophages. T cell activation during the acute stages of infection may promote parasite killing while innate effector mechanisms, such as those of NK cells, may be a by-product of other anti-parasitic responses, i.e, macrophage activation. Lymphokine-mediated activation of macrophage killing of intracellular eimerian parasites has been reported (61). Enhancement of disease resistance produced by immunomodulators such as lymphokines has the advantage over subunit recombinant vaccines since the immunity produced is nonspecifically directed against not only several species of *Eimeria* but also other diseases of economic importance.

Immunity to *Salmonella* infection has been extensively studied in mice as an experimental model for human *Salmonella typhi* and paratyphi infections. The widely used model for "enteric fever" is produced by parenteral (ip, iv, or sc) infection of mice with virulent strains of *Salmonella*, such as *S. typhimurium* or *S. enteritidis*. *S. enteritidis* is a naturally occurring pathogen of mice causing systemic symptoms pathologically resembling typhoid fever. Several experiments showed that the cellular rather than humoral response is involved in protection (26). These include: 1) lack of correlation of anti-*Salmonella* antibody levels and protection against disease, 2) live *Salmonella* infection induced delayed-type hypersensitivity (DTH) in close temporal association with the onset of antibacterial immunity in the liver and spleen, and 3) agammaglobulinemic BALB/c mice showed normal protection to infection with live virulent *Salmonella* (32). In chickens, the uptake of *S. enteritidis* and *S. thompson* across the cecal mucosa has been visualized by ultrastructural studies. Wandering macrophages containing bacteria were observed spanning the epithelial and lamina propria regions through breaks in the basement membrane suggesting a role for macrophage in the transport of bacteria to the circulation (73). *S. typhimurium* were recovered mostly from the caecal tonsil region and persisted until 33 days post infection (49). Recent studies suggest that in chickens cellular mechanisms rather than humoral antibodies play a role in defense against salmonellosis. As in mice, there is a lack of correlation between resistance to infection with the level of serum and biliary anti-*Salmonella* antibodies. Rather, resistance correlates with the development of cell-mediated immunity (18, 49). Compared to chickens receiving a killed *S. gallinarum* vaccine, chickens immunized with a live vaccine produced significantly higher macrophage inhibition factor (MIF) at 10 and 21 days post immunization after challenge with live bacteria (18).

MAJOR HISTOCOMPATIBILITY COMPLEX (MHC) GENES AND NON-MHC GENES

The immune response is a genetically controlled event and many of the responsible genes are located in the major histocompatibility complex (MHC). The MHC was originally defined by its influence on tumor rejection in mice (80). The chicken MHC, or B-complex was first described by Briles and co-workers in 1950 (12). As in mammals, the B-complex has been shown to control a wide range of immunological reactions such as resistance to autoimmune, viral, bacterial and parasitic diseases (3). Furthermore, non-immunity related traits such as reproduction, hatchability, growth rate, feed use efficiency, and body weight are also influenced (3). The B-complex is located on the 16th chromosome and is closely linked to the nucleolar organizer region (NOR) containing ribosomal transcriptional units (rDNA) (8). B-complex genes and proteins are divided into three classes designated I, II, IV. Unlike the mammalian MHC, class III genes are not found in the B-complex. Class I, II, and IV genes are also designated as B-F, B-L and B-G respectively. Molecular analysis of the chicken MHC suggests that the class I and class II genes are widely separated on the genome and each is tightly linked with non-class I, non-class II genes (34, 35). The advent of molecular biological techniques and the availability of genetically defined B-congenic and recombinant chicken lines now enable continued detailed analysis of B-complex genes and the role of B-gene products in host immunity. Avian genome manipulation to improve economically important traits as well as traits associated with immunity and disease resistance will soon be feasible.

B-Complex Genes and Proteins: Structure and Function. Molecular cloning of B-complex class I, II, IV genes has been accomplished (10, 33-35). Five independent class II genes and six independent class I genes have been identified. Sequence homology (60%) between chicken and mammalian class II but not class I genes has been shown. The chicken B complex appears very different from mammalian MHCs in that distances between genes within a region are much shorter as is the distance between class I and class II regions. The close proximity of class I and class II genes is probably responsible for the very low frequency of observed recombination between the B-F and B-L subregions. Another distinctive feature of the avian MHC is the tight association of B-complex genes with unrelated genes, such as the guanine nucleotide-binding protein-related gene (C12.3) which controls lymphocyte proliferation (34, 35).

Genes of the MHC encode polymorphic cell surface glycoproteins involved in the control of various aspects of immune responses. The class I gene products (HLA-A, B, C, in man; H-2K, D, L in mice; B-F in chickens) are the classical transplantation antigens expressed by virtually all nucleated cells and function as restricting elements for cytotoxic T lymphocytes. Molecular analysis of B-F antigens showed an alpha chain protein of 40 to 45 kDa bound to a 12 kDa beta$_2$-microglobulin (72). The class II or Ia gene products (HLA-DR, DP, DQ in man; I-E and I-A in mice; and B-L in chickens) restrict antigen presentation to T helper lymphocytes. They are composed of two non-covalently associated glycosylated proteins, an alpha chain (33 to 34 kDa) and a beta chain (28 to 29 kDa) expressed on the surface of B lymphocytes, cells of the myeloid lineage and activated T cells (28, 56). The B-G (class IV) subregion is tightly linked to the B-F and B-L subregions and encodes a family of highly polymorphic 40 to 46 KDa proteins associated with red blood cells (65) but not involved in skin-graft rejection, graft versus host reaction or mixed lymphocyte reaction (37, 50). The B-G subregion has recently been cloned and Southern blot analysis suggests that B-G antigens are encoded by a highly polymorphic multigene family (33).

Influences of MHC Genes and Non-MHC Genes on the Immune Response. MHC genes have a profound effect on the ability of animals to respond to specific antigens. Immune response (Ir) genes located in the subregion containing class II genes were first discovered in guinea pigs by Benacerraf and his coworkers (5). These genes directly influence Tcell-mediated immunity, and indirectly through helper T cells, affect humoral immune responses. Control of the immune system by Ir genes can now be better explained as a consequence of characterization of the cell surface glycoproteins encoded by the MHC. They function to bind peptide fragments produced by intracellular degradation in a manner suitable for recognition by T cells. Failure to bind and subsequently present a peptide results in selective unresponsiveness due to an apparent defect in antigen presentation. On the other hand, not all peptides that bind to MHC molecules elicit a T cell response for the simple reason of size limitations in the TCR repertoire. In chickens, the MHC has been shown to influence humoral immune responses to simple, chemically defined antigens and to complex antigens (6, 70). Furthermore, interactions between T cells and B cells in the antibody response to T-dependant antigens, macrophage presentation of antigen to T cells and splenic germinal center formation are also controlled by class II gene products of B-complex (84, 85, 87). Recent studies indicate that class I antigens are important determinants of CTL recognition of virus-infected chicken cells (62, 88).

Influence of MHC Genes and Non-MHC Genes on Disease Resistance. It is well established that a certain disease conditions are more (or less) likely to occur in a host carrying a particular MHC marker. Examples of MHC control of resistance to disease in chickens include virally induced diseases such as Marek's disease (MD), lymphoid leukosis and Rous sarcomas. The best observed correlation between disease resistance to MD virus and MHC type has been seen in B^{21} chickens (3). However, the role of non-MHC linked genes in the pathogenesis of Marek's disease needs to be further investigated since a recent study suggests that disease susceptibility to Marek's disease may be controlled by genes closely linked to the C12.3 gene (35). In Rous sarcoma (22, 75, 77), certain B-haplotypes are linked with regression as a dominantly inherited trait while others permit progressive tumor growth and metastasis. However, until the molecular structure of the chicken MHC is better defined, caution should be exercised in interpreting the reported disease associations with B-complex genes.

We have recently examined several inbred chicken strains that share the same B-haplotype as 15I5-B congeneic chickens but express different genetic backgrounds to investigate the role of MHC linked and non-MHC linked genes in controlling disease susceptibility to coccidiosis. Since coccidia multiplication *in vivo* depends upon the dose of injected inoculum, a small dose (1×10^4) inoculum was used to investigate genetically determined variations in host response. Under these conditions, the coccidia multiply in the intestine and can subsequently be detectable in feces at 5 to 10 days following inoculation. Wide variations in host responses to *E. tenella* infection were demonstrated following primary inoculation and could be segregated into three categories of disease susceptibility: susceptible, intermediate or resistant (Table 3). The inability to segregate into non-overlapping groups suggested that the susceptibility differences to coccidiosis among inbred chickens is a complex process controlled by multiple genes. Examination of the patterns of disease resistance following primary and secondary infections suggested that there are two distinct mechanisms controlling immunity, an innate mechanism involved during primary infection is different from an immune mechanism controlling acquired immunity following secondary inoculation. The results support other genetic studies that have suggested both MHC and non-MHC genes influence disease susceptibility in the host response to eimerian infections (15, 58, 59).

The cellular basis for the genetic control of disease susceptibility to coccidiosis was studied using two recombinant antigens in chickens orally inoculated with *E. acervulina* (58). The p130 recombinant antigen encoded by the cSZ-1 cDNA clone represents the p240/p160 immunodominant *E. acervulina* sporozoite surface antigen and the p150 recombinant antigen encoded by the cMZ-8 cDNA clone represents the P250 immunodominant *E. acervulina* merozoite surface antigen (40). All of the immune sera obtained from *E. acervulina* inoculated chickens contained circulating antibodies strongly reactive on immunoblots with the p150 but not the p130 antigen. In general, the cellular but not humoral response to the p150 recombinant surface merozoite antigen correlated with the degree of protection following challenge infection. The vaccine potential of the p150 recombinant merozoite protein has been investigated using two different modes of antigen delivery (56). One group of chickens were immunized intra-muscularly with the p150 antigen in complete Freund's adjuvant. Another group was inoculated with live *E. coli* transformed with the recombinant plasmid carrying the cMZ8 cDNA encoding the p150 antigen (*E. coli,* pCO5cMZ8) (44). Intramuscular injection of the p150 antigen in

CFA or oral inoculation of chickens with live *E. coli* pCO5cMZ8 induced antigen-specific systemic and intestinal immune responses. Higher levels of antigen-specific T cell responses were correlated with better protection against challenge than were lower or minimal T cell responses. In general chickens immunized by intramuscular injection of the p150 recombinant antigen in CFA showed higher antibody and T cell responses compared to chickens orally inoculated with live *E. coli* pCO5cMZ8. Analysis of protection in various chicken strains showed that protection induced with the immunization of non-viable coccidial antigen depends upon the B-haplotypes of the host and the mode of antigen delivery. Since eimerian parasites have many antigenically distinct life cycle

Table 3. Summary of host responses to *E. tenella* infection in 15I5-B congenic and inbred chickens.

Chickens (B Haplotype)	Primary Inoc. Oocyst[b]	LS[c]	PCV[d]	Secondary Inoc.[e] LS24	WisF125	Antibody[f] Serum	Bile
15I B^{15}	I	-	-	S	-	-	-
5.6-2	S	R	I	S	S	H	L
15.7-2 B^{2}	R	M	L	S	S	H	H
15.15I-5 B^{5}	I	M	L	R	S	I	-
15.C-12 B^{12}	S	M	L	I	I	I	I
15.P-13 B^{13}	S	S	I	S	I	H	I
15.P-19 B^{19}	I	M	I	R	R	I	I
15.N-21 B^{21}	R	M	L	I	-	H	I
B^{13}	I	M	I	R	-	I	I
FP $B^{15}B^{21}$	S	S	H	R	R	L	L

[a]MHC haplotypes; [b]S (Susceptible), I (Intermediate), or R (Resistant) designates the degree of disease susceptibility; [c]S (Severe), or M (Moderate) denotes the degree of severity of the caecal lesion; [d]Percent reduction in PCV levels are categorized as H (High), I (Intermediate), or L (Low); [e]Disease susceptibility/resistance following a secondary inoculation is designated as S (Susceptible), I (Intermediate), or R (Resistant); [f]Antibody responses are categorized as H (High), I (Intermediate), or L (Low) level of antibody production on the basis of ELISA titer (adapted from ref. 58).

stages, achieving complete protection by vaccination may require a better understanding of the complexities of B and T cell priming in natural infection as well as knowledge of parasite antigens that can elicit protective immunity. The finding that only certain chicken strains respond to the p150 merozoite recombinant antigen suggest that the presence of multiple T cell epitopes in vaccines could improve the efficacy of vaccines, especially with many outbred strains of chickens of unknown genetic backgrounds. Potential limitations to these approaches, however, impede their future usefulness. For instance, the high degree of antigenic drift exhibited by some species of *Eimeria* may render vaccines against them ineffective. Furthermore, since at least seven different species of *Eimeria* are known to infect chickens, the development of such subunit vaccines will depend upon combination vaccine encompassing protective antigens from all species of *Eimeria*.

Studies of intestinal effector mechanisms suggest that IEL may play an important role in mucosal defense against various *Eimeria* spp. We have recently investigated the mechanism whereby underlying genetic factors influence innate immunity in avian coccidiosis by comparing NK cell activities of two different chicken strains showing different levels of disease susceptibility to coccidiosis. The efficiency of NK-cell killing by chicken IEL is influenced by the host genetic background (16). The relatively rapid induction of killing of tumor targets from as early as 2 hr was seen by SC chicken IEL against an avian leukosis virus transformed tumor, RP9. On the other hand, the delayed and less efficient killing of RP9 was seen by FP IEL NK cells. The NK-cell activity difference seen in FP and SC IEL is intriguing for gut immunity implicated in avian coccidiosis since the sporozoites of *E. tenella* first penetrate enterocytes and then enter IEL that leave the epithelium, pass through LPL and enter the crypt where the parasites are liberated (19, 48). Whether the sporozoites can enter NK cells among the IEL population or not is as yet uncertain. However, the potential invasion of coccidia into chicken IEL is of considerable importance for the study of mucosal immunity against this protozoan infection. The genetic susceptibility of different strains of chickens to coccidiosis might be determined at this level within the gut mucosa.

Study of the genetic control of disease susceptibility to salmonellosis in chickens has been limited. However the extensive genetic analysis of the host response to murine typhoid caused by *S. typhimurium* serves as a model for the genetic control of resistance to this organism and as a probe to evaluate mechanisms of immunity to typhoid fever (reviewed in 67). In mice orally infected with *S. typhimurium*, the bacteria either multiply in the small bowel or penetrate the intestinal mucosa and enter in the PP of the small intestine. The bacteria gain access to the circulation via the lymphatic, enter the reticuloendothelial system (RES), and multiply within splenic and hepatic tissues. A secondary bacteremia ensues followed by systemic dissemination of the organism. When mice of different inbred strains are inoculated via the same route, they exhibit dose-dependent variable susceptibility to *S. typhimurium*. The parenteral LD_{50} of highly virulent *S. typhimurium* for some strains (e.g., BALB/c, C57BL/6) is less than 10 bacteria, whereas the LD_{50} for other strains (e.g., CBA, A/J) is greater than 10^4 (67). Furthermore,recent studies indicated that expression of several distinct or closely linked genes appearing to act at different phases of infection control specific and non-specific immunity involved in disease resistance to salmonellosis. The host gene that controls early replication of *S. typhimurium* in splenic and hepatic tissues after i.v. or s.c. challenge was designated *Ity* (for immunity to typhimurium) located on chromosome 1 and not linked to the MHC (66). The *Ity* gene is closely linked or identical to the gene designated *Lsh* that controls the extent to which *Leishmania donovani* replicates in the RES of mice during the first few weeks of parasitic diseases (66). Skamene *et al.* (79) found that the distribution of the *Bcg* gene regulating the early phase of resistance to *Mycobacterium bovis* bacille Calmette-Guerin (BCG) is identical to the patterns of *S. typhimurium* and *L. donovani* responses and suggested that resistance to all three microorganisms appears to be regulated by a single gene. The mode of genetic control of innate immunity to salmonellosis appears to involve resident murine macrophages (23, 66). Mice genetically resistant (Ity^{r}) behaved as if they were highly susceptible to *Salmonella* if they were also homozygous for the defective allele of the endotoxin response gene located on chromosome 4. These two traits have been linked in C3H/HeJ mice. C3H/HeJ mice were unable to restrict the net multiplication of *S. typhimurium* in their spleen and their macrophages were not activated by LPS. It has been postulated that such unactivated phagocytes are unable to restrict *Salmonella* growth. The gene which controls the late-phase of the anti-*Salmonella* response is designated *xid* (for X-linked immunodeficiency) and mice carrying this gene tended to die later in the course of murine typhoid. CBA/N and F1 male mice derived from CBA/N female parents, carrying the *xid* X-linked recessive allele, expresses a variety of functional B-lymphocyte abnormalities, including poor antibody responses to T-independent and T-dependent antigens, poor splenic B-cell proliferative responses to certain T-dependent antigens, and low levels of serum IgM and IgG3 (89). Most macrophage and T cell functions of *xid* mice appear to be normal. *Xid* mice are able to control an early net replication of *S. typhimurium* in thier RES but the initiation of an IgG anti-*Salmonella* antibody response mice is delayed. When viewed collectively, these data indicate that enhanced disease susceptibility to *S. typhimurium* seen in *xid* mice is due to their inability to make an adequate protective antibody response that is apparently required for survival of infected mice late in the course of murine typhoid. In Ity^{r} *nu/nu* mice, no significant differences were observed in the growth patterns of salmonellae in the RES organs of T-cell deficient mice and their *nu/+* littermates for up to 13 days after iv challenge. Thereafter, net salmonella multiplication was greater in *nu/nu* than *nu/+* mice suggesting that T cells are necessary for the late phase of the protective anti-*Salmonella* antibody response and/or development of acquired cellular immunity.

Limited information is available concerning a genetic control of salmonellosis in chickens. Analysis of antibody responses to *S. pullorum* in segregating $B^{1}B^{1}$ populations suggested polygenic control (71). Genetic and age influences on infectivity of the *S. gallinarum* 9R strain was investigated in three different outbred chickens (78). The 9R strain of *S. gallinarum* produced hepatic and splenic lesions without mortality in meat-type and brown-egg-producing strains of chickens, but not Leghorns. White Leghorns had fewer *Salmonella* bacteria in the spleen, lower fecal excretion, and a shorter period of systemic infection than did the brown-egg-producing strain when the birds were inoculated subcutaneously at one day of age. There were no significant differences in serological responses between White Leghorn and meat-type birds (78).

PERSPECTIVE

Much remains to be studied concerning the role of MHC genes and immunological factors controlling intestinal immune responses in poultry diseases. The GALT operates in an extremely complex milieu compared to other lymphoid tissues, and a variety of nonspecific environmental factors are likely to affect the response to enteric pathogens. Furthermore, the ontogeny and function of the GALT depends upon age, immune status and genetic background of the host. Further detailed immunological studies of isolated mucosal lymphocytes may provide insights into both the abnormalities of mucosal immune function in disease and the immune system of mucosae in normal individuals. With increasing public health concerns of salmonellosis and campylobacter infections of poultry products, better control strategies must be developed. Understanding the various factors controlling protective intestinal immunity to these pathogens will be crucial in developing immunological-based controls. With technological advances in molecular biology, future application of immunotherapeutic and immunoprophylactic uses of recombinant gene products will be facilitated. Advances in molecular biology and gene manipulation involving cloning of MHC genes and transgenic insertion into the germlines of chickens and other species should facilitate improvements in genetic stocks of poultry and livestock.

REFERENCES

1. **Arnaud-Battandier, F., B. M. Bundy, M. O'Neill, J. Bienenstock, and D. L. Nelson.** 1978. Cytotoxic activities of gut mucosal lymphoid cells in guinea pigs. J. Immunol. **121**:1059-1065.
2. **Arnaud-Battandier, F., E. C. Lawrence, and R. M. Blaese.** 1980. Lymphoid populations of gut mucosa in chickens. Digest. Dis. Sci. **25**:252-259.
3. **Bacon, L. D.** 1987. Influence of the MHC on disease resistance and productivity. Poul. Sci. **66**:802-811.
4. **Befus, A. D., N. Johnston, G. A. Leslie, and J. Bienenstock.** 1980. Gut-associated lymphoid tissue in the chicken. I. Morphology, ontogeny and some functional characteristics of Peyer's patches. J. Immunol. **125**:2626-2632.
5. **Benacerraf, B., and H. O. McDevitt.** 1972. Histocompatibility-linked immune response genes. Science **175**:273-279.
6. **Benedict, A. A., L. W. Pollard, P. R. Morrow, H. A. Abplanalp, P. H. Mauer, and W. E. Briles.** 1975. Genetic control of immune responses in chickens. I. Responses to a terpolymer of poly (Glu^{60} Ala^{30} Tyr^{10}) associated with the major histocompatibility complex. Immunogenetics. **2**:313-324.
7. **Bjorkman, P. J., M. A. Saper, B. Samraoui, W. S. Bennett, J. L. Strominger, and D. C. Wiley.** 1987. Structure of the human class I histocompatibility antigen, HLA-A2. Nature **329**:506-512.
8. **Bloom, S. E., and L. D. Bacon.** 1985. Linkage of the major histocompatibility (B) complex and the nucleolar organizer in the chicken. J. Heredity **76**:146-154.
9. **Bockman, D. E., and M. D. Cooper.** 1973. Pinocytosis by epithelium associated with lymphoid follicles in the bursa of Fabricius, appendix and peyer's patches. An electron microscopic study. Am. J. Anat. **136**:455-478.
10. **Bourlet, Y., G. Behar, F. Guillemot, N. Frechin, A. Billault, A. Chausse, R. Zoorob, and C. Auffray.** 1988. Isolation of chicken major histocompatibility complex class II (B-L) beta chain sequences: comparison with mammalian beta chains and expression in lymphoid organs. EMBO. **7**:1030-1039.
11. **Brandtzaeg, P., K. Baklien, K. Bjerke, T. O. Rognum, H. Scott, and K. Valnes.** 1987. Nature and properties of the human gastrointestinal immune system. *In* Immunology of the gastrointestinal tract (K. Miller and S. Nicklin, eds.). pp. 1-85. CRC press, Boca Raton, FL.

12. **Briles, W. E., W. H. McGibbon, and M. R. Irwin.** 1950. On multiple alleles affecting cellular antigens in chickens. Genetics **35:**633-652.
13. **Britten, V., and H.P.A. Hughes.** 1986. American trypanosomiasis, toxoplasmosis and leishmaniasis: intracellular infections with different immunological consequences. Clin. Immunol. All. **6:**189-226.
14. **Bucy, R. P., C.-L. H. Chen, J. Cihak, U. Losch, and M. D. Cooper.** 1988. Avian T cells expressing gamma delta receptors localize in the splenic sinusoids and the intestinal epithelium. J. Immunol. **141:**2200-2205.
15. **Bumstead N., and B. Millard B.** 1987. Genetics of resistance to coccidiosis: Response of inbred chicken lines to infection by *Eimeria tenella* and *Eimeria maxima*. Brit. Poul. Sci. **28:**705-715.
16. **Chai, J. Y., and H. S. Lillehoj.** 1988. Isolation and functional characterization of chicken intestinal intra-epithelial lymphocytes showing natural killer cell activity against tumour target cells. Immunology **63:**111-117.
17. **Chan, M. M., C.-L. H. Chen, L. L. Ager, and M. D. Cooper.** 1988. Identification of the avian homologue of mammalian CD4 and CD8 antigens. J. Immunol. **140:**2133-2138.
18. **Chandran, N. J. D., J. S. Moses, N. Dorairajan, and R. A. Balaprakasam.** 1983. Cell mediated immune response in chicks against salmonellosis. Cheiron **12:**6294-299.
19. **Chally, J. R., and W. C. Burns.** 1959. The invasion of cecal mucosa by *Eimeria tenella* sporozoites and their transport by macrophages. J. Parasitol. **6:**238-241.
20. **Chen, C. -L.H., J. Cihak, U. Losch, and M. D. Cooper.** 1988. Differential expression of two T cell receptors, TCR1 and TCR2, on chicken lymphocytes. Eur. J. Immunol. **18:**539-543.
21. **Chiba, M., W. Bartnik, S. G. ReMine, W. R. Thayer, and R. G. Shorter.** 1981. Human colonic intraepithelial and lamina proprial lymphocyte: cytotoxicity *in vitro* and the potential effects of the isolation method on their functional properties. Gut **22:**177-180.
22. **Collins, W. M., W. E. Briles, R. M. Zsigray, W. R. Dunlop, A. C. Corbett., K. K. Clark., J. L. Marks, and T. P. McGrail.** 1977. The B locus (MHC) in the chicken: Association with the fate of RSV-induced tumors. Immunogenetics 333-343.
23. **Colwell, D. E., S. M. Michalek, and J. R. McGhee.** 1986. Lps gene regulation of mucosal immunity and susceptibility to *Salmonella* infection in mice. Curr. Topics Microbiol. Immunol. **124:**121-147.
24. **Crocker P. R., J. M. Blackwell, and D. J. Bradly.** 1984. Expression of the natural resistance gene Lsh in resident liver macrophages. Infect. Immunity **43:**1033-1040.
25. **Dillon, S.B., and T.T. MacDonald.** 1984. Functional properties of lymphocytes isolated from murine small intestinal epithelium. Immunology **52:**501-505.
26. **Eisenstein T. K., and B. M. Sultzer.** Immunity to *Salmonella* infection. Advan. Exp. Med. Biol. **162:**261-296.
27. **Ernst, P., A.D. Befus, and J. Bienenstock.** 1985. Leukocytes in the intestinal epithelium: a unique and heterogeneous compartment. Immunol. Today **6:**50-55.
28. **Ewert, D. L., M. S. Munchus, C.-L. H. Chen, and M. D. Cooper.** 1984. Analysis of structural properties and cellular distribution of avian Ia antigen by using monoclonal antibody to monomorphic determinants. J. Immunol. **132:**2524-2530.
29. **Ferguson, A.** 1980. Intraepithelial lymphocytes of the small intestine. Gut **18:**921-925.

30. **Fleischer, B.** 1980. Effector cells in avian spontaneous and antibody-dependant cell-mediated cytotoxicity. J. Immunol. **125:**1161- 1166.
31. **Flexman, J. P., Shellam, G. R., and G. Mayrhofer.** 1983. Natural cytotoxicity, responsiveness to interferon and morphology of intraepithelial lymphocytes from the small intestine of the rat. Immunology **48:**733-744.
32. **Fultz, M. J., F. D. Finkelman, and E. S. Metcalf.** 1989. Altered expression of the *Salmonella typhimurium*-specific B-cell repertoire in mice chronically treated with antibodies to immunoglobulin D. Infect. Immunity **57:**432-437.
33. **Goto, R., C. G. Miyada, S. Young, R. B. Wallace, H. Abplanalp, E. Bloom, W. E. Briles, and M. M. Miller.** 1988. Isolation of a cDNA clone from the B-G subregion of the chicken histocompatibility (B) complex. Immunogenetics **27:**102-109.
34. **Guillemot, F., A. Billault, O. Pourquie, B. Ghislaine, A. Chausse, R. Zoorob, G. Kreibich, and C. Auffray.** 1988. A molecular map of the chicken major histocompatibility complex: the class II beta genes are closely linked to the class I gene and the nucleolar organizer. EMBO. **7:**2775-2785.
35. **Guillemot, F., A. Billault, and C. Auffray.** 1989. Physical linkage of a guanine nucleotide-binding protein-related gene to the chicken major histocompatibility complex. Proc. Natl. Acd. Sci. USA **86:**4594-4598.
36. **Guy-Grand D., C. Griscelli, and P. Vassalli.** 1978. The mouse gut T lymphocyte, a novel type of T cell. Nature, origin and traffic in mice in normal and graft-versus-host conditions. J. Exp. Med. **148:**1661-1667.
37. **Hala, K., M. Vilhelmova, and J. Hartmanova.** 1976. Probable crossing-over in the B blood group system of chickens. Immunogenetics **3:**97-103.
38. **Herberman, R. B., and H. T. Holden.** 1978. Natural cell-mediated immunity. Adv. Cancer Res. **27:**305-377.
39. **Isobe, T., M. Kohno, K. Suzuki, and S. Yoshihara.** 1989. *Leucocytozoon caulleryi* infection in bursectomized chickens. *In* Proc. of the Vth International Coccidiosis Conference. pp. 79-85. INRA, Paris.
40. **Jenkins M. C., H. S. Lillehoj, and J. B. Dame. 1988.** *Eimeria acervulina*: DNA cloning and characterization of recombinant sporozoite and merozoite antigens. Exp. Parasitol. **66:**96-107.
41. **Jeurissen, S. H. M., E. M. Janse, G. Koch, and G. F. De Boer.** 1989. Postnatal development of mucosa-associated lymphoid tissue in chickens. Cell Tissue Res. **258:**119-124.
42. **Kasai, M., M. Iwamori, Y. Nagai, K. Okumura, and T. Tada.** 1980. A glycolipid on the surface of mouse natural killer cells. Eur. J. Immunol. **10:**175-180.
43. **Keller, R., T. Bachi, and K. Okumura.** 1983. Discrimination between macrophages and NK-type tumoricidal activities via anti-asialo GM1 antibody. Exp. Cell. Biol. **51:**158-164.
44. **Kim K. S., M. C. Jenkins, and H. S. Lillehoj.** 1989. Immunization of chickens with live *Escherichia coli* expressing *Eimeria acervulina* merozoite recombinant antigen induces partial protection against coccidiosis. Infect. Immunity **57:**2434-2440.
45. **Kiyono, H., L. Mosteller-Barnum, A. Pitts, S. Williamson, S.Michalek, and J. McGhee.** 1985. Isotype specific immunoregulation. J. Exp. Med. **161:**731.

46. **Kiyono, H., T. Kurita, I. Suzuki, J. H. Eldridge, J. F. Morrison, and J. R. McGhee.** 1988. IgA-specific regulatory T cells in the mucosal system. *In* Mucosal Immunity and Infections at mucosal surfaces. (W. Strober, M. E. Lamm, J. R. McGhee and S. P. James, eds.). pp. 63-73. Oxford Univ. Press, Inc., New York, NY.
47. **Lam, K. M., and T. J. Linna.** 1980. Transfer of natural resistance to Marek's disease (JMV) with nonimmune spleen cells. II. Further characterization of protecting cell population. J. Immunol. **125:**715-718.
48. **Lawn, A.M., and M.E. Rose.** 1982. Mucosal transport of *Eimeria tenella* in the cecum of the chickens. J. Parasitol. **68:**1117-1123.
49. **Lee, G. M., G. D. F. Jackson, and G. N. Cooper.** 1983. Infection and immune responses in chickens exposed to *Salmonella typhimurium*. Avian Dis. **27:**577-583.
50. **Lee, R. W. H., and A. W. Nordskog.** 1981. Role of the immune-response region of the B-complex in the control of the graft-vs-host reaction in chickens. Immunogenetics **13:**85-92.
51. **Leibold, W. G. Janotte, and H. H. Peter.** 1980. Spontaneous cell mediated cytotoxicity (SCMC) in various mammalian species and chickens: selective reaction pattern and different mechanisms. Scand. J. Immunol. **11:**203-222.
52. **Lillehoj, H. S.** 1987. Effects of immunosuppression on avian coccidiosis. Cyclosporin A but not hormoanal bursectomy abrogates host protective immunity. Infect. Immunity **55:**1616-1621.
53. **Lillehoj, H. S., and J. Y. Chai.** 1988. Comparative natural killer cell activities of thymic, bursal, splenic and intestinal intraepithelial lymphocytes. Dev. Comp. Immunol. **12:**629-643.
54. **Lillehoj, H. S., and K. S. Cheung.** 1990. Postnatal development of T lymphocyte subpopulations in the intraepithelium and lamina propia. Immunology. (Submitted).
55. **Lillehoj, H. S., M. C. Jenkins, and L. D. Bacon.** 1990. The effects of major histocompatibility genes in antigen delivery on induction of protective mucosal immunity to *Eimeria acervulina* following immunization with recombinant merozoite antigen. Immunology. (In Press).
56. **Lillehoj, H. S., S. Kim, E. P. Lillehoj, and L. D. Bacon.** 1988. Quantitative differences in Ia antigen expression in the spleens of 15I5-B congenic and inbred chickens as defined by a new monoclonal antibody. Poul. Sci. **67:**1525-1535.
57. **Lillehoj, H. S., E. P. Lillehoj, D. Weinstock, and K. Schat.** 1988. Functional and biochemical characterization of chicken T lymphocyte antigens. Eur. J. Immunol. **18:**2059-2065.
58. **Lillehoj H. S., M. C. Jenkins, L. D. Bacon, R. H. Fetterer, and W. E. Briles.** 1988. *Eimeria acervulina*: Evaluation of the cellular and antibody responses to the recombinant coccidial antigens in B-congenic chickens. Exp. Parasitol. **67:**148-158.
59. **Lillehoj H. S., M. D. Ruff, L. D. Bacon, S. Lamont, and T. Jeffers.** 1989. Genetic control of immunity to *Eimeria tenella*. Interaction of MHC genes and non-MHC genes influence levels of disease susceptibility. Vet. Immunol. Immunopathol. **20:**135-148.
60. **Lillehoj, H. S.** 1989. Intestinal intraepithelial and splenic natural killer cell responses to eimerian infections in inbred chickens. Infect. Immunity **57:**1879-1884.

61. **Lillehoj, H. S., S. Y. Kang, L. Keller, and M. Sevoian.** 1989. *Eimeria tenella* and *E. acervulina*: Lymphokines secreted by an avian T cell lymphoma or by Sporozoite-stimulated immune T lymphocytes protect chickens against avian coccidiosis. Exp. parasitol. **69:**54-64.
62. **Maccubbin, D. A., and L. Schierman.** 1986. MHC-restricted cytotoxic response of chicken T cells: expression, augumentation, and clonal characterization. J. Immunol. **136:**12-16.
63. **McGhee, J. R., H. Kiyono, and C. D. Alley.** 1984. Gut bacterial endotoxin: influence on gut-associated lymphoreticular tissue and host immune function. Sur. Immunol. Res. **3:**241-252.
64. **Mestecky, J., and J. R. McGhee.** 1987. Immunoglobulin A (IgA): Molecular and cellular interactions involved in IgA biosynthesis and immune response. Adv. Immunol. **40:**153-229.
65. **Miller, M. M., R. Goto, and W. E. Briles.** 1988. Biochemical confirmation of recombination within the B-G subregion of the chicken major histocompatibility complex. Immunogenetics **27:**127-132.
66. **O'Brien A. D., D. L. Rosenstreich, and B. A. Taylor.** 1980. Control of natural resistance to *Salmonella typhimurium* and *Leishmania donovani* in mice by closely linked but distinct genetic loci. Nature **287:**440-442.
67. **O'Brien, A. D.** 1986. Influence of host genes on resistance of inbred mice to lethal infection with *Salmonella typhimurium*. Curr. Topics Microbiol. Immunol. **124:**37-48.
68. **Owen, R. L., and A. L. Jones.** 1974. Epithelial cell specialization within human Peyer's patches: an ultrastructural study of intestinal lymphoid follicles. Gastroenterology. **66:**189-203.
69. **Peppard, J. V., M. E. Rose, and P. Hesketh.** 1983. A functional homologue of mammalian SC exists in chickens. Eur. J. Immunol. **13:**566-570.
70. **Pevzner, I. Y., C. L. Trowbridge, and A. W. Nordskog.** 1978. Recombination between genes coding for immune response and the serologically determined antigens in the chicken B system. Immunogenetics. **7:**25-33.
71. **Pevzner, I. Y., H. A. Stone, and A. W. Nordskog.** 1981. Immune response and disease resistance in chickens. I. Selection for high and low titers to *Salmonella pullorum* antigen. Poul. Sci. **60:**920-926.
72. **Pink, J. R. L., M. W. Kieran, A. M. Rijnbeck, and B. M. Longenecker.** 1985. A monoclonal antibody against chicken MHC class I (B-F) antigens. Immunogenetics **21:**293-297.
73. **Popiel, I, and P. C. Turnbull.** 1985. Passage of *Salmonella enteritidis* and *Salmonella thompson* through chick illeocecal mucosa. Infect. Immunity **47:**786-792.
74. **Sanders, B. G., and W. L. Case.** 1977. Chicken secretory immunoglobulin:chemical and immunological characterization of chicken IgA. Comp. Biochem. Physiol. **56B:**273-278.
75. **Schierman, L. W., and W. M. Collins.** 1987. Influence of the major histocompatibility complex on tumor regression and immunity in chickens. Poul. Sci. **66:**871.
76. **Sharma, J. M., and B. D. Coulson.** 1981. Natural killer cell activity in specific pathogen-free chickens. J. Nat. Cancer. Inst. **63:**527-531.

77. **Schierman, L. W., D. H. Watanabe, and R. A. McBride.** 1977. Genetic control of Rous sarcoma regression in chicken:Linkage with the major histocompatibility complex. Immunogenetics **5:**325-332.
78. **Silva, E. N., G. H. Snoeyenbos, O. M. Weinack, and C. F. Smyser.** 1981. Studies on the use of 9R strain of *Salmonella gallinarum* as a vaccine in chickens. Avian Dis. **25:**38-52.
79. **Skamene E. P., A. Gros, P. A. L. Forget, C. Kongshavn, St. Charles, and B. A. Taylor.** 1982. Genetic regulation of resistance to intracellular pathogens. Nature **297:**506-509.
80. **Snell, G. D.** 1953. The genetics of transplantation. J. Nat. Cancer Inst.**14:**691-700.
81. **Sowder, J. T., C.-L. H. Chen, L. L. Ager, M. M. Chan, and M. D. Cooper.** 1988. A large subpopulation of avian T cells express a homologue of the mammalian gamma/delta receptor. J. Exp. Med. **167:**315-322.
82. **Suzuki, I., K. Kitamura, H. Kiyono, T. Kurita, D. R. Green, and J. R. McGhee.** 1986. Isotype-specific immunoregulation. Evidence for a distinct subset of T contra-suppressor cells for IgA responses in murine Peyer's patches. J. Exp. Med. **164:**501-506.
83. **Tagliabue, A., D. Befus, D.A. Clark, and J. Binenstock.** 1982. Characteristics of natural killer cells in the murine intestinal epithelium and lamina propria. J. Exp. Med. **155:**1785.
84. **Vainio, O. C. Koch, and A. Toivanen.** 1984. B-L antigens(class II) of the chicken major histocompatibility complex control T-B cell interaction. Immunogenetics **19:**131.
85. **Vainio, O., P. Toivanen, and A. Toivanen.** 1987. Major histocompatibility complex and cell cooperation. Poul. Sci. **66:**795-801.
86. **Vainio, O., and O. Lassila.** 1989. Chicken T cells: Differentiation antigens and cell-cell interactions. Crit. Rev. Poul. Biol. **2:**97-102.
87. **Vainio, O, T. Veroma, E. Ecrola, P. Toivanen, and M. J. H. Ratcliffe.** 1988. Antigen-presenting cell-T cell interaction in the chicken is MHC class II antigen restricted. J. Immunol **140:**2864-2868.
88. **Weinstock, D., and D. Schat.** 1987. Virus-specific syngeneic killing of reticuloendotheliosis virus transformed cell line target cells by spleen cells. *In* Avian Immunology II. W. T. Weber and D. L. Ewert., eds. pp. 253-263. Alan R. Liss. Inc., New York, NY.
89. **Wicker, L. S., and I. Scher.** 1986. X-linked immune deficiency (*xid*) of CBA/N mice. Curr. Topics Microbiol. Immunol. **124:**87-102.
90. **Wilson, A. D., C. R. Stokes, and F. J. Bourne.** 1986. Morphology and functional characteristics of isolated porcine intraepithelial lymphocytes. Immunology **59:**109.

DISCUSSION

R. MEINERSMANN: Using surface markers on the lymphocytes in chickens, can you distinguish between cytotoxic T lymphocytes and supressor cells?

H. LILLEHOJ: I don't know. Even though the phenomenon of supressor cells has been described in chickens, we don't have any markers that can distinguish these T cells as yet.

R. MEINERSMANN: Is CTL response necessarily accompanied by an inflammatory response, or, to put it the other way around, if there is no inflammatory response, as is the case of most intestinal colonization by *Salmonella* and *Campylobacter* of chickens, can we assume that CTL is not active if there is no inflammation?

H. LILLEHOJ: I wouldn't know until I tested it. There is no correlation of inflammation and cytotoxic T cell generation.

R. MEINERSMANN: In chickens that are germfree, or totally bacteria-free, the immune system takes longer to develop. If we were to give competitive exclusion flora so that you have a much more rapid introduction of bacteria, can we assume that the immune system develops more quickly in those animals?

H. LILLEHOJ: Yes, actually, that has been demonstrated in murine systems. In nude mice or with germfree mice, intraluminal antigen such as bacteria early on can enhance the development of the immune system.

N. STERN: Are there intrinsic differences among MHC differentiated lines of chickens to respond to foreign particles on first exposure, or are the different lines of chickens going to have to be stimulated first and immunologically respond and then likely to respond differently to challenge by the specific organism? Do they need to be stimulated first?

H. LILLEHOJ: I think that immune response can have influences at both levels such as were measured in pigs. In pigs you can measure an *E. coli* pilin receptor. Their generation is controlled by the host, but not necessarily by the immune response, so that, if receptors are not expressed there will be no colonization. Secondly, the immune response is indeed expressed at the levels of specific generation of T cells and B cells. Much experimental data show in mammalian systems that MHC can actually influence T cell receptors so that different mouse strains will have different numbers of T cells that can respond to different antigen and that is influenced by the thymus and the expression of the MHC antigen.

K. NAGARAJA: As in mammals, how much of recirculation of lymphocytes do we see in poultry?

H. LILLEHOJ: I do not have any data on that point, although in sheep and other animals, by thoracic cannulation, you can actually measure the number of T cells or B cells that recirculate through lymphatic. I do not have any further information on chickens.

K. NAGARAJA: What is your feeling on that?

H. LILLEHOJ: I do not know because it depends on the size of the chickens and the size of the strain or genetic strain and age. I don't know that you can actually do that kind of experiment.

J. HASSAN: The relationship of the bura (of Fabricius), should it go along with the GALT? What is your opinion about that?

H. LILLEHOJ: The bursa is clearly identified as part of GALT in chickens. But, if you look at the composition of the bursa, most cells that are there are B cells that express IgM. So you have to say that they are immature type of cells and these type of cells leave the bursa and go to the spleen or peripheral lymphoid organ. Under the influence of local T cells that secrete specific cytotoxins, they can differentiate into IgA or IgG type of cells.

J. HASSAN: For the monoclonal antibody you used, you use CD nomenclature. Is that synonomous with the one in human beings?

H. LILLEHOJ: Yes, biochemically they are identical. This has been published by Max Cooper and our group. The CD4 and CD8 antigens are very similar.

R. CURTISS: Is there any information to indicate whether antigen processing in terms of MHC restriction is the same leading to a secretory response as to a humoral IgG response?

H. LILLEHOJ: I think, although I can only refer to mammalian systems where people have been looking at different types of antigen presentation in the skin and the liver, in the skin ther are Langerhans' cells that can present antigens. In the intestine in human celiac disease, the epithelial cells in the gut express IA and usually IA antigen expression is correlated to antigen presentation. We can only assume that this kind of thing can happen in chickens in their gut.

R. MEINERSMANN: Have any T cells been indentified in chickens that have Fc receptors for any immunoglobulin?

H. LILLEHOJ: I don't think so. We have described the first Fc receptor in chickens and they are not present on T cells. They are present on non-T non-B cells. I am assuming that they are Fc receptors equivalent to CD16 that are located on natural killer cells (in mammals).

D. CORRIER: As far as the bursa and its relation to B cells and the regulation of IgM response, does the bursa play any role in regualting B cells or plasma cells that ultimately secrete IgA?

H. LILLEHOJ: The level of IgA expressing B cells in bursa as well as IgG is very low. There is about 5-10% IgG expressing cells and less than 5% expressing IgA. We found great numbers in IEL and LP. I think that bursa generally plays a role in replenishing B cells in peripheral lymphoid organs. Interestingly, we find a high level of B cells in the cecal tonsil. I don't know if anybody has looked at the role of B cells in the cecal tonsil in local immune response.

D. CORRIER: If you take a bird and bursectomize it on day-of-hatch, say you use Bruce Glick's method, what does that do to secretory IgA?

H. LILLEHOJ: One day of age may be too late because the B cells leave the bursa to go to peripheral lymphoid tissue early on. You have to treat with drugs such as cyclophosphamide about 18 days (old embryo), and then at hatching you may have to treat more to eliminate circulating B cells. We have done that type of experiment, and if you screen carefully they have no IgA or IgM B cells. It doesn't work every time.

D. CORRIER: No B cells. So, if you do a good job, you will eliminate IgA?

H. LILLEHOJ: Yes, you do.

IS VACCINATION A FEASIBLE APPROACH FOR THE CONTROL OF *SALMONELLA*?

K. V. Nagaraja
C. J. Kim

College of Veterinary Medicine
University of Minnesota
St. Paul, Minnesota 55108

M. C. Kumar

E. B. Olson Farms
Division of Jennie-o Foods, Inc.
Willmar, Minnesota 56201

B. S. Pomeroy

College of Veterinary Medicine
University of Minnesota
St. Paul, Minnesota 55108

ABSTRACT

Prophylactic vaccination is a possible method of preventing vertical transmission of *Salmonella*. There are two fundamental requirements for the large scale use of prophylactic vaccines. The vaccines should be both safe and effective. The method of preparation will influence the effectiveness of the vaccine. Adjuvants of many types have been used with vaccines with varying success. Research work on the use of vaccines for the control of salmonellosis in domestic poultry has been ongoing at the Avian Disease Research program for quite some time. Laboratory and field studies indicate encouraging results on the use of vaccines for the control of salmonellosis.

Outer membrane proteins of *Salmonella* have been shown to stimulate antibody response and induce protective immunity in mammals. We have attempted to examine the use of outer membrane proteins from *Salmonella* for protection against *Salmonella enteritidis* in chickens. Preliminary results are very encouraging. The results suggested

Colonization Control of Human Bacterial Enteropathogens in Poultry

that outer membrane proteins of the organism give better protection. Vaccination is a practical approach for the control of *Salmonella* in domestic poultry at the present time.

INTRODUCTION

Salmonellosis in poultry is caused by either the host adapted serotypes such as *S. gallinarum* and *S. pullorum* or by non-host adapted types which are very often pathogenic for man. Eradication programs for *S. gallinarum* and *S. pullorum* have been successful and on the whole they have resulted in the elimination of these serotypes from the poultry industry in the U. S. *Salmonella* have been isolated from almost all animal species, including poultry, cows, pigs, pets, other birds, sheep, seals, donkeys, lizards, and snakes (8). In the vast majority of cases, human beings acquire *Salmonella* by the ingestion of contaminated food or water. Poultry (chickens, turkeys, ducks) and poultry products (primarily eggs) are the most important sources of human infection and are estimated to be responsible for about one-half of the common vehicle epidemics (3). The development of vaccines to prevent human and poultry diseases has been a major accomplishment in the field of immunology. Some diseases have been eliminated through the use of vaccines. However, while existing vaccines have certainly diminished the incidence, morbidity, and mortality of a large number of infectious diseases, salmonellosis is still very difficult to prevent and control through immuno-prophylaxis in domestic animals.

In the age of antibiotics why should anyone attempt to develop a bacterial vaccine? Vaccines are needed even with available antibiotics because antibiotic treatment of many bacterial infections cannot prevent serious sequelae. Evolution of drug resistant bacteria reduces the effects of antibiotics. Salmonellosis by non-host adapted serotypes is a major public health problem in the U. S. and many other countries. The use of bacterins and attenuated live cultures as vaccines in the prevention of avian salmonellosis has been examined experimentally, but has never had wide application under field conditions in the U. S. until recently.

Prophylactic vaccination is also a possible method of preventing vertical transmission of *Salmonella*. For the large scale use of prophylactic vaccines, they must be both safe and effective. The effectiveness of these vaccines may vary with the method of its preparation. Adjuvants of many types from alum through oil emulsions to polynucleotides have been used in a variety of diseases with prophylactic vaccines. An ideal vaccine should mimic the immunological stimulation associated with the natural infection, evoke minimal side effects, be readily available, cheap, stable, and easily administered. Bacterins and attenuated live cultures for use as vaccines in the prevention of avian paratyphoid infections have been studied experimentally, but have never been applied widely in the field conditions (55). Furthermore, there are considerable deficiencies in current vaccines, including unwanted side-effects, contaminating materials, etc. New strategies are therefore urgently required for development of a new generation of vaccines.

Live vaccines may invade host cells. The efficacy of live bacterial vaccines may be due to their particular distribution within the body as well as to their capability of stimulating cell mediated immunity (16). Live vaccines require fewer vaccinations, last longer and can be administered in drinking water (30). Nevertheless, live vaccines may present hazards as a result of residual virulence (53), caused sometimes by insufficient attenuation (2).

To illustrate the dilemma of using a live vaccine, the use of a rough strain of *S. gallinarum*, called 9R strain will be given as an example. As a result of the work of Smith (52) on the use of live attenuated vaccines, a notable advance was made in the control of fowl typhoid (FT). Live avirulent stable rough strain 9R of *S. gallinarum* was used extensively to protect poultry against FT. But, its use has been debated by many authors. Gordon and Luke (22) used the 9R FT vaccine in poultry breeding flocks in the field. In two such flocks, blood agglutinins developed. However, the vaccinal strain was isolated 11 months after vaccination, and there was presumptive evidence that vaccination of adult birds with the rough strain may induce pathological changes in the ovary of some birds.

The existing and developing modes of vaccination includes the use of killed or live attenuated microbial agents, purified microbial proteins or their subunits, recombinant vaccine, internal image anti-idiotypes and synthetic peptides. Various surface components of bacteria other than enterobacteria have been used as vaccines. Capsular polysaccharide vaccines are now established (5), and vaccines prepared from outer membrane proteins of meningococci have been promising in experimental infections (20). Extracts of *Salmonella* that are not well characterized have also been observed to afford protection in animal models (6, 44). The envelope of gram-negative bacteria is a complex structure which consists of an outer membrane, an intermediate layer composed of peptidoglycan and an inner, cytoplasmic membrane. Each of these surface layers are morphologically, chemically, and immunologically distinct. The outer and inner membrane contains the usual constituents of membrane, protein, and phospholipid. The outer membrane contains essentially all of the lipopolysaccharides (LPS) of the cell envelope as well. Purified LPS preparations are poor immunogens and provide little (4) or no protection, and phospholipids are not immunogens.

In recent years, interest in the development of an effective vaccine has led researchers to consider OMP of gram-negative bacteria for their potential use as a vaccine. The OMP of *E. coli* and *Salmonella* have been the subject of numerous studies (19). Some of the so called major proteins of these bacteria form pores for the diffusion of small hydrophilic compounds and are therefore called porins (42). The porins are probably exposed on the surface of the bacteria because they serve as bacteriophage receptors (51). This suggests that they might also be reached by antibodies and furthermore, that it might be possible to influence, or prevent, enterobacterial infections by suitable vaccination, provoking the synthesis of antiporins. The OMP of other gram-negative bacteria such as *Bordetella pertussis* (48, 49), *Vibrio cholerae* (50), *Proteus* (25), *Neisseria meningitides* (18), and *Pseudomonas aeruginosa* (23) were found to be protective. There are few studies on the use of OMP of *Salmonella* as antigens in humans and chickens. Most of the works on *Salmonella* OMP as potential vaccines have been evaluated in mice or rabbits (43). Kuusi *et al.* (28) reported that OMP preparation extracted from a rough mutant of *S. typhimurium* proved to be good immunogens in mice and rabbits (20). They also demonstrated that antisera raised against porin, a major outer membrane protein, could protect the mice even after the removal of anti-LPS antibodies by specific absorption. These observations suggested the importance of protein antigens in protective immunity.

Research conducted at the Avian Health Program at the College of Veterinary Medicine, University of Minnesota, has successfully demonstrated that vaccines can be used for the control of salmonellosis (32-40, 46-47, 54). To begin with, autogenous bacterins in an oil emulsion preparation were shown to be very effective in breaking the *Salmonella* cycle in turkey breeder-hatchery operations. The following will summarize some of the highlights of these experiments.

The first experiment was designed to investigate two main objectives. They were: 1) to determine whether by vaccination it is possible to prevent *Salmonella* infection of breeder flocks from their contaminated environment, and 2) to examine whether it is possible to obtain *Salmonella*-free progeny from *Salmonella*-infected turkey breeder flocks.

A breeder/hatchery operation was selected where *Salmonella* infection of the particular serotype *S. san-diego*, was self-perpetuating in a cycle. The breeder flocks in this facility were composed of grandparent and parent birds. The yearly operation involved approximately 26,500 breeder replacements and 20,700 breeders. The breeders were housed in three different locations. The breeder replacements were brooded for four months and then selected and moved to three breeding farms. A comprehensive study of this breeder/hatchery showed evidence of cycling of *Salmonella* infection. This was determined by the isolation of the same serotype from live birds, their environment, hatchery debris and ten-day mortality. Breeder replacements produced from the operations own parent breeding stock served as the source of the infection. An autogenous mineral oil adjuvant vaccine was prepared from the *Salmonella* serotype *S. san-diego* isolated from the breeder facility. The organism was grown in veal infusion broth at 37 ° C for 48 hours and inactivated with formalin. Using a continuous flow Sorval centrifuge, the bacteria were harvested and the broth discarded. Resuspension of bacteria was done in 0.85% normal saline and concentration was adjusted to 0.25% transmission at a wavelength of 540 mu. After the sterility of the suspension was checked, it was then mixed with mineral oil and Arlacel. This mineral oil adjuvant vaccine of *S. san diego* was used to vaccinate the breeders. Breeder replacements at 18 weeks of age were vaccinated following the official NPIP *Salmonella* testing program. One hundred percent of the breeder replacements (#22,230) were vaccinated. They were then housed in the breeder barns where there was a consistent history of the presence of *S. san-diego*. The birds were given feed without animal by-products throughout the study. The vaccination was repeated once more, ten weeks later. An extensive monitoring program on the vaccinated flock and their eggs and progeny was instituted. Cloacal swabs from 10% of birds randomly selected were collected twice a month and examined for *Salmonella*. During the laying cycle, hatchery debris, which included cull poults, infertiles, dead in shells, cloacal squeezings, and fluff samples from each hatch, were monitored. The ten-day mortality in poults from the vaccinated flock was examined. Fifty random blood samples per flock vaccinated were examined for their serological response to the vaccine.

The hatchery debris and progeny from the vaccinated flock remained negative for *S. san-diego* throughout the observation period. The breeder replacements were found negative for any *Salmonella* on official test of the NPIP during the year. The height of antibody response was considerably greater in vaccinated breeders for *S. san-diego*. The enhanced antibody response to the mineral oil adjuvant vaccine seen in breeders in this experiment was concluded to result in an increased degree of protection. The high antibody levels persisted at protective levels for considerably longer periods of time.

The second experiment was designed to examine the use of killed vaccine against *Salmonella arizonae* infection in turkeys. Infection of turkey flocks with *S. arizonae* is of economic significance to the turkey industry. Adverse effects on hatchability and reduced productivity have been noted in turkeys infected with *S. arizonae*. Vaccination for *S. arizonae* with a mineral oil adjuvented vaccine was examined in turkey breeder flocks (38). The objectives were: 1) to determine whether by vaccination it is possible to prevent *S. arizonae* infection in breeder flocks from their contaminated environment and 2) to examine whether it is possible to obtain *S. arizonae*-free progeny from turkey breeders maintained in a contaminated environment. Several laboratory and field studies conducted by us at the College of Veterinary Medicine demonstrated the use of a vaccine for the prevention of egg transmission of *S. arizonae*. These result eventually led to the availability of a federally licensed commercial vaccine against *S. arizonae* infection in turkey breeders.

Outer membrane proteins (OMPs) of *Salmonella* have been shown to stimulate antibody response and produce protective immunity in mammals. Current research at the University of Minnesota has focussed on the use of outer membrane protein from *Salmonella* (11-14, 41) for protection against homologous and heterologous challenge in turkeys (26, 33). Turkeys vaccinated with OMPs from a *Salmonella* serotype that belonged to Serogroup "B" were challenged with homologous and heterologous *Salmonella* serotypes. The heterologous serotypes included C1, C2, and D1 groups. Birds were vaccinated at six weeks of age. They were challenged at ten weeks of age. They were posted at three weeks post challenge to examine the clearance of challenged *Salmonella*. Internal organs (cecal junction, spleen, liver and bone marrow) were collected for isolation of challenged *Salmonella*. This study demonstrated that OMPs induce immune response which cleared the organism of challenge strains from tissues. Currently, use of a vaccine against *S. enteritidis* infection in chickens is being explored. Experiments have been conducted both with whole cell bacterin and a vaccine made from OMP's from *S. enteritidis*. The results have been very encouraging. The following report presents this particular study in detail.

MATERIALS AND METHODS

Forty two-week-old purebred Leghorn chickens ascertained to be free from *S. enteritidis* infection were used.

Feed. *Samonellae*-free feed containing no antibiotics was used. Clean water was provided *ad libitum*.

Bacteria. For challenge studies, extraction of proteins and bacterin preparation, *S. enteritidis* #87-4390 provided by National Veterinary Services Laboratory (NVSL, Ames, IA) was used. Before it was used, biochemical tests and serotyping were performed to confirm the isolate for *S. enteritidis*.

Isolation of Outer Membrane (OM) Protein. *S. enteritidis* was grown on tryptic soy agar (DifCo Laboratories, Detroit, MI) in Roux flasks at 37 ° C for 48 hours. At the end of the incubation, the bacterial cells were harvested from the flasks with 10 mM HEPES buffer (pH 7.4) (Sigma Chemical Co., St. Louis, MO). The bacterial cell suspension was centrifuged at 12,000 x g for 30 minutes. The resulting bacterial pellet was collected and resuspended in 10 mM HEPES buffer. The bacterial cells in the suspension were disrupted by passing them through a French press (Wabash Metal Production Company, Inc., Wabash, IN) at 15,000 to 20,000 lb/in^2. Intact cells and large debris were removed by centrifugation at 4,000 x g for 20 minutes. Total membrane preparation was harvested from the supernatant by centrifugation at 100,000 x g for 60 minutes at 4 ° C. The gel-like pellet was resuspended in 10 mM HEPES buffer and extracted with an equal volume of a detergent solution (2% sodium lauryl sarcosinate in 10 mM HEPES buffer, pH 7.4) for overnight at 4 ° C. The detergent insoluble fraction was harvested by centrifugation of the suspension at 100,000 x g for 60 minutes at 4 ° C and the pellet was resuspended in distilled water. The protein concentration of the pellet was determined by Bradford protein assay method (15) and final concentration was adjusted to 4 mg/ml.

Bacterin. Formalin-killed mineral oil adjuvanted bacterin was prepared for use. In brief, *S. enteritidis* organism was grown on tryptic soy agar (Difco) in Roux flasks for 48 hours at 37 ° C. Bacterial cells were harvested in normal saline. The concentration of the bacterial suspension was adjusted to contain 10^{11} organisms/ml. The purity of the culture was examined by inoculating brilliant green agar plates (Difco). The bacterial culture was killed by adding 0.3 % formalin with agitation. To prepare the oil emulsion of the bacterin, equal volumes of formalin-killed aqueous bacterial suspension and mineral oil with an emulsifier Arlacel A (Sigma) were mixed well in a Waring blender (VWR Scientific, Chicago, IL) to obtain a homogeneous oil emulsion of the bacterin.

Experimental Design. The birds were individually wingbanded and randomly distributed. They were divided into three groups designated as 1, 2, and 3. Each group contained 30 birds. All birds were ascertained to be *S. enteritidis*-free by cloacal swabs and serological testing for a week before use. Birds in group 1 were vaccinated with OM protein extracts in oil emulsion at 2 mg/bird and birds in group 2 were vaccinated with 0.5 ml of the oil emulsion of the bacterin. Vaccinations in group 1 and 2 were performed subcutaneously. The birds in group 3 were kept as unvaccinated controls. Each group of birds was further divided into two subgroups designated as A and B. Birds in subgroup A were challenged orally at four weeks post-vaccination with 1 ml of broth culture containing 10^7 colony forming unit/ml (CFU/ml) of live *S. enteritidis,* and birds in subgroup B were kept as vaccinated but not challenged controls. Cloacal swab samples were collected from all birds at weekly intervals until the end of the experiment. After challenge, daily mortality was checked and any birds found dead were cultured for *S. enteritidis*. All birds were sacrificed four weeks post-challenge. All internal organs were examined for lesions. Tissues from liver, spleen, ovary, oviduct, bone marrow, and ileo-cecal junction were cultured for *S. enteritidis*.

Serology. Blood samples from all birds were collected at weekly intervals post-vaccination. They were tested by microagglutination test for the presence of antibodies to *S. enteritidis*. The whole cell antigen of *S. enteritidis* was used as described by Williams and Whittemore (56).

Bacteriological Methods. All culture samples from cloacal swabs and internal organs were enriched in tetrathionate broth (Difco) for 24 hours at 42 ° C followed by streaking on brilliant green agar plates and incubating for 24 hours at 37 ° C. Colonies resembling *Salmonella* were transferred to triple sugar iron agar (Difco). The cultures giving typical *Salmonella* reaction were confirmed serologically.

RESULTS

Serology. Table 1 shows the results of the microagglutination tests. Chickens in group 1 and 2 vaccinated with OM protein and killed bacterin, respectively, showed a positive seroconversion. Antibody titers increased post challenge in the vaccinated birds. In general, the geometric mean antibody titers in birds injected with OM protein vaccine were much higher than the bacterin injected group.

Table 1. Geometric mean antibody titers using microagglutination (MA) test.

Group	Treatment	Subgroup	MA Titers on Weeks Post-Vaccination							
			1	2	3	4*	5	6	7	8
1	Protein (2 mg/bird)	A	1.6	4.1	5.3	3.7	7.0	4.5	4.0	3.3
		B	1.7	3.9	6.0	3.5	3.4	2.7	2.4	2.0
2	Bacterin	A	1.2	2.5	3.6	5.0	5.7	3.9	3.6	2.8
		B	1.5	2.8	3.8	4.2	3.0	2.5	2.4	2.0
3	Control	A	0.0	0.0	0.0	0.0	2.6	2.5	2.9	3.0
		B	0.0	0.0	0.0	0.0	0.0	0.0	0.0	0.0

*Birds were challenged orally at four weeks post-vaccination with live *S. enteritidis*.
Subgroup A: Challenged birds
B: Unchallenged birds

Bacteriological Results. Table 2 shows the isolation of *S. enteritidis* from cloacal swab cultures. Birds in group 1 and 2 showed very low isolation rate from the cloacal swab culture after challenge. In the OM protein injected group, only one bird yielded *S. enteritidis*. In control unvaccinated but challenged group, very high percentage of birds yielded *S. enteritidis* in their cloacal swabs for two weeks post-challenge.

The results of the isolation of *S. enteritidis* from tissues are summarized in Table 3. Two birds from the protein vaccinated group and three birds in the group given killed bacterin were positive for *S. enteritidis*. In control unvaccinated but challenged group, six out of thirteen birds yielded *S. enteritidis* from the tissues.

DISCUSSION

It is evident from previous observation (12) that protein antigens are also important for protection in salmonellosis, other than LPS. Most of the studies (17, 21, 43) about induction of protective immunity by such protein antigens has been demonstrated in mice and rabbit. The conclusions that emerge from these studies are that OM proteins elicit an antibody response in both animal and human (3) and that OM proteins hold much promise for safer, more useful vaccines against a number of gram-negative bacterial infections (1, 43).

Table 2. Isolation of *Salmonella enteritidis* from cloacal swab culture.

Group	Treatment	Subgroup	No. positive/No. total on Weeks Post-Vaccination							
			1	2	3	4*	5	6	7	8
1	Protein (2 mg/bird)	A	0/15	0/15	0/15	0/15	0/15	1/15	0/15	0/15
		B	0/15	0/15	0/15	0/15	0/15	0/15	0/15	0/14
2	Bacterin	A	0/15	0/15	0/15	0/15	2/15	2/15	1/14	0/14
		B	0/15	0/15	0/15	0/15	0/15	0/15	0/15	0/15
3	Control	A	0/15	0/15	0/15	0/15	9/15	8/13	3/13	3/13
		B	0/15	0/15	0/15	0/15	0/15	0/15	0/15	0/15

*Birds were challenged orally at four weeks post-vaccination with live *S. enteritidis*.
Subgroup A: Challenged birds
B: Unchallenged birds

In the present study, sarkosyl was used to prepare detergent insoluble fractions from *S. enteritidis* outer membranes. They were enriched greatly with outer membrane proteins. The sarcosinate extraction procedure described in the present study is less time consuming and technically simple. This procedure will result in porins extraction very efficiently (7). The porins form the basis of the outer membrane structure in gram-negative bacteria and are very important in immunity. However, one has to realize that this procedure may not result in the extraction of minor protein material.

This study also demonstrated that OM protein extract, at certain levels, induced an immune response which cleared the organism following challenge. The results of culture of ileocecal junction indicated the clearance of the organism following challenge in vaccinated groups.

The OM protein preparation used in the present study is suspected to contain some LPS. Some LPS core types have been considered to provoke antibody production. However, in previous studies, LPS or anti-LPS antibodies in OM protein preparation were considered unlikely to be responsible for the protection (27, 29).

The OM protein preparation was found to be superior to killed bacterins in terms of protection and clearance of *S. enteritidis* after challenge. These results support earlier finding (9, 10) which demonstrated the protective activity of protein extract from salmonellae (24, 28). Besides the promising protective ability of the OM preparation, it is necessary to eliminate the LPS from the preparation to reduce endotoxic activity.

The results of the present work are encouraging to justify future work for the development of more effective and less toxic *Salmonella* vaccine.

Table 3. Isolation rate of *Salmonella enteritidis* from tissues.

Group	Treatment	Subgroup	No. positive/No. birds — Tissues: Liver	Spleen	Ovary	Oviduct	Bone Marrow	Cecal Junction	Total* Positive	Percent Negative
1	Protein (2 mg/bird)	A	1/15	2/15	1/15	0/15	0/15	2/15	2	86.7
		B	0/14	0/14	0/14	0/14	0/14	0/14	0	100.0
2	Bacterin	A	2/14	2/14	2/14	1/14	0/14	2/14	3	78.6
		B	0/15	0/15	0/15	0/15	0/15	0/15	0	100.0
3	Control	A	5/13	5/13	4/13	4/13	0/13	6/13	6	53.8
		B	0/15	0/15	0/15	0/15	0/15	0/15	0	100.0

*Total positive for at least one tissue.
Subgroup A: Challenged birds
B: Unchallenged birds

REFERENCES

1. **Adamus, G., M. Mulczyk, D. Witkowska, *et al*.** 1980. Protection against keratoconjunctivitis shigellosa induced by immunization with outer membrane proteins of *Shigella* spp. Infect. Immun. **30:**321-324.
2. **Arnon, R., M. Shapira, and C.O. Jacob.** 1983. Synthetic vaccines. J. Immun. Methods. **61:**261-273.
3. **Aserkoff, B., S.A. Schroeder, and P.S. Brechman.** 1970. Salmonellosis in the United States - a five-year review. Am. J. Epidemiol. **92:**13.
4. **Augerman, C.R., and T.K. Einenstein.** 1978. Comparative efficacy and toxicity of a ribosomal vaccine, acetone killed cells, lipopolysaccharide, and a live vaccine prepared from *Salmonella typhimurium*. Infect. Immun. **19:**575-582.
5. **Austrian, R.** 1977. Pneumococcal infection and pneumococcal vaccine. N. Engl. J. Med. **297:**938-939.
6. **Barber, C., Eylan, E., and Heiber, R.** 1972. The protection role of proteins from *Salmonella typhimurium* in infection of mice with their natural pathogen. Rev. Immun. **36:**77-81.

7. **Barenkamp, S.J, R.S. Munson Jr., and D.M. Granoff.** 1981. Subtyping isolates of *Haemophilus influenzae* type b by outer membrane protein profiles. J. Infect. Dis. **143:**668-676.
8. **Bennett, I.L., Jr., and E.W. Hook.** 1959. Infectious diseases some aspects of salmonellosis. Annu. Rev. Med **10:**1.
9. **Bouzoubaa, K., K.V. Nagaraja, B.S. Pomeroy, and J.A.Newman.** 1990. Use of outer membrane proteins from *Salmonella gallinarum* for the prevention of egg transmission in fowl typhoid. Am. J. Vet. Res. (In Press).
10. **Bouzoubaa, K., K.V. Nagaraja, J.A. Newman, et al.** 1987. Use of cellular proteins from *Salmonella gallinarum* for prevention of fowl typhoid infection in chickens. Avian Dis. **32:**699-704.
11. **Bouzoubaa, K., K.V. Nagaraja, F.Z. Kabbaj, J.A. Newman, and B.S. Pomeroy.** 1989. Feasibility of using proteins from *Salmonella gallinarum* vs 9R live vaccine for the prevention of fowl typhoid in chickens. Avian Diseases. **33:**385-391.
12. **Bouzoubaa, K., K.V. Nagaraja, and J.A. Newman.** 1985. Cellular proteins from *Salmonella gallinarum* for prophylactic immunization. Proc. 66th Conf. Research Workers in Animal Disease. p. 65.
13. **Bouzoubaa, K., K.V. Nagaraja, J.A. Newman, and B.S. Pomeroy.** 1987. Use of Cellular Proteins from *Salmonella gallinarum* for prevention of fowl typhoid infection in chickens. Avian Diseases. **314:**699-704.
14. **Bouzoubaa, K., K.V. Nagaraja, and J.A. Newman.** 1985. Isolation and purification of cellular proteins from *Salmonella* for prophylactic immunization. Proc. 36th North Central Avian Dis. Conf. p. 47.
15. **Bradford M.M.** 1976. A rapid and sensitive method for the quantitation of microgram quantities of protein utilizing the principle of protein-dye-binding. Anal. Biochem. **72:**248-254.
16. **Cameron, C.M., O.L. Brett, and W.J.P. Fuls.** 1974. The effect of immunosuppression on the development of immunity to fowl typhoid. Ondersterpoort. J. Vet. Res. **41:**15-22.
17. **Collins F.M.** 1974.Vaccines and cell-mediated immunity. Bacteriol. Rev. **38:**371-402.
18. **Craven, D.E. and C.E. Frasch.** 1979. Protection against group B meningococcal disease: Evaluation of serotype 2 protein vaccine in a mouse bacteremia model. Infect. Immun. *26:*110-117.
19. **DiRienzo, J.M., K. Nakamura, and M. Inouye.** 1978. The outer membrane proteins of gram-negative bacteria: biosynthesis, assembly and functions. Annu. Rev. Biochem. **47:**481-532.
20. **Frasch, C.E., L. Parkes, R.M. McNeils, *et al.*** 1976. Protection against group B meningococcal disease. I. Comparison of group-specific and type-specific protection in chicken embryo model. J. Exp. Med. **144:**319-329.
21. **Gilleland, H.E., Jr., M.G. Parker, J.M. Matthews, *et al.*** 1984. Use of purified outer membrane protein F (porin) preparation of *Pseudomonas aeruginosa* as a protective vaccine in mice. Infect. Immun. **44:**49-54.
22. **Gordon, W.A.M., and D. Luke.** 1959. A note on the use of the 9R fowl typhoid vaccine in poultry breeding flocks. Vet. Rec. **71:**926-927.
23. **Hedstrom, R.C., O.R. Pavlovskis, and D.R. Galloway.** 1984. Antibody response of infected mice to outer membrane proteins of *Pseudomonas aeruqinosa.* Infect. Immun.**43:**4953.

24. **Johnson, W.** 1972. Ribosomal vaccine, I. Immunogenicity of ribosomal fractions isolated from *Salmonella typhimurium* and *Yersinia pestis.* Infect. Immun. **5**:947-952.
25. **Karch, H., and K. Nixdorff.** 1981. Antibody-producing cell responses to an isolated outer membrane protein and to complexes of this antigen with lipopolysaccharides or with vesicles of phospholipids from *Proteus mirabilis*. Infect. Immun. **31**:862-867.
26. **Kim, C.J., and K.V. Nagaraja.** 1987. Control of *Salmonella* in turkeys by immunization. Proc. 68th Conf. Research Workers in Animal Diseases. p. 245.
27. **Kleid, D.G., D. Yansura, B. Small, *et al.*** 1981. Cloned viral protein vaccine for foot-and-mouth disease:Responses in cattle and swine. Science **214**:1125.
28. **Kuusi, N., M. Nurminen, H. Saxen, *et al.*** 1979. Immunization with major outer membrane proteins in experimental salmonellosis of mice. Infect. Immun. **25**:857-862.
29. **Kuusi, N., M. Nurminen, H. Saxen, *et al.*** 1981. Immunization with major outer membrane protein (Porin) preparations in experimental murine salmonellosis. Effect of lipopolysaccharide. Infect. Immun. **34**:328-332.
30. **Levi, B., I. Witz, M. Malkinson, N. Singer, Y. Wiseman, and E.Z. Ron.** 1980. Development of a multivalent live vaccine active against a wide range of Enterobacteriaceae. *In* New development with human and veterinary vaccines. Alan (eds). Alan R. Liss Inc. NY. pp. 119-123.
31. **Loeb, M.R., and D.H. Smith.** 1982. Human antibody response to individual outer membrane proteins of *Haemophilus influenzae* type B. Infect. Immun. **37**:1032-1036.
32. **Nagaraja, K.V., B.S. Pomeroy, L.T. Ausherman, and K.A. Friendshuh.** 1987. *Salmonella* control programs in Minnesota Turkey Industry. Proc. of the XXIII World Veterinary Congress. Montreal, Canada p. 317.
33. **Nagaraja, K.V., C.J.Kim, and B.S.Pomeroy.** 1988. Outermembrane proteins in prophylactic vaccines for *Salmonella*. J. Amer. Vet. Med. Assoc. **192**:1784.
34. **Nagaraja, K.V., 5. Nivas, B.S. Pomeroy, J.A. Newman, and I. Peterson.** 1982. A three year study of *Salmonella arizonae*-free parent turkey breeding flocks. J. Amer. Vet. Med. Assoc. **181**:284.
35. **Nagaraja, K.V., S. Nivas, B.S.Pomeroy, J.A. Newman, and I. Peterson.** 1981. *Salmonella* feasibility studies in turkeys. Minnesota Turkey Research. University of Minnesota. **179**:112-115.
36. **Nagaraja, K.V., D.A. Emery, L.F. Sherlock, J.A. Newman, and B.S. Pomeroy.** 1984. "Control of *Salmonella* by Immunization". MN Turkey Res., U. of MN Agric. Expt. Stat. Pub. **30**:61-63.
37. **Nagaraja, K.V., D.A. Emery, J.A. Newman, and B.S. Pomeroy.** 1985. *Salmonella* control in turkeys. Ohio State University Dept. Series. **90**:84-87.
38. **Nagaraja, K.V., M.C. Kumar, J.A. Newman, and B.S. Pomeroy.** 1985. Control of *Salmonella arizonae* infection in turkey breeder flocks by immunization. J. Amer. Vet. Med. Assoc. **187**:309.
39. **Nagaraja, K.V., B.S. Pomeroy, J.A. Newman, and I. Peterson.** 1982. Vaccination of turkeys with killed oil adjuvant *Samonella san-diego* vaccine. J. Amer. Vet. Med. Assoc. **181**:284.
40. **Nagaraja, K.V., J.A.Newman, and B.S. Pomeroy.** 1984. Use of oil adjuvant vaccines for the control of *Salmonella* infections in Turkeys. Proc. International Symposium on *Salmonella*. pp. 374-375.

41. **Nagaraja, K.V., K. Bouzoubaa, and B.S. Pomeroy.** 1988. Prophylactic immunization with outer membrane proteins from *Salmonella gallinarum* for the prevention of fowl typhoid. Proc. 37th Western Poultry Disease Conf. pp. 121-122.
42. **Nakae, T.** 1976. Outer membrane of *Salmonella* isolation of protein complex that produces transmembrane channels. J. Biol. Chem. **251**:2176-2178.
43. **Nataraja, M., V. Udhayakumar, K. Krishnaraju, *et al.*** 1985. Role of outer membrane proteins in immunity against murine salmonellosis-1. Antibody response to crude outer membrane proteins of *Salmonella typhimurium*. Comp. Immun. Microbiol. Infect. Dis. **8**:9-16.
44. **Plant, J., A.A. Glynn, and B.M. Wilson.** 1978. Protective effects of a supernatant factor from *Salmonella typhimurium* infection of inbred mice. Infect. Immun. **22**:125-131.
45. **Pomeroy, B.S., S.C. Nivas and K.V. Nagaraja.** 1979. *Salmonella* Feasibility Studies in Turkeys. Proc. 22nd North Central Poultry Conf. p. 53.
46. **Pomeroy, B.S., K.V. Nagaraja, L.T. Ausherman, I.L. Peterson, and K.A. Friendshuh.** 1989. Studies on feasibility of producing *Salmonella*-free turkeys. Avian Diseases. **33**:1-7.
47. **Pomeroy, B.S., K.V. Nagaraja, L.T. Ausherman, I.L. Peterson, and K.A. Friendshuh.** 1987. Studies on feasibility of producing *Salmonella*-free turkeys. Proc. of the XXIII World Veterinary Congress. Montreal, Canada. p. 317.
48. **Redhead, K.** 1984. Serum antibody responses to the outer membrane proteins of *Bordetella pertussis*. Infect. Immun. **44**:724-729.
49. **Robinson, A., and K. Hawkins.** 1983. Structure and biological properties of solubilized envelope proteins of *Bordetella pertussis*. Infect. Immun. **39**:590-598.
50. **Sears, S. D., K. Richardson, C. Young, *et al.*** 1984. Evaluation of the human immune response to outer membrane proteins of *Vibrio cholerae*. Infect. Immun. **44**:439-444.
51. **Siitonen, A., V. Johansson, M. Nurminen, *et al.*** 1977. *Salmonella* bacteriophage that uses outer membrane protein receptors. FEMS Microbiol Letter **1**:141-144.
52. **Smith, H.W.** 1956. The use of live vaccines in experimental *Salmonella gallinarum* infection in chickens with observations on their interference effect. J. Hyg. **54**:419-432.
53. **Tizard, I.** 1982. An introduction to veterinary immunology. W.B. Saunder company. p. 363.
54. **Wallapa, Santivar, K.V. Nagaraja, B.S. Pomeroy, and J.A. Newman.** 1983. Studies on the control of salmonellosis in turkeys by their vaccination and competitive exclusion. J. Amer. Vet. Med. Assoc. **183**:357.
55. **Willams, J. E.** 1978. Paratyphoid infections. *In* Disease of poultry. Hofstad, M. S., Barnes, H. J., Clanek, B. W., Reid, W. H. and Yoder Jr, H. W., eds., 7th ed. Iowa State University Press, Ames, Iowa. pp. 117-167.
56. **Williams, J.E., and A.D. Whittemore.** 1971. Serological diagnosis of pullorum disease with the microagglutination system. Appl. Microbiol. **21**:394-399.

DISCUSSION

H. HUONG: In many of your experiments, if you look at the control, a lot of them have only 60% positive, while your vaccinated strains show 100% protection. I was wondering, if you adjust the challenge dose to have your control to 100% positive, then what is the outcome of your vaccination?

K. NAGARAJA: If you increase the challenge dose you can break the immunity. With a heavy challenge you can break the immunity with any vaccination trial. I'm sure there would have been a higher percent of positive birds, even with the vaccinated birds. But, remember how much challenge dose we get in the field. You have to imitate exactly what is happening in the field and you have to be realistic with experimental trials. We can increase the challenge dose and can show vaccine breaks.

H. HUONG: How did you determine the challenge dose?

K. NAGARAGA: By LD_{50}.

H. HOUNG: How many LD_{50}?

K. NAGARAJA: It is in that paper. I cannot remember the exact number.

H. HUONG: Another question is the cost. Which would be cheaper, the live vaccine or the outer membrane prep?

K. NAGARAJA: I don't know. The outer membrane proteins takes enough resources to produce. There are methods to maximize production so it could be effectively produced at a cheaper cost. Perhaps immunochemicals methods. Remember, these are all laboratory trials and take a lot of money. If one is interested, there are immunochemical proedures to use to effectively produce large amounts at a cheaper cost.

R. CURTISS: I have a question in terms of how effective an outer membrane protein preparation or bacterin can be. Depending on how one grew the bacteria, with and increasing number of pathogens now, *Vibrio cholera, Bordettella, Yersinia, Shigella, Salmonella*, it is very clear that a number of genes that are critical for colonization and invasion and overcoming host defense are environmentally regulated or regulated turned on in response to the host. So, to get those antigens expressed ought to be important for processing the microorganism to be used as an outermembrane protein or bacterin prep. But it seems to me in the system you describe it would be ideal to test whether that is really true or not. It seems to be a crazy idea that the most important antigens are the ones there all the time. I don't know how you fiddled the bugs prior to extraction.

K. NAGARAJA: No, we haven't examined if any procedure is better to produce different concentrations of protein. It is worth examining. Maybe we can change the media conditions to express quantities we want.

C. GYLES: Do you know what the outer membrane protein content of the bacterin is?

K. NAGARAJA: Yes.

C. GYLES: How does it compare in quality and quantity with the outer membrane protein vaccine preparation?

K. NAGARAJA: The outer membrane protein, it was much more than would have been in the bacterin.

C. GYLES: Would it be feasible to increase the amount of killed bacterin to give the same antigens challenge?

K. NAGARAJA: I think it is.

R. MEINERSMANN: In the trials in which you use the autogenous vaccine, essentially treating the problem flocks, did you institute any other control measures other than the vaccine?

K. NAGARAJA: All thses experiments preceded with the laboratory work. I did not show the results of the laboratory studies. This experiment was examined under laboratory controlled conditions and we were able to abort the infection. Infect the birds and vaccinate and you can abort the infection. Only after the laboratory study we looked at the problem flocks and went with the field trials.

C. GYLES: The two turkey flocks that you reported on with a persistent *Salmonella* problem. Those represent a killed vaccine superimposed on a live natural vaccine. Have you ever tried this on a clean flock, that is to see the effect on subsequent challenge because, as you know, there is a big impact on how killed vaccines work if they represent a booster to live vaccine?

K. NAGARAJA: I have not done that but I know some of my colleagues have done that.

J. HASSAN: I know you have used the urease system of extraction, was there any difference between urease and sarkosyl extraction?

K. NAGARAJA: We have examined urease extraction and in fact it is in the first paper we published on *Salmonella gallinarum* last year in Avian Diseases. With the French press you don't use any detergents and it is much easier to avoid use of detergents. We also examined the use of isothiocynate to extract the proteins. We tried all these procedures and ended up with the French press because you don't have to use detergents.

A. FRASER: How would you visualize vaccinating millions of birds with outer membrane protein preparation?

K. NAGARAJA: From the point of production, you mean?

A. FRASER: Yes, what strategy would you use to vaccinate the birds?

K. NAGARAJA: By subcutaneous vaccination. We have been using the vaccines by intramuscular or intraperitoneal administration. There are so many vaccines, this is one more addition.

SPEAKER X: What type of tissue reactions did you see when using outer membrane proteins compared to bacterin?

K. NAGARAJA: With the bacterin you see some reaction but not with outer membrane protein because of the volume. The volume is much less with the outer membrane protein, it is O.25 ml. So you don't see any reaction with the outer membrane protein. With the bacterin I have seeen some but it is not widespread, however people have reported to me some granulomas here and there.

EXTENDED ABSTRACTS

THE PREVENTION AND TREATMENT OF *SALMONELLA* INFECTIONS IN CHICKENS WITH CHLORTETRACYCLINE

M. Hinton

Department of Veterinary Medicine
University of Bristol
Langford House, Langford
Avon BS18 7DU, United Kingdom

ABSTRACT

It is essential to be able to evaluate antibacterial agents experimentally under controlled conditions and in a realistic manner. A model is described in which the value of an antibiotic, in this case chlortetracycline, was assessed for either preventing the colonization of the young chicks gastrointestinal tract with salmonellae, following oral challenge via the feed, or for treating an established infection. It was possible to obtain meaningful results when low levels of challenge were used and the birds only became infected subclinically. This has two obvious interrelated advantages: 1) the birds do not suffer illness unnecessarily, and 2) the experimental design does not become unbalanced because of birds dying of salmonellosis during the course of the experiment.

Colonization Control of Human Bacterial Enteropathogens in Poultry

INTRODUCTION

Salmonellae infections in young chicks are rarely associated with clinical illness but are of importance because carrier birds act as a reservoir of salmonellae for people who either handle or consume contaminated poultry meat.

It is essential, therefore, to be able to study the factors which may influence the colonization of the chicken's gastrointestinal tract with salmonellae in a way which simulates conditions on commercial farms. Accordingly, a model has been developed for the study of feed-borne infections in which feed is contaminated with small numbers of salmonellas using a two-stage mixing process (2, 3). The model has been used to follow the sequence of colonization of the birds' organs with invasive and non-invasive serovars (6, 9) and also to investigate the effect of both growth enhancing antibiotics and acid treatment of contaminated feed on salmonellae colonization (4, 5).

This paper describes two experiments in which the influence that an antibiotic, at two concentrations, had on salmonellae colonization in young chickens was assessed.

MATERIALS AND METHODS

Birds and Their Management. The chicks were purchased as unsexed 'day-olds' from a commercial hatchery and kept in groups of 10-15 in cardbcard boxes with wood shavings as bedding. Feed and water were available *ad libitum*.

Bacterial Strains. Two nalidixic acid-resistant *Salmonella* strains were used, namely, *Salmonella kedougou* (strain 131a/1) (7) and *S. typhimurium* DT14 (strain F98)(1). The *S. typhimurium* strain was invasive while the *S. kedougou* was not (9). Both were sensitive to ampicillin (25 μg), chloramphenicol (50 μg), nitrofurantoin (200 μg), streptocmycin (25 μg), sulphafurazole (500 μg), and tetracycline (50 μg)(8).

Preparation of Contaminated Feed. A 10^{-3} or a 10^{-4} dilution of an overnight BHI broth culture of the respective *Salmonella* strain was mixed into a broiler mash by using a two stage mixing process (2). Briefly, the diluted broth was dispersed (1:100 v/w) in the coconut and then the contaminated coconut was mixed (1:50 w/w) in the feed, which contained no grcwth promoting antibiotic or coccidiostat.

Enumeration of *Salmonella*. The method used for counting the *Salmonella* in the caecal contents and liver parenchyma had a limit of detection of 10 *Salmonella*/g (3). The plating medium (brilliant green agar supplemented with 30 μg/ml nalidixic acid, (BGNA) was incubated overnight at 37 ° C in air. The identity of the challenge strains was confirmed from time to time by using the slide agglutination test.

Experiment 1. The experiment was repeated three times. There were 10 birds/group in replicate 1 and 15/group in replicates 2 and 3. Feed, supplemented with either 50 or 300 ppm chlortetracycline (CTC; Aurofac 100, Cyanamid) was contaminated with *S. kedougou* at the rate of 12, 50 and 11 organisms/g respectively in the 3 replicates. Unsupplemented contaminated feed served as the control. These diets were given from the time the birds arrived from the hatchery. Swabs of cloacal faeces were obtained from the birds on days 1, 2, 3, 6, 7, and 8 after purchase and then used to inoculate BGNA agar. Birds were killed humanely on days 2, 6, or 8 and the number of *Salmonella* in the caecal contents determined.

Experiment 2. The experiment was repeated four times. The birds (12/group) were given feed contaminated with 1 of 2 levels of *Salmonella* in each replicate (Table 2). Two groups were given diet T3 (high inoculum) and T4 (low inoculum). At day 7, one group given the high and low inoculum was given uncontaminated feed containing 300 ppm CTC for seven days while the other group received uncontaminated unsupplemented feed.

Six of the 12 birds in each group were killed at seven days of age before any CTC treatment and six after treatment was completed at 14 days of age. The numbers of *Salmonella* in the caecal contents and liver parenchyma were determined. Negative cultures were scored as l/g and the counts transformed to $\log_{10}$. The results were evaluated using an analysis of variance with the model formula Fl*(F2/F3). Fl was the challenge dose in the feed (T3 or T4), F2 the age of the bird in weeks and F3 whether or not the birds were given feed supplemented with CTC. The model was necessarily incomplete since no birds were dosed with CTC during the first week. Accordingly, 'treatment' (F3) was nested within 'age' (F2) in order to produce a balanced design for analysis.

RESULTS

No 'wild-type' salmonellae were isolated from samples of feed prior to contamination with the 'test' strains and none of the paper sheets in the boxes used to transport the birds fram the hatchery were positive. No birds developed salmonellosis although 2 two-week-old birds given the T3 diet in the first replicate of experiment 1 were unwell when killed. In addition, one bird in that experiment died within two days of purchase. It was not positive for salmonellae.

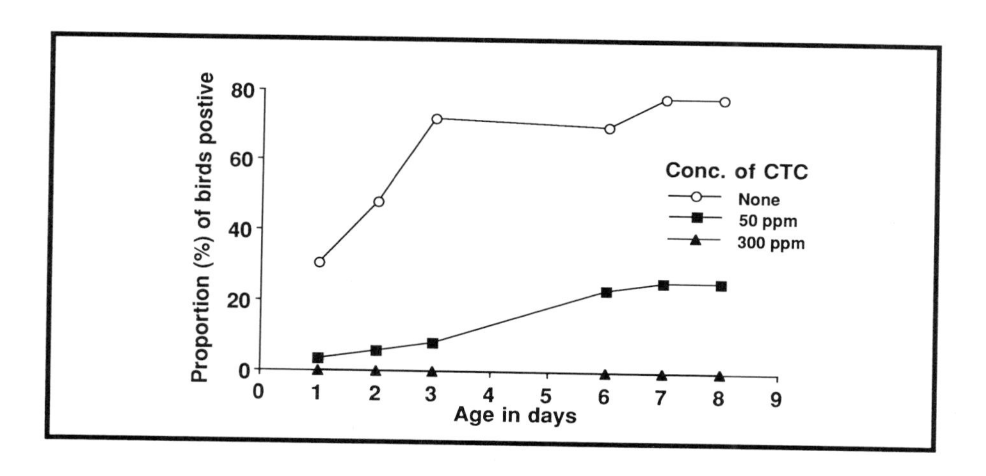

Fig. 1. Isolation rate of *S. kedougou* from the cloacal faeces.

Experiment 1. The isolation rate of *S. kedougou* from the cloacal faeces is illustrated in Fig. 1. The proportion of excretors among birds given no antibiotics increased from 25% on day 1 to 64% on day 8. No birds given feed containing 300 ppm CTC became infected, while the inclusion of 50 ppm CTC was associated with a reduction in the rate of colonization in the challenged birds.

The distribution of birds according to the number of *S. kedougou* in their caecal contents is given in Table 1. The proportion of infected birds given unmedicated feed was similar at 2, 6, and 8 days after challenge commenced. A comparison of all positive birds aged 6 and 8 days, using a one-way analysis of variance, indicated that the geometric mean *Salmonella* count in the caecal contents of the control birds and those given feed containing 50 ppm CTC was not significantly different ($\underline{P} > 0.05$) at $\log_{10}$ 6.85 and 6.10/g, respectively.

Table 1. The distribution of chicks in experiment 1 according to the numbers of *Salmonella kedougou* in their caecal contents.

Age (Days)	Concentration of of Chlortetracycline in Feed (ppm)	No. of birds with *Salmonella* Counts ($\log_{10}$ cfu/g) in Range of:				Proportion % Positive	Geometric Mean *Salmonella* Count in Positive Birds
		0	1-2	3-5	6-8		
2	0	4	-	5	4	69	5.97
	50	13	-	-	-	-	-
	300	13	-	-	-	-	-
6	0	5	-	1	7	62	7.11
	50	11	-	1	1	15	5.94
	300	13	-	-	-	-	-
8	0	4	-	4	6	71	6.64
	50	11	-	1	2	21	6.70
	300	14	-	-	-	-	-

Experiment lB. The distribution of birds according to the numbers of salmonellae in the caecal contents and liver parenchyma are given in Table 2, and the results of the analysis of variance summarized in Table 3. Overall, the geometric mean *Salmonella* count in the caecal contents was significantly lower ($\underline{P} < 0.00l$) in birds aged two weeks (F2), but this was due to the beneficial effect of the CTC treatment (Table 4) since the average counts in the unmedicated birds when aged 1 and 2 weeks were not significantly different at $\log_{10}$ 6.97 and 6.51, respectively.

In all birds the numbers of *S. typhimurium*/g of liver was lower than that recorded in the caecal contents. Overall the number of *Salmonella* in the liver was significantly influenced (P = 0.04) by the challenge dose (FI) although the differences between the average counts was not very large ($\log_{10}$ 2.78 and 2.45 for birds given the T3 and T4 diets respectively). The numbers of *Salmonella* fell significantly (P <0.001) during the week the control birds were given uncontaminated feed (Table 4). However, the inclusion of CTC in feed was associated with a significant (P *ca*. 0.01) additional reduction compared with the controls. The average numbers of *Salmonella*/g for these two groups were $\log_{10}$ 1.68 and 2.27, respectively.

Table 2. Distribution of birds in experiment 2 according to the numbers of *Salmonella typhimurium* in the caecal contents and liver parenchyma.

Organ	Diet Code*	Age (Weeks)	Medicated Feed Given in Second Week +	No. of Birds with *Salmonella* Counts ($\log_{10}$ cfu/g) in Range of:				Geometric Mean *Salmonella* Count
				0	1-2	3-5	6-8	
Caecum	T3	1		-	-	6	41	7.03
		2	No	-	-	3	21	6.69
		2	Yes	1	3	5	14	5.27
	T4	1		-	-	12	36	6.91
		2	No	-	-	4	24	6.34
		2	Yes	3	2	13	6	4.49
Liver	T3	1		-	9	39	-	3.55
		2	No	4	16	3	1	2.39
		2	Yes	5	19	-	-	1.63
	T4	1		9	6	33	-	2.95
		2	No	1	20	3	-	2.16
		2	Yes	4	20	-	-	1.74

*Diets T3 contained 196 and 137, and 333 *S. typhimurium* organisms/g in replicates 1-4. The corresponding numbers for the T4 diets were 19.6, 13.7, 23.5, and 33.3 organisms/g, respectively.

Table 3. Summary of the analysis of variance tables in experiment 2.

	Source of Variation*	Variance Ratio	P
Caecal Contents	F1	3.0	0.181
	F2	39.6	<0.001
	F1.F2	1.2	0.271
	F2.F3	32.8	<0.001
	F1.F2.F3	0.6	0.450
Liver	F1	12.2	0.040
	F2	58.9	<0.001
	F1.F2	2.6	0.106
	F2.F3	6.3	0.013
	F1.F2.F3	0.5	0.470

*See text.

Table 4. The geometric mean ($\log_{10}$ cfu/g) *Salmonella typhimurium* count in the caecal contents and liver parenchyma of birds in experiment 2 given unmedicated and medicated feed.

Age (Weeks)	Caecal Contents		Liver Parenchyma	
	Unmedicated Feed	Medicated Feed*	Unmedicated Feed	Medicated Feed*
1	6.97		3.25	
2	6.51	4.88	2.27	1.68
P	<0.001		0.013	

*The medicated feed contained 300 ppm chlortetracycline.

DISCUSSION

The results of experiment 1 demonstrated that the inclusion of 300 ppm CTC in the feed prevented the birds becoming colonized with *Salmonella*, while at the lower rate of inclusion (50 ppm) fewer birds became infected although interestingly, the average number of *Salmonella* in the caecal contents of the birds that did so did not differ signficantly fram the controls. Similar findings to this have been demonstrated in birds given feed containing short chain acids, but which became infected (7); observations which together suggest that antibacterial agents affect the absolute number of animals that became colonized by reducing the chances of an individual consuming an 'infective dose', but do not influence the numbers of *Salmonella* in the intestinal contents of the fewer animals that do become infected.

Experiment 2 satisfactorily assessed the therapeutic value of CTC. The assessment of the clinical efficiency of antibacterial agents presents several practical problems. Case control studies are time consuming and expensive while animals challenged with large numbers of pathogenic organisms experimentally may become severely ill and die which is clearly unacceptable on welfare grounds. It also unbalances the experimental design and this complicates subsequent analysis. On the other hand if low challenge doses are used, and the animals fail to develop clinical signs, the experiment may be deemed by same to have failed. In this study two challenge doses were used with the higher one (T3) causing illness in a few of the birds. However, from a bacteriological point of view there was little to choose between the birds on the two challenge regimes and hence the value of the agent was proven adequately by using a low challenge which only induced a subclinical infection.

ACKNOWLEDGEMENTS

I am grateful to Drs. R.H. Gustafson and H. Smith for helpful discussions and to Mrs. V. Allen, Mr. R. Cornish, Mrs. E. Coombs, Mrs. R. Hinton, Mr. M.D. Mohammad, and Mrs. P. Parsons for technical assistance.

REFERENCES

1. **Barrow, P.A., and J.F. Tucker.** 1986. Journal of Hygiene **96:**161-169.
2. **Hinton, M.** 1986. The artificial contamination of poultry feed with *Salmonella* and its infectivity for young chickens. Letters in Applied Microbiology. **3:**97-99.
3. **Hinton, M.** 1988. *Salmonella* infection in chicks following the consumption of artifically contaminated feed. Epidemiology and Infection **100(2):**247-256.
4. **Hinton, M.** 1988. *Salmonella* colonization in young chickens given feed supplemented with the growth promoting antibiotic avilamycin. Journal of Veterinary Pharmacology and Therapeutics **11(3):**269-275
5. **Hinton, M., and A.H. Linton.** 1988. Control of *Salmonella* infections in broiler chickens by the acid tretment of their feed. Veterinary Record **123(16):**416-421.

6. **Hinton, M., G.R. Pearson, E.J. Threlfall, B. Rowe, M. Woodward, and C. Wray.** 1989. Experimental *Salmonella enteritidis* infection in chicks. Veterinary Record **124(9)**:223.
7. **Impey, C.S., G.C. Mead, and M. Hinton.** 1987. Influence of continuous challenge via the feed on competitive exclusion of salmonellas for broiler chicks. Journal of Applied Bacteriology **63(2)**:139-146.
8. **Linton, A.H., K. Howe, and A.D. Osborne.** 1975. The effects of feeding tetracycline, Nitrovin, and Quindoxin on the drug-resistance of coli-aerogenes bacteria from calves and pigs. Journal of Applied Bacteriology **38(3)**:255-275.
9. **Xu, Y.M., G.R. Pearson, and M. Hinton.** 1988. The colonization of the alimentary tract and visceral organs of chicks with salmonellas following challenge via the feed: bacteriological findings. British Veterinary Journal **144(4)**:403-410.

STUDIES ON PREVENTION OF *SALMONELLA* CARRIAGE BY ORAL IMMUNIZATION OF POULTRY

T. S. Agin
A. Morris-Hooke

Miami University
Oxford, Ohio 45056

ABSTRACT

Temperature-sensitive (ts) mutants of *S. enteritidis* serovar *heidelberg* were isolated following chemical mutagenesis with nitrosoguanidine and enrichment with penicillin and D-cycloserine. Two different phenotypes were obtained, coasters (which continue replication for one or more generations at 42 ° C) and tights (which cease replication immediately after shift to 42 ° C). Biochemical profiles of ten mutants of each phenotype chosen for further study were identical to that of the parental wild-type (wt), and reversion rates ranged from 5×10^{-7} to 1×10^{-8}.

Four-to-six day-old, specific-pathogen-free, White Leghorn chicks were immunized by oral inoculation with varying doses of tight and coasting ts mutants. Immunized and control chicks were challenged with a streptomycin-resistant derivative of the parental wt one month later and shedding of the wt was monitored by plating cloacal samples on Brilliant Green agar. At selected times after challenge with the wt, the birds were sacrificed and spleens, livers, gall bladders, ovaries, and intestines were sampled for the presence of the challenge strain.

Immunization had no effect on the weight gain of the chicks. Colonization and invasion occurred in all groups following oral challenge with as few as 10^2 colony forming units (CFU) of the wild-type. *Salmonella*-specific serum ELISA antibodies were induced in one immunized group and they were also found in all groups following challenge with the wt, but none could be detected in bile and intestinal washes.

Colonization Control of Human Bacterial Enteropathogens in Poultry

INTRODUCTION

Purified component or killed whole cell vaccines which induce humoral immune responses have had only limited success in preventing *Salmonella* colonization of poultry. Live attenuated vaccines, which usually induce both humoral and cellular immunity, include *gal*E, *aro*A, *pur*A, and ts mutants (1, 2, 7).

The ts vaccine we propose to develop should offer several advantages over other types of attenuated *Salmonella*. First, all the surface antigens will remain intact. Second, ts mutants with lesions in essential genes are unable to sustain replication in any nutritional environment. Third, coasting ts mutants capable of limited replication in the vaccinee should allow the expression of antigenic determinants seen only during replication *in vivo*. Coasting may also enhance immunogenicity by prolonging stimulation of the immune system. Fourth, through genetic manipulation, multiple ts mutations of identical phenotype can be incorporated into a single strain reducing reversion to negligible levels (4, 5).

The research presented here focused on the isolation and characterization of single ts mutants of *S. enteritidis* and preliminary immunological evaluation of two of the ts mutants in preventing *Salmonella* colonization in White Leghorn chickens.

MATERIALS AND METHODS

Bacteria and Culture Media. *Salmonella enteritidis* 141 serovar *heidelberg* (from Carl Weston, Hubbard Farms, NH), was cultured on tryptic soy agar (TSA) or in tryptic soy broth (TSB) unless otherwise indicated.

Mutagenesis, Enrichment, and ts Mutant Isolation. *S. enteritidis* was subjected to mutagenesis with nitrosoguanidine (30 μg/ml) and enrichment with penicillin (1,000 to 3,000 units/ml) and D-cycloserine (30mM to 100mM) according to a protocol described previously (3), except that addition of the antibiotics was delayed for 90 or 360 min after transfer to the nonpermissive temperature (42 ° C). Samples of the surviving cells were diluted appropriately, plated on TSA, and incubated at 29 ° C for 24 h. The colonies were then replica-plated and incubated at 42 ° C and 29 ° C to detect mutants incapable of sustained growth at 42 ° C.

Characterization of ts Mutants. Three different types of ts mutants were isolated: tight mutants, which cease replication immediately at 42 ° C; coasting mutants, which continue replication at 42 ° C for one to three generations; and leaky mutants, which have impaired growth rates at 42 ° C, and were discarded. Reversion rates were determined by incubating 1 x 10^8 CFU on TSA at 42 ° C. Biochemical profiles and antibiograms of the ts mutants were compared with those of the wt using the Sceptor diagnostic system (Johnston Laboratories, Towson, MD).

Antibiotic-resistant strains. Streptomycin-resistant (Sm^R) mutants of the wt and coasting ts mutant A/2/7, and a rifamycin-resistant (Rif^R) mutant of tight ts mutant F/2/16 were derived by plating $>10^9$ CFU on TSA containing 1,000 μg/ml Sm or 300 μg/ml Rif. The antibiotic-resistances were used as markers to distinguish the ts mutants and the wt during immunization and challenge experiments.

Animals. Specific-pathogen-free, White Leghorn chicks were hatched on the premises, housed in wire-floored temperature controlled brooder boxes, and transferred to large wire-floored cages after one month. Sterile, antibiotic-free, chicken-starter food and water were provided *ad libitum*.

Immunization Schedule. The effects of immunization were studied in four-to-six day-old chicks. Groups of 10-20 chicks were immunized orally with 10^5 or 10^6 CFU of the tight or coasting mutants. Control groups received 100 μl sterile saline. Weight gain and general demeanor of the birds were monitored. Presence of the immunizing strains in the intestine and the cecal contents was determined up to 84 hours after immunization by plating the contents onto Brilliant Green agar (BG) and BG containing 300 ug/ml Rif (BGR) or 1,000 μg/ml Sm (BGS). Shedding of the ts mutants was monitored by plating cloacal swabs onto BG, BGS, and BGR, both after culture in TSB and by direct plating.

Challenge Schedule. Immunized and control birds were challenged with selected doses of Sm^R-wt, 30 days after immunization. Weight gain and general demeanor of the birds were monitored. Shedding of the challenge strain was detected by plating cloacal swabs onto BGS. At selected times after challenge, groups of chickens were sacrificed by $C0_2$ asphyxiation and the livers, gall bladders, spleens, kidneys, ovaries and cecal material removed and homogenized in sterile saline using a Tekmar Stomacher. The homogenates were plated on BGS, incubated at 37 ° C and scored for the presence of *Salmonella*.

Serum, Biliary, and Intestinal Antibodies. One-to-two-day-old chicks were separated into two groups of 25 experimental birds and one group of 26 control birds. The experimental birds were immunized orally with 10^6 CFU of F/2/16 Rif^R or A/2/7 daily for three days; control birds received saline. Blood was collected from six one-day-old birds by cardiac puncture, the sera were pooled and stored at -20 ° C. Sera were also collected from all birds 6, 18, and 31 days post-immunization and 12, 15, and 25 days after challenge (day 35). Three birds from each group were sacrificed on the same days, bile samples were collected, the small intestines removed and washed internally with 3-5 ml ice-cold saline which was clarified by centrifugation and stored at -20 ° C.

Enzyme Linked Immunosorbent Assay (ELISA). Whole cell antigen was prepared by treating Sm^R-*S. enteritidis* 141 with gentamicin sulfate at a final concentration of 100 μg/ml for 60 min. Penicillin was then added (1,000 units/ml) for 2 hours. The cells were washed three times in Dulbecco's phosphate buffered saline (PBS), suspended in 1.5 ml PBS, 0.02% sodium azide, and stored at 4 ° C. The antigen stock contained approx. 1×10^{10} cells/ml. The ELISA was performed as described previously (8).

RESULTS

Isolation and Characterization of ts Mutants. The number and phenotypes of ts mutants isolated after mutagenesis and enrichment are given in Table 1. Delaying the addition of the antibiotics for 3 h (instead of 90 min) increased the frequency of coasting mutants recovered by 30%. More than 200 ts mutants were characterized with respect to cut-off temperatures, mean generation times (MGTs) at both permissive and nonpermissive temperatures, and reversion rates. Two mutants, F/2/16 and A/2/7, were chosen for the immunization and challenge experiments on the basis of their MGTs at 29 ° C (30 min, close to that of wt) and low reversion frequencies (1×10^{-8}). The ts mutant A/2/7 is a coaster which continues replication for three generations at 42 ° C. F/2/16 is a tight ts mutant which stops replication immediately after transfer to 42 ° C (Fig. 1).

Cut-off temperatures for both mutants were identical (42 ° C), and their biochemical profiles were similar to that of the weight (data not shown).

Table 1. Isolation of ts mutants of *S. enteritidis*.

Experiment[a]	Survivors	TS (%)	Tight (%)	Coasters (%)
One[b]	805	171 (21)	27 (16)	55 (32)
Two[c]	692	222 (32)	36 (16)	94 (42)

[a]Mutants were isolated following treatment with 30 μg/ml nitrosoguanidine and two cycles of enrichment with penicillin and D-cycloserine as described in Materials and Methods.

[b]Penicillin (1,000 units/ml) and 30mM D-cycloserine were used for the enrichment steps.

[c]Penicillin (10,000 units/ml) and 100 mM D-cycloserine were used for the enrichment steps, and their addition was delayed for three hours after the cultures were shifted to the nonpermissive temperature.

Effect of Oral Immunization on Weight Gain. Groups of four-to-six day-old chickens were inoculated orally with either the tight or coasting mutants and the weights of both immunized and control birds were monitored. No differences in weight gain were observed for up to eight days after immunization (Fig. 2) and shedding of the immunizing strain was not detected.

Effect of Immunization on Shedding of the Challenge Strain. In order to determine whether immunization with the ts mutants could protect chickens against colonization and shedding of the challenge strain, chickens immunized with the tight or coasting mutants and control birds were challenged orally with 5×10^3 or 5×10^2 CFU and the presence of the challenge strain was determined in fecal samples. No distinct shedding pattern was observed in any of the three groups, even when the challenge dose was lowered from 5×10^3 to 5×10^2 CFU, and when fecal samples were taken over 35 days (Fig. 3).

Serum, Biliary, and Intestinal Antibodies. Sera were collected from chicks inoculated with 10^6 CFU of the tight or coasting mutants or saline, and bile and intestinal washes were also collected. No ELISA activity could be detected in the bile or intestinal washes. Low levels of antibody activity were detected in the sera of newly hatched chicks; the specificity of the activity was confirmed by absorption experiments with whole cells of *S. enteritidis* 141 (data not shown). The serum ELISA antibody activity in the samples from the tight-immunized group increased significantly by day nine ($p = 0.029$) (Fig. 4). Serum ELISA activity also increased, in all groups, after challenge with 10^2 CFU of the wild-type. It appeared, moreover, that the increase in activity in the immunized birds occurred before it did in the control birds, suggesting an anamnestic response, but the numbers are clearly too small for valid statistical analysis.

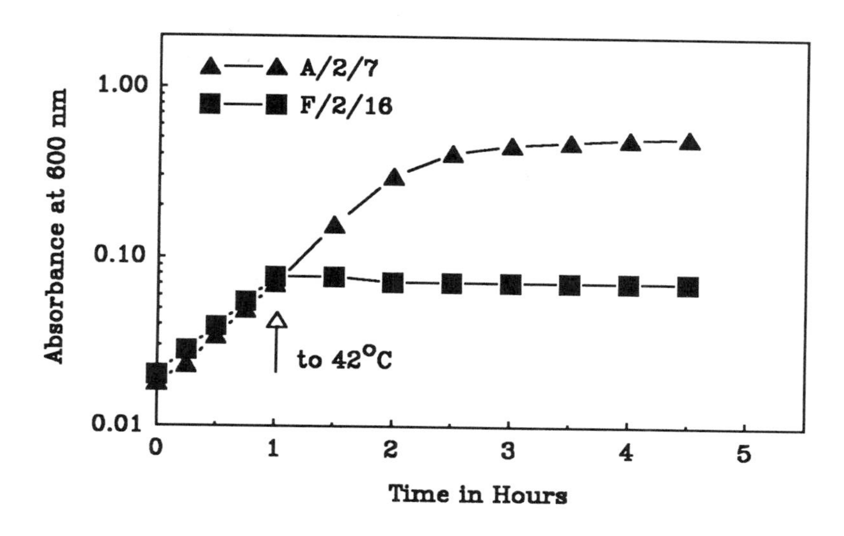

Fig. 1. Growth profiles of tight and coasting ts mutants of *S. enteritidis* 141 at permissive and nonpermissive temperatures. Cultures of ts mutants F/2/16 and A/2/7 were inoculated into TSB and incubated at 200 rpm, 29 ° C. At the time indicated by the arrow, the cultures were shifted to 42 ° C.

DISCUSSION

The purpose of this study was to isolate and characterize ts mutants of *S. enteritidis* and to explore the potential of the ts mutants as vaccines to prevent colonization of White Leghorn chickens by *Salmonella*. More than 200 ts mutants were isolated from *S. enteritidis* 141, and two were chosen for immunological evaluation because of their growth profiles, reversion frequencies and biochemical characteristics.

Immunization with either the tight or coasting mutant, using several dosing and challenge schedules, had no apparent effect on colonization or invasion by the challenge strain. No differences were detected among the three groups of birds following challenge with the wt *Salmonella*. All three groups shed the challenge strain for up to 35 days and, furthermore, it was found in all organs tested from birds in all three groups. The only possible difference among the three groups that could be discerned was that a smaller percentage of birds immunized with the tight mutant shed the challenge strain, but this was not consistent throughout the experiment.

It is possible that immunity is induced, but it is overcome by this challenge strain which is remarkably efficient at colonization: just 10^2 CFU of *S. enteritidis* 141 can colonize 100% of the inoculated birds. This is in sharp contrast with the results of experiments with *S. enteritidis* serovar typhimurium, in which the challenge dose required for colonization was 10^8 CFU for birds as young as two weeks (9).

It is clear that the ts mutants of *S. enteritidis* 141 used in these studies did not elicit the immune response expected under the conditions tested. The reasons why no responses were detected must be explored, and the problem overcome, perhaps by delaying immunization for one to two weeks, increasing the immunizing dose, increasing the number of immunizing doses, or combining immunization with other methods, such as competitive exclusion or the addition of lactose to the drinking water (6).

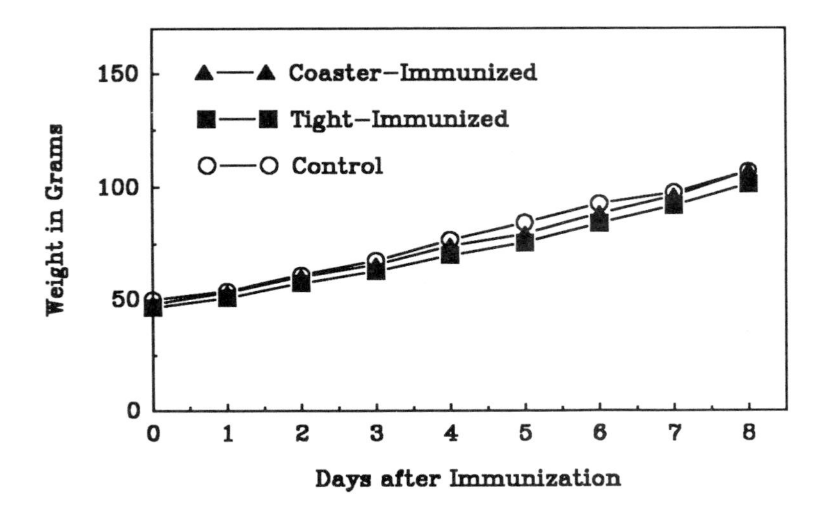

Fig. 2. Effect of oral immunization with ts mutants on weight gain. Groups of ten four-to-six day-old *Salmonella*-free chickens were immunized with either 10^5 F/2/16 or 10^5 A/2/7 in 100 μl saline. Control chicks were fed 100 μl saline. Immunized and control chickens were weighed daily.

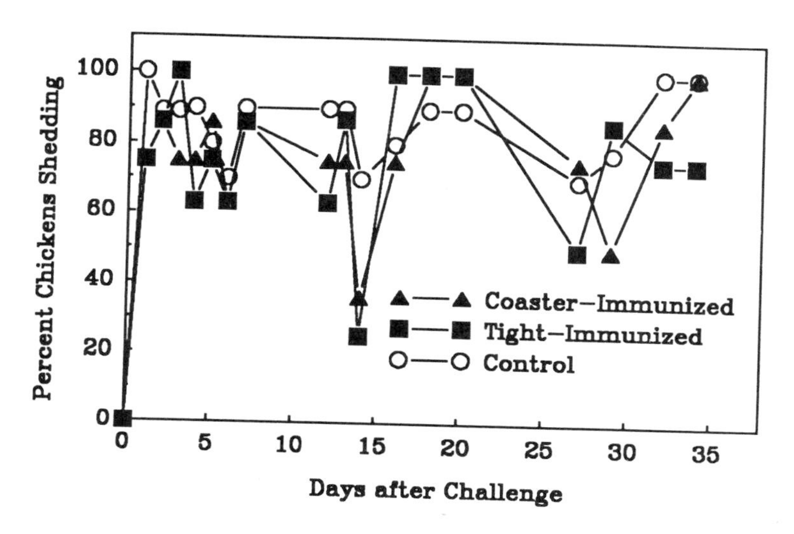

Fig 3. Effect of oral immunization on colonization after challenge with *S. enteritidis* 141. Groups of ten four-to-six day-old chickens were immunized with 10^5 CFU of F/2/16 or A/2/7 in 100 μl saline. Control birds were fed 100 μl saline. Thirty days later, immunized and control birds werer challenged with 5 x 10^2 CFU *S. enteritidis* 141 in 100 μl saline. Shedding was monitored by culturing cloacal samples in TSB at 37 ° C for 3 h and plating on Brilliant Green agar.

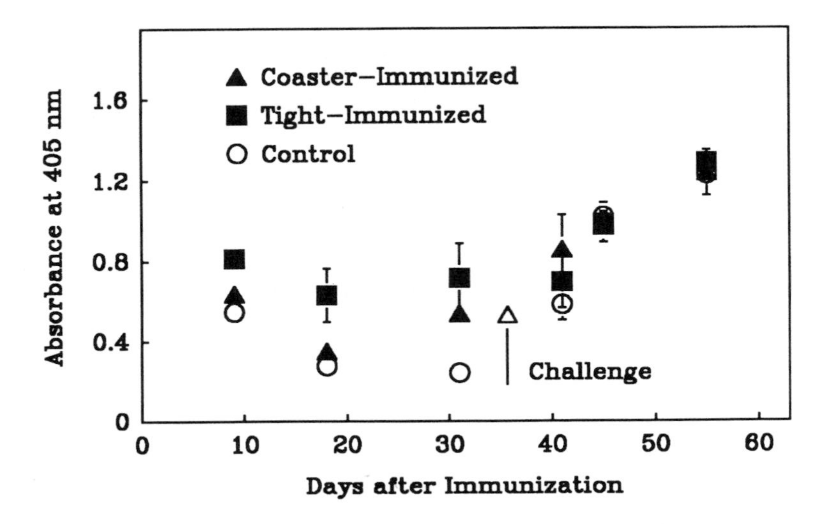

Fig. 4. *Salmonella*-specific ELISA antibodies in the sera of immunized and control chickens. Four-to-six day-old chicks were immunized orally on three consecutive days with 10^6 CFU of the tight or coasting mutants or saline. Sera were collected on days 9, 18, and 31 post-immunization. Immunized and control chickens were challenged orally with the wild-type on day 35 and sera collected on days 41, 45, and 55. Serum antibody to*Salmonella* was measured using an enzyme-linked immunosorbent assay.

REFERENCES

1. **Fahey, R. J., and G. N. Cooper.** 1970. Oral immunization against experimental salmonellosis: I. Development of temperature-sensitive mutant vaccines. Infect. Immun. **1:**263-270.
2. **Germanier, R.** 1984. Oral vaccination against enteric bacterial infections: an overview. Infection **12:**138-142.
3. **Morris Hooke, A., M. P. Oeschger, B. J. Zeligs, and J. A. Bellanti.** 1978. Ideal target organism for quantitative bactericidal assays. Infect. Immun. **20:**406-411.
4. **Morris Hooke, A., J. A. Bellanti, and M. P. Oeschger.** 1985. Live attenuated bacterial vaccines: new approaches for safety and efficacy. Lancet **1:**1472-1474.
5. **Morris Hooke, A., D. 0. Bordelli, M. C. Cerquetti, and J. A. Bellanti.** 1987. Differential growth characteristics and immunogenicity of tight and coasting temperature-sensitive mutants of *Pseudomonas aeruginosa*. Infect. Immun. **55:**99-103.
6. **Oyofo, B. A., J. R. DeLoach, D. E. Corrier, J. 0. Norman, R. L. Ziprin, and H. H. Mollenhauer.** 1989. Effect of carbohydrates on *Salmonella typhimurium* colonization in broiler chickens. Avian Dis. **33:**531-534.
7. **Sigwart, D. F., B. A. D. Stocker, and J. D. Clements.** 1989. Effect of a *pur*A mutation on efficacy of *Salmonella* live vaccine vectors. Infect. Immun. **57:**1858-1861.
8. **Sordelli, D. 0., R. A. Rojas, M. C. Cerquetti, A. Morris Hooke, P. J. Degnan, and J. A. Bellanti.** 1985. Enzyme-linked immunosorbent assay antibody responses to a temperature-sensitive mutant of *Pseudomonas aeruginosa*. Infect. Immun. **50:**324-327.
9. **Suphabphant, W., M. D. York, and B. S. Pomeroy.** 1982. Use of two vaccines (live G30D or killed RW16) in the prevention of *S. typhimurium* infections in chickens. Avian Dis. **27:**602-615.

THE IMMUNOLOGICAL BASIS OF CHICKEN'S PROTECTION AGAINST *SALMONELLA TYPHIMURIUM* F98

J. O. Hassan

Department of Biology
Washington University
St. Louis, Missouri 63130

ABSTRACT

S. typhimurium F98 parent strain and three of its mutants: 85kbp virulence plasmid cured (LP-), rough and LP- (0-*), and non flagellate (H-), were used to investigate their ability to colonize, invade, and to induce humoral and cellular responses in chickens by oral vaccination. Faecal excretion of *S. typhimurium* was used to determine the level of protection against secondary challenge. Quantification of *S. typhimurium* in the gastro-intestinal tract and visceral organs was used to determine colonization and invasiveness. The strains and their extracted surface antigens were characterized on SDS-PAGE. The antigens were used to detect IgM, IgA, and IgG, level in serum, intestinal washings, and bile, using ELISA. IgM, IgA, and IgG, secreting cells was detected in the spleen. Cell mediated immune responses was detected by delated type hypersensitivity (DTH).

A drastic reduction in faecal excretion of *S. typhimurium* was observed in the immunized chickens after secondary challenge. The rough mutant gave the highest level of protection. All the *Salmonella* antigens induced IgM, IgA, and IgG production. IgG was the predominant antibody in serum, while IgA was the most predominant in the intestinal washing and bile. The antibody secreting cell assay detected the peak of IgM, IgA, and IgG secretions a week earlier than their respective peaks in the serum. Cell mediated immune responses was induced by all the antigens except LPS.

The ability of the chicken immune system to handle the rough mutant may be related to its lack of LPS. The OMP antigen gave the highest DTH response,especially in chickens infected with the rough mutant. The lack of LPS in this mutant exposed the OMP for better processing and recognition. A combination of the efficiency of the rough mutant with a sensitive faecal *Salmonella* detection assay can be used to create a surveillance system that may lead to the establishment of *Salmonella*-free stock in the poultry industry.

Colonization Control of Human Bacterial Enteropathogens in Poultry

INTRODUCTION

Salmonellosis is a zoonotic disease with a complex epizootiological cycle. Poultry accounts for a greater proportion of human *Salmonella* infections. One of the most perplexing problems in the control of salmonellosis is the role played by carriers. The control of *Salmonella* infection in poultry with live vaccine will most likely provoke a strong objection from the consumer, due to its public health hazard. It is therefore necessary to investigate the immunogenicity of *Salmonella* antigens in detail, with the aim of gaining more information that may lead to the production of genetically modified vaccine.

In this work, *S. typhimurium* F98 wild type, and three of its mutants; large plasmid minus, rough, and non-flagellate were used to infect one or four-days old chicks, to determine their ability to colonize, invade, and to protect chickens against a secondary challenge with the parent strain. Lipopolysacharide (LPS), flagella, and outer membrane protein (OMP) were extracted and used to detect specific IgM, IgA, and IgG response in infected chickens. Specific anti-*Salmonella* IgM, IgA and IgG secreting cells were enumerated in the spleen of chickens infected with the wild type. The ability of LPS, flagella, and OMP to induce delayed type hypersensitivity were investigated. Faecal excretion of the challenge strain was used to determine the level of protection induced by immunization with the wild type and its mutants. The relationship between immune responses and protection will be analyzed.

MATERIALS AND METHODS

***Salmonella* Strains.** The *S. typhimurium* F98 phage type 14 has been described previously, (Smith and Tucker, 1975, 1980; Barrow *et al.*, 1987). The *S. typhimurium* mutants have been described before (Barrow *et al.*, 1988).

Chickens. Specific pathogen free (SPF) Light Sussex flock, housed at 21 ° C and fed *ad libitum* (Smith and Turker, 1975).

Bacteria Culture and Inoculation. The *Salmonella* strains were maintained at the Institute of Animal Health, Houghton Laboratory, Department of Microbiology. Broth cultures were made in 10 ml nutrient broth, incubated at 37 ° C overnight in a shaking water bath. Chickens were inoculated orally into the crop with 0.1 lml (one-day-old) or 0.3 ml (four-days-old) undiluted broth culture of *Salmonella* strain.

PREPARATION OF ANTIGENS

Flagella. 5 g of wet *S. typhimurium* parent strain or its mutants were homogenized for 1 min., centrifuged at 5,000 g x 25 min., at 4 ° C, and the supernatant at 75,000 g x 45 min. The colorless deposit was resuspended in phosphate buffered saline (PBS), acidified with 1:20 vol. lM HCl. for 30 min. centrifuged at 80,000 g x 1 hr. The supernatant was neutralized with 1M NaOH. The flagella was polymerised with 2 vol. cold saturated 70% ammonium sulphate, left overnight at 4 ° C, and centrifuged at 20,000 g x 15 min. The obtained flagella was dialysed, and stored in formalin at -20 ° C.

Lipopolysacharide. LPS was extracted by the phenol water method (Galanos *et al.*, 1977) from *S. typhimurium* parent strain and its mutants.

Outer Membrane Protein. *Salmonella* cells were harvested and washed 3ce in PBS at 4 ° C. The suspended bacteria were sonicated in an ice bath for 4 x 1 min. spell, with 30min. rest between each sonication. Centrifuged at 10,000 g x 30 min. at 4 ° C. Pellet was resuspended in 100 μl PBS and 10 μl of 20% N-lauroylsarcosine sodium salt. (Sarcosyl; Sigma) was added at room temperature. The mixture was centrifuged at 100,000 g x 30 min. at 4 ° C. The obtained OMP was dissolved in PBS containing sodium azide and stored at -20 ° C.

EXPERIMENTAL PROTOCOL

Groups of one-day-old or four-day-old chicks were infected orally with *S. typhimurium* parent strain or any of its mutants. Intestinal content, intestinal wall, and visceral organs were collected from a group of chickens infected at one-day-old at determined hours and from another group infected at four-days-old, on a weekly basis for bacterial quantification. Blood, bile, and intestinal washings were collected weekly and used for the determination of serum, bile, and intestinal immunoglobulin level with an enzyme linked immunosorbent assay (ELISA). Spleen was taken from a group of chickens infected at four-days- old, weekly and used for the detection of antibody secreting cells. At four weeks post infection, some of the chickens infected at four-days were injected intradermally with 100 μl of *Salmonella* antigen in the footpad, and cellular infiltration was determined by histological analysis. Faecal excretion of groups of chickens infected at four-days-old were monitored, and immunized chickens and unimmunized group were challenged with the wild type and excretion of the wild type was monitored.

RESULTS

***S.typhimurium* F98 Mutants as Vaccine.** As illustrated in Table 1, all the immunized strains were excreted for a considerable length of time by the chickens. Following secondary challenge, the chickens immunized with the rough mutant and the nonflagellate mutant showed a better level of protection against secondary challenge than the unimmunized chickens.

Colonization and Invasiveness of *S. typhimurium* F98. Figure l (a-d) shows that all the strains can colonize the intestine, and invade the intestinal wall. All strains except the rough mutant was able to actively invade the yolk sac. The rough mutant was isolated later than others from the liver and spleen and at a lower count.

Figure 2 (a-b) illustrates the isolation pattern over many weeks. The highest level of colonization was in the caecal content. All the strains invade and proliferate in the bursa and caecal tonsil. Only the rough mutant was not isolated from the liver and spleen after one week. Following challenge with the wild type, isolation of the wild type was mostly from the unimmunized chickens.

Humoral Responses. Figure 3 (a-d) depicts the observed IgM, IgA, and IgG responses in the serum, intestine, and bile. All the antigens detect the same pattern of antibody response, but the LPS gave a lower titre. IgG is the predominant isotype in the serum, while IgA is more dominant in the intestine and bile. IgM was detected earlier in the serum; it peaked by the third week, while serum IgA and IgG peaked in the fifth week. IgA gave the lowest titre in the serum. IgM and IgA were detected in the first week in the intestine. IgM peaked in the third week, a week earlier than IgG and IgA. IgG has

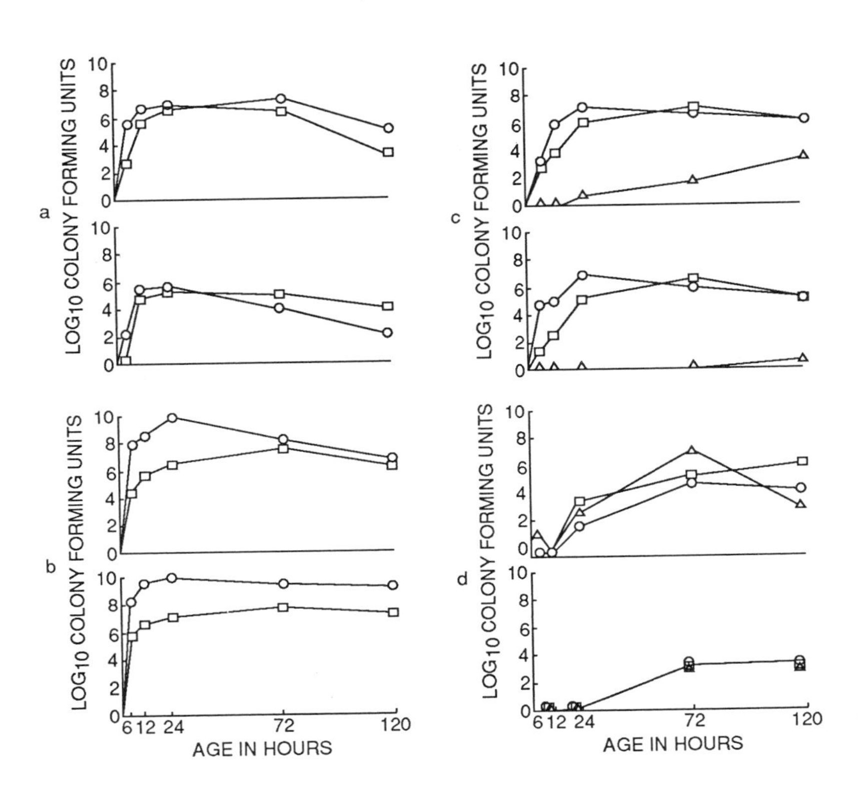

Figure 1. (a) Isolation of *S. typhimurium* from the (○) ileum content and (□) ileum wall of chickens infected at one day old with *S. typhimurium* F98 Nal^r. (b) Isolation of *S. typhimurium* from the (○) caecal content and (□) caecal wall of chickens infected at one day old with *S. typhimurium* F98 Nal^r. (c) Isolation of *S. typhimurium* from the (○) caecal tonsil, (□) bursa of Fabricius, and (Δ) thymus of chickens infected at one day old with *S. typhimurium* F98 Nal^r. (d) Isolation of *S. typhimurium* from the (○) liver, (□) spleen, and (Δ) yolk sac of chickens infected at one day old with *S. typhimurium* F98 Nal^r. In all parts the bottom graph represents the rough mutant and the top graph all others.

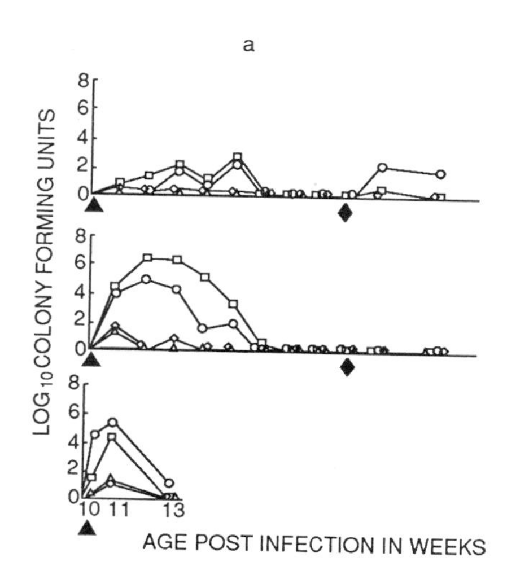

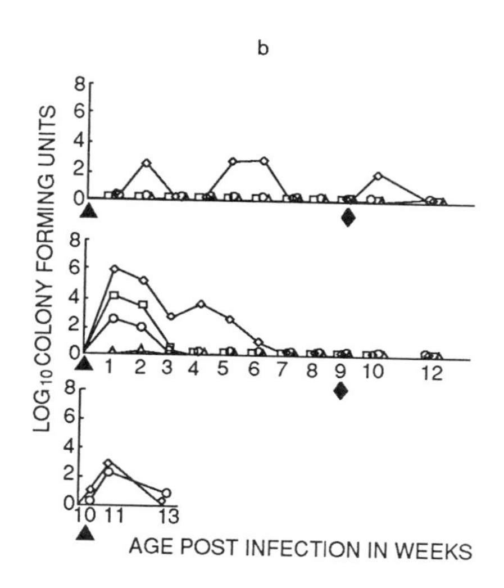

Figure 2. (a) Isolation of *S. typhimurium* from the (○) caecal tonsil, (□) caecal content, (Δ) ileum content, and (◊) ileum wall of chickens infected at four days old with *S. typhimurium* F98 Nal^r. (b) Isolation of *S. typhimurium* from the (○) liver, (□) spleen, (Δ) thymus, and (◊) bursa of Fabricius of chickens infected at four days old with *S. typhimurium* F98 Nal^r. In both parts the top graph represents the rough mutant, the middle graph all others, and the bottom graph the controls.

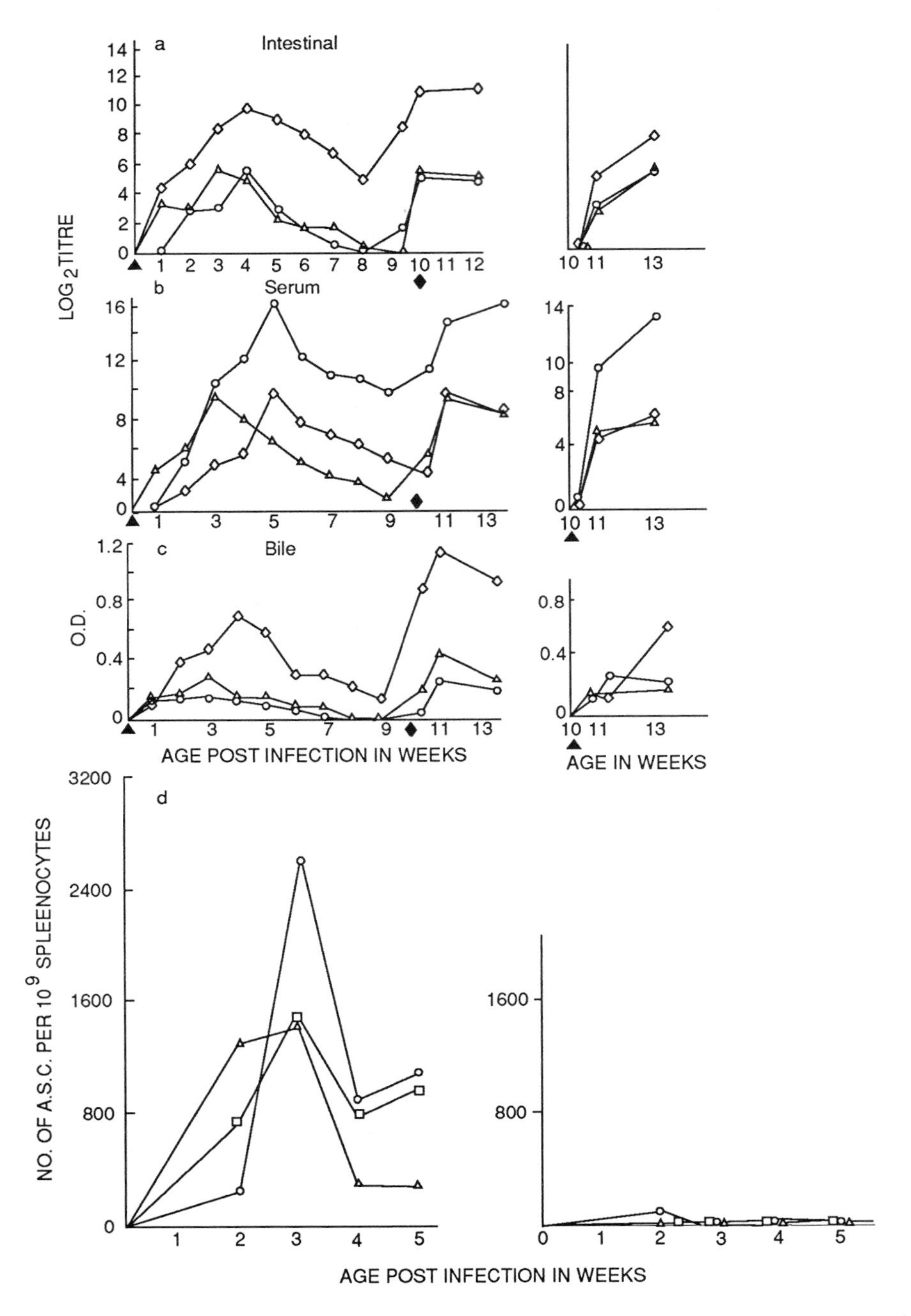

Figure 3. (a–c) (Δ) IgM, (◊) IgA, and (○) IgG response as detected by *Salmonella* antigen. Right panels represent controls. (d) (Δ) IgM, (□) IgA, and (○) IgG *Salmonella*-specific antibody-secreting cells in the spleen of chickens infected with *S. typhimurium* F98 Nal^r: (Left) at four days old and an uninfected chicken; (right) with an Elisa using *Salmonella* sonicated cell antigen.

the lowest titre in the intestine. IgA was the major antibody in the bile. Its pattern is similar to serum IgA than intestinal IgA. Traceable amounts of IgM were detected in the third week, but there was no convincing evidence for IgG in the bile. Following secondary challenge, a high titre of IgG was detectable by the third day after challenge in the serum while there was little or no detectable antibody in the unimmunized chickens. A similar observation was seen for IgA in the intestine and bile. This phenomenon did not accompany IgM responses after secondary challenge.

Table 1. Percentage of chickens excreting *Salmonella typhimurium* strain and mutant following initial infection.

Faecel Excretion of	Age	Rough Mutant	Non-Flagellate (H^-) Mutant	Control	Parent Strain	Virulence Plasmid Cured Mutant (LP^-)
Primary Inoculation	1	70	87	0	97	100
with F98 Nal^r	2	632	77	0	80	97
(Immunizing Mutants)	3	50	77	--	80	73
	4	43	80	--	57	50
	5	37	37	--	33	30
	6	30	13	--	10	17
	7	27	10	--	20	10
	8	17	10	0	10	10
	9	17	3	0	3	17
Secondary Inoculation	10	17	7	63	47	30
with F98 spc^r	11	0	3	33	13	3
(Challenge Strain)	12	0	3	17	3	7

Cellular Responses. No cellular response was observed in the uninfected chickens challenged with *S. typhimurium* antigens. As shown in Table 2, less than ten cells were seen in response to LPS. The flagella antigen induces a moderate response. The OMP induced a high cellular infiltration by heterophils (mostly eosinophils with pink granules) in all chicken groups, but the highest observed response was with chickens immunized with the rough mutant.

Table 2. Delayed typed hypersensitivity test (cellular infiltration of foot pad section) in chickens.

Age of Birds at Challenge	Antigen Inoculated	Parent Strain	Non-Flagellate Mutant	Rough Mutant
Four Weeks	SSC Antigen	+ + +	+ + +	+ + +
	H Antigen	+ +	NT	+ +
	LPS Antigen	< +	< +	NT
	CMP Antigen	+ + +	+ + + +	+ + + +

+ = 10 - 100 Cells
+ + = 100 - 500 Cells
+ + + = 500 - 1000 Cells
+ + + + = above 1000 Cells
NT = Not Tested

DISCUSSION

The *S.typhimurium* strains used in this work are good colonizers. The rough mutant gave 100% protection rate in the first week post challenge. All the strains were good tissue invaders, except the rough mutant which was unable to invade the yolk sac. The invasion of the yolk sac requires active penetration by the bacteria. The exact mode of invasion of the intestinal wall and visceral organs is not fully understood in chickens. Active uptake by the epithelium cells has been suggested (Turnbull and Richmond, 1977). Phagocytic cell involvement has also been implicated (Popiel and Turnbull, 1985). The ability of the rough mutant to invade the tissue and not the yolk sac suggests active uptake of the bacteria cells from the lumen. The demonstration of *Salmonella* by peroxidase anti-peroxidase technique in phagocytes within the laminal propial of caecum, in the spleen, and the bursa of Fabricius of chicken infected with *S. typhimurium* (Hassan, 1989) confirms this suggestion.

The migration of the *Salmonella*-laiden phagocytes to the liver, spleen, caecal tonsil, and the bursa of Fabricius must be through local circulatory system, since *Salmonella* was not isolated from the blood of chickens infected with *S. typhimurium*. (Barrow *et al.*, 1988). The lack of isolation of the rough mutant from the liver and spleen before 24 hrs. post infection and one week after infection, despite the high level of intestinal and caecal titre, may be due to the ability of the phagocytes to kill the rough mutant better than the other strains. Resistance by *S. typhimurium* to host defense mechanism in mice increases with the LPS chain length (Shaio and Rowland, 1985; Modrzakowski and

Spitznagel, 1979) showed that susceptibility of the rough mutant to granule cationic proteins of human neutrophils increases as the sugar residue decreases in LPS. The removal of LPS from the rough mutant exposed the OMP, enhancing their presentation to T-cells for recognition, resulting in the production of antibodies with high affinity for OMP, as against the low induction of IgA and cellular responses by the LPS (Hassan, 1989).

The clearance of *S. typhimurium* from the ileum, liver, and spleen coincided with the peak of humoral and cellular responses. The clearance of the challenge strain within one week in the chickens immunized with the rough mutants can be associated with the activation of memory cells within the first three days of challenge leading to the production of highly efficient humoral and cellular response. The observation associated with the nonflagellate mutant should be interpreted with caution because it was produced with a very strong mutagen N-methyl-N-nitrosoguanidine (NTG). (Barrow *et al.*, 1988) which could have produced other unrecognised changes to the phenotype of the organism, in addition to the loss of flagella.

The rough mutant, as demonstrated in this presentation is an effective vaccine for the control of *S. typhimurium* infection in chicken. Its effect on faecal excretion of *S. typhimurium* by chicken will no doubt reduce the risk of contaminated poultry carcass to man. This mutant is also unique in that it has been shown to be avirulent in man. (Barrow *et al.*, In press). The ability of the rough mutant to protect against other *Salmonella* serotypes was not investigated, but the OMP which appears to be the dominant antigen in this mutant is antigenically related to all Enterobacteriaceae.

CONCLUSION

S. typhimurium strains colonize and invade the intestine and visceral organs. Immunization with *S. typhimurium* strains reduced the faecal shedding of challenged strain. All the extracted antigens were immunogenic, but LPS gave low response. The antibody secreting cell assay detected active production of immunoglobulins by plasma cells. The lack of LPS on the rough mutant makes it more susceptible to phagocytes. The exposure of the OMP on the rough mutant makes it more immunogenic. *S. typhimurium* F98 rough mutant is a good vaccine strain.

ACKNOWLEDGMENTS

This work was done at the Department of Microbiology, Institute of Animal Health, Houghton Laboratory, Houghton, U. K.

REFERENCES

1. **Barrow, P.A., J.M. Simpson, and M.A. Lovell.** 1988. Intestinal colonisation in the chicken by food-poisoning *Salmonella* serotypes; microbial characteristics associated with faecal excretion. Avian Pathol. **17(3)**:571-588.

2. **Barrow, P.A., M.B. Huggins, M.A. Lovell, and J.M. Simpson.** 1987. Observations on the pathogenesis of experimental *Salmonella typhimurium* infection in chickens. Res. in Vet. Sci. **42(2)**:194-199.
3. **Gallanos, C., O. Lüderitz, E.T. Rietschel, E.T., and O. Westphal.** 1977. Biochemistry of Lipids II. Int. Rev. Biochem. **14**:239-335.
4. **Hassan, J.O.** 1989. Ph.D. Thesis, University of Birmingham, U.K.
5. **Modrzakowski, M.C., and J.K. Spitznagel.** 1979. Bactericidal activity of fractionated granule contents from human polymorphonuclear leukocytes: Antagonism of granule cationic proteins by lipopolysaccharide. Infect. and Immun. **25(2)**:597-602.
6. **Popiel, I., and P.C.B. Turnbull.** 1985. Passage of *Salmonella enteritidis* and *Salmonella thompson* through chick ileocecal mucosa. Infect. and Immun. **47(3)**:786-792.
7. **Shaio, M-F., and H. Rowland.** 1985. Bactericidal and opsonizing effects of normal serum on mutant strains of *Salmonella typhimurium*. Infect. and Immun. **49(3)**:647-653.
8. **Smith, H.W., and J.F. Tucker.** 1975. The effect of antibiotic therapy on the faecal excretion of *Salmonella typhimurium* by experimentally infected chickens. J. Hyg., Cambridge. **75(2)**:275-292.
9. **Turnbull, P.C.B., and J.E. Richmond.** 1977. The *in vivo* behaviour of salmonellas in the intestinal tract of the chick: an animal model of pathological and public health importance. J. Clin. Pathol. **30**:785.

PRODUCTION OF POULTRY FEED FREE OF *SALMONELLA* AND *ESCHERICHIA COLI*

H. E. Ekperigin

Poultry Consultant
Davis, California 95616

R. H. McCapes

University of California
Davis, California 95616

K.V. Nagaraja

University of Minnesota
St. Paul, Minnesota 55108

R. Redus

VE Corporation
Arlington, Texas 76011

W. L. Ritchie

Farm Service Elevator Company
Willmar, Minnesota 56201

W. J. Cameron

Willmar Poultry Company Inc.
Willmar, Minnesota 56201

Colonization Control of Human Bacterial Enteropathogens in Poultry

ABSTRACT

Efficiency of a new pelleting process in eliminating *Escherichia coli*, salmonellae, and non-lactose-fermenting organisms other than salmonellae (non-lactose-fermenters), was evaluated in eleven trials. The new process consisted of an APC™ System (VE® Corporation, Arlington, TX) and a pellet mill. Instead of being conditioned by steam produced in a boiler, poultry mash pelleted by the new process was first thoroughly mixed in an Original Vertical Conditioner™ with steam and other hot gases generated by direct combustion in a Vaporator®. Efficiency was assessed by comparing the microbial loads in raw and processed mash.

From raw mash, *E. coli* was isolated from 72.0 -100.0% of samples in all eleven trials, salmonellae from 5.0 - 10.0% of samples in three trials, and non-lactose-fermenters from 100.0% of samples in the four trials in which they were cultured. Among trials, the mean number x10^5 of colony-forming units (cfu) of microorganisms per gram of raw mash was 70.0 ± 41.2 and 92.0 ± 56.6 for *E. coli* and non-lactose-fermenters, respectively.

From processed mash, *E. coli* was isolated from samples in four of the eleven trials. Samples were negative for salmonellae and non-lactose-fermenters. Within limitations of the sampling and analytical tests utilized, the new pelleting process appeared to be 100.0% effective against *E. coli*, salmonellae, and non-lactose-fermenters when mash entering the pellet mill from the conditioner had a temperature of 82.9 ± 2.1 ° C, a moisture content of 16.3 ± 1.3%, and had been thoroughly mixed with steam and hot gases for 4.3 ± 0.4 minutes.

INTRODUCTION

Diseases caused by some species of salmonellae can have adverse effects on the economy of the poultry industry. In humans, they are of great public health significance (Bryan *et al.*, 1979; National Academy of Sciences, 1985; Williams *et al.*, 1984). Poultry and other livestock are considered to be important sources of human salmonellae poisoning (Bryan *et al.*, 1979; National Academy of Sciences, 1985).

Egg, environmental, and feed contamination are the main sources of salmonellae infection in poultry. Technology has been available, and utilized, to eliminate the egg and environment as sources of infection (Ghazikhanian *et al.*, 1984; Poss, 1984). On the other hand, technology to eliminate salmonellae from feed has been lacking. Pelleting results in significant reductions in levels of salmonellae in feeds (Bryan *et al.*, 1979; Cox *et al.*, 1986; Harvey, 1972). However, merely reducing the levels of salmonellae in feeds will not ensure elimination of the infection from animals because only as few as one colony-forming unit (cfu) of salmonellae per gram of feed is required to initiate infection (Gangarosa, 1978; Schleifer *et al.*, 1984).

Williams (1978) considered research into methods of producing and feeding salmonellae-free feeds an essential first step in any program to eliminate salmonellae from animals. According to the United States Advisory Committee on *Salmonella* (Dubbert, 1986), production of salmonellae-free feed is dependent on the implementation of research to identify optimum combinations of temperature, heating time, and moisture (optimum TTM) that will kill salmonellae contained in feed. Liu *et al.* (1959) defined optimum TTMs for salmonellae, but lamented the lack of equipment that could achieve the desired temperatures under practical conditions. Cox *et al.* (1986) described attempts to utilize a boiler-conditioner and a pellet mill to eliminate salmonellae from poultry feed. Unfortunately, the steam pressure which was high enough to ensure a consistent elimination of salmonellae was not compatible with the pelleting process because it made mash too moist for pelleting. Recently, an equipment configuration called the APC™ System (VE® Corporation, Arlington, TX) was introduced to the feed industry (Beaumont, 1986; Emnett, 1986). It was claimed to be capable of permitting the attainment of high mash temperatures without causing mash to become too moist for pelleting. The APC™ System consists of a Vaporator® and an Original Vertical Conditioner™. Fuel is ignited and combined directly with water in the Vaporator®, resulting in the generation of steam, nitrogen, carbon dioxide, and a trace of carbon monoxide. The hot gases are channeled directly into the Original Vertical Conditioner™ through an opening in its lower end, and thoroughly mixed by paddles with feed introduced through the top (Fig. 1). This direct utilization of all the hot products of combustion is said to make it possible to control mash temperature independent of moisture levels.

This is a report of experiments conducted to determine the effects of the new pelleting process on the level of contamination of untreated (raw) poultry mash by *Escherichia coli*, salmonellae, and non-lactose-fermenting organisms other than salmonellae.

MATERIALS AND METHODS

Effects of the new pelleting process on microbial load were determined by collecting samples of raw and processed mash during processing and comparing the levels of their contamination by *E. coli*, salmonellae, and other non-lactose-fermenters (*Shigella, Proteus, Edwardsiella*, and *Serratia* species). Two experiments were conducted. The first consisted of seven trials; the second of four trials.

Sampling. Samples of raw mash were collected following batch-mixing, but before entry into one of the 25-ton storage bins above the conditioner (Location A, Fig. 1).

The number of samples collected in each of the eleven trials of the first experiment varied from 14 - 20. Samples of processed mash were collected after the crumbler (Location D, Fig. 1) and consisted of mash which had been conditioned, pelleted, cooled, and sometimes crumbled by the new process.

In the second experiment, ten sets of samples consisting of three subsamples per set were collected at regular intervals during feed processing in each of the four trials. The first and second subsamples were utilized for microbiological analyses, while the third was analyzed for moisture. Samples were collected of raw mash as described earlier (Location A, Fig. 1), conditioned mash (Location B, Fig. 1), and pelleted mash as it exited the pellet mill, but before entering the cooler (Location C, Fig. 1). All samples were placed on ice after collection.

Microbiology. In the first experiment, samples were analyzed for salmonellae and *E. coli* as described earlier (McCapes *et al.*, 1989).

The first sub-samples of feed collected during the second experiment were examined for *E. coli*, salmonellae, and non-lactose-fermenters using the enrichment procedure with tetrathionate brilliant green (42 ° C overnight) and brilliant green agar (37° C for 24 hours) as the enrichment and selective medium, respectively. The population of

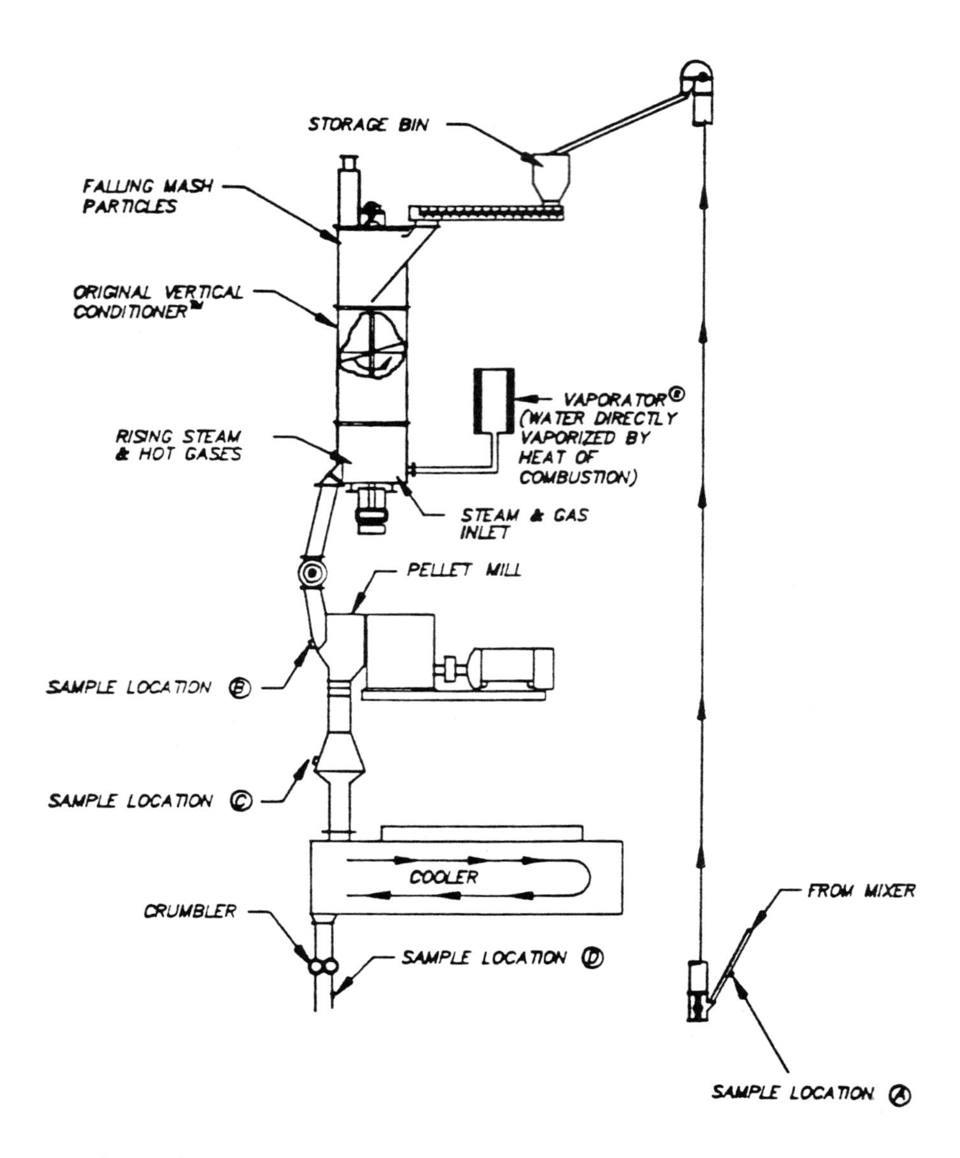

Fig. 1. Diagram of a pelleting process in which the conventional boiler and conditioner have been replaced respectively by a Vaporator® and an Original Vertical Conditioner™.

microorganisms present in feed samples was determined using the standard dilution technique.

The level of contamination of feed by *E. coli*, salmonellae, or non-lactose fermenters was expressed either as the number of cfu's of each type of microorganism per gram, or as the percentage of all examined feed samples that tested positive for the organism. Levels of contamination, before and after conditioning and pelleting were compared. The new pelleting process was regarded as being 100.0% effective against a microorganism if the organism could not be isolated from pellets produced from mash known to be contaminated.

Heating Time, Moisture, Temperature, pH. Heating time was estimated and moisture and temperature of mash determined as described earlier (McCapes *et al.*, 1986). pH was determined using an electronic pH meter.

Statistical Analysis. Data from the first experiment were analyzed by multiple regression (Snedecor and Cochran, 1973) using Stat View 512+ (Brain Power Inc., Calabasas, CA) . The statistical tool utilized in the second experiment included analysis of variance (Snedecor and Cochran, 1973) and Duncan's multiple range test (Steel and Torrie, 1960).

RESULTS

Experiment 1. Results of the first experiment are summarized in Tables 1 and 2. *E. coli* was isolated from 72.0 - 100.0% of raw mash samples in all seven trials. *Salmonella* was isolated from raw mash samples in three of the seven trials, and the rate of isolation was 5.0 - 10.0%. *Salmonella* species isolated included *S. senftenberg* in trials 2 and 6, *S. bredeney* in trial 6, and *S. mbadaka* in trial 7.

Salmonella was not isolated from pelleted mash in any of the seven trials. The new pelleting process was 100.0% efficient in eliminating *Salmonella* from feed. Conditions which ensured the elimination of *Salmonella* from mash used in trials 2 and 6, were inadequate against *E. coli* present in the same mash.

E. coli was isolated from 12.5 - 56.0% of pelleted mash samples in four of the seven trials. Efficiency of the new pelleting process in eliminating *E. coli* from mash used in those trials ranged from 22.0 - 87. 5%. No *E. coli* was isolated from mash pelleted in the other three trials, and efficiency was 100.0%. The heating times, temperatures, and nutrient compositions of mash used in trials in which efficiency was 100.0% are compared in Table 2 to those of mash used in trials in which efficiency was less than 100.0%. Among trials in which efficiency was 100.0%, mean heating time was 4.0 ± 0.7 min and conditioned-mash temperature, moisture, protein, and fat were 84.4 ± 1.3 ° C, 15.4 ± 0.8%, 26.7 ± 11.0%, and 4.9 ± 2.1%, respectively. The corresponding figures for trials in which efficiency was less than 100.0% were 3.7 ± 0.3 min, 86.3 ± 2.6 ° C, 13.4 ± 2.6%, 34.5 ± 7.0%, and 6.0 ± 1.6%. Efficiency thus appeared to vary with changes in levels of moisture, temperature, heating time, or fat and protein contents. Analysis of data by multiple regression showed that the impact of moisture level on efficiency of kill was significant ($P = 0.14$). The regression equation relating efficiency (y) to temperature (x_1), heating time (x_2) moisture (x_3), protein (x_4), and fat (x_5) was:

$$y = 239.42 - 2.56x_1 + 20.47x_2 + 14.74x_3 - 2.87x_4 + 20.17x_5$$

Table 1. Effect of a new pelleting process[1] on the *Salmonella*[2] and *Escherichia coli* contents of poultry mash.

Trial No.	*Salmonella*-Positive Samples		Efficiency of Process[3] Against *Salmonella*	*E. coli*-Positive Samples		Efficiency of Process Against *E. coli*
	Raw	Pelleted		Raw	Pelleted	
1	0	0	NA[4]	72.0	56.0	22.0
2	5.0	0	100.0	85.0	55.0	35.0
3	0	0	NA	100.0	12.5	87.5
4	0	0	NA	100.0	0	100.0
5	0	0	NA	100.0	0	100.0
6	10.0	0	100.0	100.0	20.0	80.0
7	10.0	0	100.0	95.0	0	100.0

[1]New pelleting process consisted an Anaerobic Pasteurizing Conditioning System APC™ System (VE® Corporation, Arlington, TX) and a pellet mill (Buhler-Miag, Minneapolis, MN).

[2]*Salmonella* species isolated included *S. senftenberg* in trials 2 and 6, *S. bredeney* in trial 6, and *S. mbadaka* in trial 7.

[3]Efficiency of the new pelleting process in eliminating the microorganism from mash was expressed as percentage of *Salmonella*- or *E. coli*-positive mash samples that was rendered negative by the process.

[4]Not applicable.

Experiment 2. Results of the second experiment are summarized in Tables 3 and 4. *E. coli* and non-lactose-fermenters were isolated from raw mash in all four trials. The non-lactose-fermenters were confirmed not to be salmonellae. Between trials, the mean number of colony-forming units ($x10^5$) of microbes per gram of raw mash was 70.0 ± 41.2 and 92.0 ± 56.6 for *E. coli*, and non-lactose-fermenters, respectively.

No *E. coli* or non-lactose-fermenters were isolated from conditioned or pelleted mash samples. Microcidal efficiency of the APC™ System by itself or in conjunction with a pellet mill, was estimated to be 100.0% against *E. coli* and non-lactose-fermenters.

Table 2. Experiment 1: Effect of temperature, heating, and moisture, protein, and fat contents of mash on the efficiency of a new pelleting process in eliminating *Escherichia coli* from poultry mash.

Parameter Measured	Efficiency of New Process Against *E. coli*[1]	
	Less than 100%	100%
Number of Trials	4	3
Mean Number of Samples per Trial	18	17
Mean Efficiency (%)	56.1 ± 32.5	100.0 ± 0.0
Retention (heating) Time (min.)	3.7 ± 0.3	4.0 ± 0.7
Temperature of Conditioned Mash (° C)	86.3 ± 2.6	84.4 ± 1.3
Moisture in Conditioned Mash (%)	13.4 ± 2.6	15.4 ± 0.8
Protein Content of Mash(%)	34.5 ± 7.0	26.7 ± 11.0
Fat content of Mash (%)	6.0 ± 1.6	4.9 ± 2.0

[1]Efficiency was expressed as percentage of *E. coli*-positive raw samples that was rendered negative by the new pelleting process. The new process consisted of an APC™ System (VE® Corporation, Arlington, TX) and a pellet mill (Buhler-Miag, Minneapolis, MN). The equation relating efficiency to the physical variables was:
$y = 239.42 - 2.56x_1 + 20.47x_2 + 14.7x_3 - 2.87x_4 + 20.17x_5$, where y is the efficiency of the new pelleting process against *E. coli*, and x_1, x_2,x_3,x_4 and x_5 represent temperature, heating time, moisture, protein, and fat, respectively.

The pH, temperature, and moisture of raw, conditioned, and pelleted mash are compared in Table 4. Differences in pH were not significant. Temperature of raw mash was significantly increased ($P < 0.001$) by the APC™ System, alone or in conjunction with the pellet mill. However, temperatures of conditioned and pelleted mash did not significantly differ from each other. Moisture content of mash was significantly increased ($P < 0.001$) by the APC™ System. Among trials, mean heating time was 4.6 ± 0.5 min, and temperature and moisture content of conditioned mash were 82.9 ± 2.4 ° C and 17.2 ± 1.0%, respectively. The quality of pellets produced in all four trials was good.

Table 3. Experiment 2: Effect of a new pelleting process[1] on the level of contamination of poultry mash by *Escherichia coli* and non-lactose-frmenting bacteria.

		E. coli	Lactose Negative[2]	*Salmonella*
Number (cfu x 10^5) of microorganisms per gram mash[3]	Raw mash	70.0 ± 41.2	92.0 ± 56.6	0
	Conditioned mash	0	0	0
	Pelleted mash	0	0	0
Microcidal Efficiency (%)[4]	APCS™ Alone	100.0	100.0	NA[5]
	APCS™ and Pelleter	100.0	100.0	NA

[1]The new pelleting process consisted of an Anaerobic Pasteurizing Conditioning System (APCS™; VE® Corporation, Arlington, TX) and a pellet mill (Buhler-Miag, Minneapolis, MN).

[2]Lactose-negative organisms. Includes non-lactose-fermenters like *Salmonella, Shigella, Edwardsiella,* and *Serratia* species.

[3]Mean of four trials.

[4]Efficiency of new pelleting process in eliminating a microorganism was expressed as the percentage of the population of that microbe which was killed by processing.

[5]Not applicable.

Table 4. Experiment 2: Effect of an Anaerobic Pasteurizing Conditioning System[1] and pellet mill[2] on the pH, temperature, and moisture content of raw poultry mash.

Parameter Measure[3]	Raw Mash	Conditioned Mash	Pellets
pH	6.14 0.11[a]	6.13 0.14[a]	6.04 0.10[a]
Temperature (C)	18.3 2.3[b]	82.9 2.4[c]	84.3 1.7[c]
Moisture (%)	13.5 0.6[d]	17.2 1.0[e]	13.4 0.4[d]

[1]VE® corporation, Arlington, TX.

[2] Buhler-Miag, Minneapolis, MN.

[3]Mean of four trials. Within rows, figures bearing different superscripts differ significantly (P).

DISCUSSION

The present study describes a new pelleting procedure in which the boiler-conditioner was replaced by an APC™ System, and evaluates its microcidal efficiency. The APC™ System, by itself, or in conjuction with a pellet mill, appeared to be 100.0% efficient in its ability to eliminate *E. coli*, and salmonellae, and other non-lactose-fermenting bacteria from poultry mash when optimum TTM was 82.9 ± 2.1 ° C, 4.3 ± 0.4 min., and 16.3 ± 1.3% moisture. Subsequent to the second experiment, it was speculated that placing hot samples on ice might not have caused the samples to cool rapidly enough even though that portion of our experimental protocol was based on previous reports using cold water (Rasmussen *et al.*, 1964), or an ice water bath (Liu *et al.*, 1969) to cool heated samples. Estimates of cooling rate of mash samples collected on ice were therefore determined in the laboratory and, later on, in the feedmill. In the laboratory, temperature of sample of autoclaved food collected on ice decreased from 79.4 ° C to 42.8 ° C in 49 minutes. In the feedmill, temperature of sample of conditioned mash immersed in ice dropped from 83.9 ° C to 41.7 ° C in 15 minutes. The temperature at, or below, which heat is deemed to have no lethal effect on salmonellae is 46.7 ° C (Liu *et al.*, 1969). Continued exposure to lethal levels of heat for 15-49 minutes after sample collection might have caused the death of any surviving salmonellae and resulted in an overestimation of the efficiency of kill of the new pelleting process. On the other hand, such exposure would have had no impact on estimated efficiency if all salmonellae had been eliminated during processing. The optimum TTM identified under field conditions in our study is quite compatible with those of Liu *et al.* (1969) who worked in a more controlled environment and achieved near-instantaneous cooling of samples. The optimum TTM they identified as being lethal to *S. senftenberg*, the most heat-resistant *Salmonella*, was 87.8 ° C, 1.5 min., and 15.0% moisture, or 89.4 ° C, 0.5 min. and 15.0% moisture. It is, therefore, unlikely that "residual heat" associated with slow cooling of samples would have contributed to the microcidal efficiencies observed in the present study. All the same, concerns raised in the present study about the possible enhancement of microcidal efficiency by "residual heat" associated with slowly cooled samples, indicate a need for standardization of the procedure for handling hot feed samples after collection.

A uniformly high percentage of mash samples from all eleven trials were contaminated by *E. coli*, a common component of the commensal flora of intestinal tracts of animals. However, certain serotypes of *E. coli* are quite pathogenic and associated with diseases that are of great economic importance to the poultry industry (Gross, 1984). They are also of some public health importance (National Academy of Sciences, 1985). The serotypes of *E. coli* isolated in the present study were not determined. However, it is not unreasonable to assume that feed so pervasively contaminated by *E. coli* represents a significant potential health hazard to birds consuming the feed. Since the general cycle of infection of *E. coli* is similar to that of salmonellae, utilization of the new pelleting process to eliminate both microorganisms from poultry feed could enable the production of poultry which are simultaneously free of salmonellae and *E. coli*.

Except for salmonellae, efforts were not made to identify the non-lactose-fermenting microrganisms which were isolated from poultry feed utilized in the present study.

REFERENCES

1. **Association of Official analytical Chemists, Washington, DC.** 1975. Official methods of analysis. 12th ed. W. Horowitz, ed.
2. **Beaumont, T.** 1986. New feed processing system may control *Salmonella*. Poultry and Egg Marketing, p. 3.
3. **Bryan, F. L., M. J. Fanelli, and H. Riemann.** 1979. *Salmonella* infections. *In* Food-borne infections and intoxications. 2nd ed. H. Riemann and F. L. Bryan, eds. pp. 73-130. Academic Press, NY.
4. **Cox, N. A., D. Burdick, J. S. Bailey, and J. E. Thomson.** 1986. Effect of the steam and conditioning and pelleting process on the microbiology and quality of commercial-type poultry feeds. Poultry Sci. **65**:704-709.
5. **Dubbert, W. H.** 1978. Control of *Salmonella* in processing. Proceedings: National Salmonellosis Seminar. U. S. Animal Health Association, Washington, DC.
6. **Emnett, J.** 1986. Poultry feedmills turn to alternative boiler system. Equipment World, p. 1.
7. **Gangarosa, E. J.** 1978. What have we learned from 15 years of *Salmonella* surveillance? Proceedings: National Salmonellosis Seminar. U. S. Animal Health Association, Washington, DC.
8. **Ghazikhanian, G. Y., B. J. Kelly, and W. M. Dungan.** 1984. *Salmonella arizonae* control program. *In* Proceedings: International symposium on *Salmonella*. G. H. Snoeyenbos, ed. American Association of Avian Pathologists, Inc. Kennett Square, P A., p. 142-149
9. **Gross, W. B.** 1984. Collibacillosis. *In* Diseases of Poultry. 8th ed. M. S. Hofstad, H. J. Barnes, B. W. Calnek, W. M. Reid, and H. W. Yoder, Jr., eds., pp. 270-278. Iowa State University Press, Ames, IA.
10. **Harvey, R. W. S.** 1972. *Salmonella* contaminated animal feed in relation to infection in animals and man. *In* The Microbiological Safety of Foods: A Symposium. B. C. Hobbs, and J. H. B. Christian, eds., pp. 9-17. Academic Press, NY.
11. **Liu, T. S., G. H. Snoeyenbos, and V. L. Carlson.** 1969. Thermal resistance of *Salmonella senftenberg* 775W in dry animal feeds. Avian Dis. **13**:611-631.
12. **McCapes, R. H., H. E. Ekperigin, W. J. Cameron, W. L. Ritchie, J. Slagter, V. Stangeland, and K. V. Nagaraja.** 1989. Effect of a new pelleting process on the level of contamination of poultry mash by *Escherichia coli* and *Salmonella*. Avian Dis. **33**:103-111.
13. **National Academy of Sciences.** 1985. Meat and poultry inspection: The scientific basis of the nation's program. pp. 21-42, 68-79. National Academy Press, Washington, DC.
14. **Pomeroy, B. S., K. V. Nagaraja, L. T. Ausherman, I. L. Peterson, and K. A. Friendshuh.** 1989. Studies on feasibility of producing *Salmonella*-free turkeys. Avian Dis. **33**:1-7.
15. **Poss, P. E.** 1984. Cleaning and disinfection programs in the turkey breeder industry. *In* Proceedings: International Symposium on *Salmonella*. G. H. Snoeyenbos, ed. pp. 134-141. American Association of Avian Pathologists, Inc. Kennett Square, PA.
16. **Rasmussen, O. G., R. Hansen, N. J. Jacobs, and O. H. M. Wilder.** 1964. Dry heat resistance of *Salmonella* in rendered animal products. Poultry Sci. **43**:1151-1157.

17. **Schleifer, J. H., B. J. Juven, C. W. Beard, and N. A. Cox.** 1984. The susceptibility of chicks to *Salmonella montivideo* in artificially contaminated poultry feed. Avian Dis. **28:**407-503.
18. **Snedecor, G. W., and W. G. Cochran.** 1973. Statistical methods. 6th ed. pp. 258-338. Iowa State University Press, Ames, IA.
19. **Steel, R. G. D., and J. H. Torrie.** 1960. Principles and procedures of statistics; special reference to the biological sciences. pp. 107-109. McGraw-Hill Book Co., NY.
20. **Williams, J. E., G. H. Snoeyenbos, and B. S. Pomeroy.** 1984. Avian Salmonellosis. *In* Diseases of Poultry. 8th ed. M. S. Hofstad, H. J. Barnes, B. W. Calnek, W. M. Reid, and H. W. Yoder, Jr., eds. pp. 65-140. Iowa State University Press, Ames, IA.

EFFECT OF DIETARY LACTOSE AND ANAEROBIC CULTURES OF CECAL FLORA ON *SALMONELLA* COLONIZATION OF BROILER CHICKS

D. E. Corrier
A. Hinton, Jr.
R. L. Ziprin
R. C. Beier
J. R. DeLoach

Veterinary Toxicology & Entomology Research Laboratory
USDA, Agricultural Research Service
College Station, Texas 77840

ABSTRACT

The effect of dietary lactose and anaerobic cultures of cecal flora on *Salmonella* colonization was evaluated in ten-day-old broiler chicks. One-day-old chicks were divided into groups of 15 chicks each and provided no lactose, 2.5% lactose in water, or 5% lactose in the feed for ten days. All groups were inoculated orally on the day of hatch with anaerobic cultures (AC) of volatile fatty acid producing flora from adult chickens. All chicks were challenged per os at three days of age with 10^6 novobiocin-naladixic-acid-resistant *S. typhimurium*. On day 10, cecal content was evaluated for pH and *S. typhimurium* colonization. The mean $\log_{10}$ number of *Salmonella*/g cecal content decreased from 5.74 to 4.12 in the AC-inoculated chicks. The mean number of *Salmonella* decreased significantly ($P < 0.01$) from 5.74 to 0.36 in AC-inoculated chicks provided 2.5% lactose in water and from 5.36 to 1.82% in AC-inoculated chicks provided 5% lactose in feed. Lactose in the feed or water decreased mean cecal pH by 1.5 units. The results indicate that providing dietary lactose and inoculation with AC of cecal flora reduces cecal pH and significantly decreases cecal colonization by *S. typhimurium*.

INTRODUCTION

The beneficial effects of providing lactose and milk sugar products in poultry diets were observed early in this century (1, 6, 7, 8, 27-30). Dietary lactose was reported to increase the acidity of the cecum (1, 6, 7), help control avian coccidiosis (7, 8) and *Salmonella pullorum* infection (27), and influence the growth, character, and fermentation of normal intestinal flora (1, 29). The importance of normal intestinal flora in the control of invading enteropathogens has been documented in animals (9-13, 19, 20) and by the Nurmi concept of competitive exclusion in poultry (2-4, 16, 17, 21, 24-26, 31, 32). The mechanism by which normal flora decrease intestinal colonization by enteropathogens is not clear (2, 3, 9-12, 19, 20). Among the suggested mechanisms is the production of bacteriostatic short chain volatile fatty acids (VFA), particularly acetic, butyric, and propionic acids, by normal anaerobic flora present in the cecum and colon (3, 9, 10, 19, 20). The bacteriostatic action of VFA's is exerted in the undissociated state and progressively increases as the pH of the cecum decreases below 6.0 (3, 4, 9, 10, 20). The beneficial effects of dietary lactose to control avian cocciciosis and bacillary white diarrhea was originally attributed to increased acidity of the cecal content resulting from lactose fermentation (6, 8). The purpose of the present studies was: 1) to evaluate the effect of dietary lactose and anaerobic cultures of cecal flora on cecal colonization by *S. typhimurium*, and 2) to determine the effect of dietary lactose on cecal pH, VFA, and lactic acid concentrations.

MATERIALS AND METHODS

Salmonella. *Salmonella typhimurium* was obtained from the National Veterinary Services Laboratory, Ames, IA. A novobiocin (NO)-naladixic acid (NA)-resistant strain was selected and maintained on nutrient agar. All culture media used for *Salmonella* was subsequently prepared to contain 25 μg/ml NO and 20 μg/ml NA. Inocula for challenge-exposure was prepared from 18 hr tryptic soy broth cultures. The viable cell concentration of the inocula were determined by colony counts on brilliant green agar.

Anaerobic Cultures of Cecal Flora. The cecal content from adult broiler chickens were combined, immediately transferred into an anaerobic chamber (Coy Laboratory Products, Ann Arbor, MI) mixed with five parts sterile glycerine, and stored at -70 ° C. Chicks were inoculated by crop gavage on the day of hatch with 0.25 ml of 24 hr anaerobic cultures grown in VL broth as previously described (2).

Determination of Cecal pH. Cecal content was suspended in sterile glass distilled water. The pH was determined by direct insertion of glass pH electrode (Orion Research Inc., Cambridge, MA) into the cecal suspension.

Volatile Fatty Acid and Lactic Acid Concentrations of Cecal Contents. Volatile fatty acid and lactic acid concentrations were determined with a gas chromatograph equipped with a flame ionization detector (Shimadzu GC-9A, Shimadzu Corp., Columbia, MO). Peak profiles, integration, and quantification relative to interal standard were obtained with a Shimadzu chromatopac C-R3A integrator.

The concentration of undissociated VFA was determined using the Henderson-Hasselbalch equation: $pH - pK_a = \log_{10} [A^-] / [HA]$, the mean pH of each treatment group, and respective pK_a of each VFA.

Cecal Colonization by *S. typhimurium*. Cecal contents were collected aseptically, serially diluted, and spread-plated on brilliant green agar plates. The plates were incubated for 24 hr at 37 ° C in air, and the number of colony forming units (CFU) *Salmonella*/g cecal content was determined on an automatic colony counter (Biotran III, Automatic Count/Area Totalizer, New Brunswick Scientific Co., Edison, NJ). Suspect *Salmonella* colonies were confirmed by biochemical and serological tests.

To identify *Salmonella* culture-positive chicks, one cecum from each chick was removed aseptically and placed in a stomacher bag (Stomacher Lab-Blender, Tekmar Co., Cincinnati, OH) containing 10 ml sterile distilled water. The bag was blended for 30 sec, 90 ml of selenite-cysteine broth was added and incubated at 37 ° C for 24 hr. After incubation, the broth was streaked on brilliant green agar, incubated at 37 ° C for 24 hr, and the presence of *Salmonella* colonies determined as described above.

Experiment #1. One-day-old male broiler chicks (Hubbard x Hubbard) were purchased from a commercial hatchery, placed in Petersime brooder pens, and provided an unmedicated corn-soybean-meal-based ration and water *ad libitum*. The chicks were randomly assigned into four treatment groups of 15 chicks per group and received: 1) no treatment (Controls), 2) anaerobic cultures of cecal flora, 3) lactose in drinking water, or 4) anaerobic cultures and lactose in water per group, respectively. Groups 2 and 4 were inoculated *per os* on the day of hatch with 0.25 ml of anaerobic culture prepared as described. Lactose was provided to Groups 3 and 4 as 2.5% (w/v) in the drinking water from day of hatch until ten days of age. All chicks were challenged with 10^6 CFU NO-NA resistant *S. typhimurium* by crop inoculation at three days of age. At ten days of age, cecal contents were evaluated for *Salmonella* colonization, pH, VFA (acetic, propionic, butyric), and lactic acid concentrations. The experiment was repeated three times.

Experiment #2. One-day-old male broiler chicks were placed in heated floor pens on pine-shaving litter and provided feed and water as per Experiment #1. All chicks were inoculated *per os* on the day of hatch with 0.25 ml of anaerobic culture of cecal flora prepared as described. The chicks were randomly assigned into four treatment groups of 15 chicks per group and were provided: 1) no lactose (Controls), 2) lactose, 2.5% in water, 3) lactose, 5% (w/w) in feed, or 4) lactose, 10% in feed per group, respectively. Lactose was provided from day of hatch until ten days of age. All chicks were challenged with 10^6 CFU NO-NA resistant *S. typhimurium* by crop inoculation at three days of age. At ten days of age, cecal contents were evaluated for *Salmonella* colonization, pH, VFA, and lactic acid concentrations. The experiment was repeated twice.

Data Analyses and Statistics. Colony plate count data were transformed to logarithmic values and expressed as $\log_{10}$ *Salmonella* per gram of cecal content. Differences among treatment groups were examined for significance with the General Linear Models Procedure and either Duncan's multiple range test or Tukey's studentized range test (18) using commercial statistical analysis software (PCSAS release 6.02, SAS Institute, Cary, NC).

RESULTS

Cecal Colonization by *S. typhimurium*. The mean $\log_{10}$ number of *Salmonella* per gram of cecal content significantly decreased ($P < 0.05$) in the groups of chicks in Experiment #1 that were either inoculated with anaerobic cultures or provided only lactose in the water compared with the control group (Table 1). The mean number of *Salmonella* in the chicks provided lactose in water and inoculated with cecal flora was significantly decreased (P < 0.01) compared with other treatment groups and decreased by more than 5 log units (99.99%) compared with the controls. The percentage of *Salmomella* cecal culture positive chicks decreased from 100% in the control group to 14% in the chicks provided lactose in water and inoculated with cecal flora (Experiment #1, Table 1).

Table 1. Effect of providing lactose in the drinking water and inoculation of anaerobic cultures of cecal flora on *Salmonella typhimurium* colonization of broiler chicks[a].

Treatment Group	$\log_{10}$ *Salmonella*/g Cecal Content	*Salmonella* Positive Cultures/Total (%)
1. Controls	5.74 ± 1.16	10/10 (100)
2. Anaerobic Cultures	4.12 ± 1.55[b]	13/14 (93)
3. Lactose 2.5% in Water	3.96 ± 2.22[b]	9/11 (82)
4. Lactose 2.5% in Water and Anaerobic Cultures	0.36 ± 0.95[c]	2/14 (14)

[a]Lactose was provided from Day 1 to 10 days of age. Chicks that received anaerobic cultures were inoculated orally on day of hatch. All groups were challenged with 10^6 *S. typhimurium* at three days of age.
[b]Significantly different (P < 0.05) from controls.
[c]Significantly different (P < 0.01) from other groups.

In Experiment #2, the mean number of *Salmonella* per gram of cecal content significantly decreased (P < 0.01) by more than 3.5 log unit (99.9%) in the groups of chicks provided lactose in water or feed compared with the control chicks (Table 2). There was no apparent difference in the mean number of colonizing *Salmonella* between the chicks provided lactose in water and those chicks provided lactose in feed. The number of chicks cecal culture positive for *Salmonella* decreased from 100% in the controls to 40% and 53% in the chicks provided lactose (Table 2).

Effect of Dietary Lactose on Cecal pH and Undissociated Volatile Fatty Acid Concentrations. The mean pH of cecal contents decreased by one unit or more in all of the treatment groups provided dietary lactose compared to the control groups in Experiments #1 and #2 (Table 3). The mean cecal pH ranged from 6.1 to 6.5 in the nonlactose control groups and from 4.8 to 5.5 in the groups provided lactose in water or feed. Inoculation of Group 2 (Experiment #1) with anaerobic cultures alone had no apparent effect on cecal pH as compared with the noninoculated controls.

Table 2. Effect of providing lactose in the water or feed and inoculation of anaerobic cultures of cecal flora on *Salmonella typhimurium* colonization of broiler chicks[a].

Treatment Group	Log_{10} *Salmonella*/g Cecal Content	*Salmonella* Positive Cultures/Total (%)
1. Controls	5.36 ± 1.05	15/15 (100)
2. Lactose 2.5% in Water	1.42 ± 1.62[b]	6/15 (40)
3. Lactose 5% in Feed	1.82 ± 2.62[b]	6/15 (40)
4. Lactose 10% in Feed	1.59 ± 1.63[b]	8/15 (53)

[a]Lactose was provided from Day 1 to 10 days of age. All chicks were inoculated orally with anaerobic cultures of cecal flora on day of hatch. All groups were challenged with 10^6 *S. typhimurium* at three days of age.
[b]Significantly different ($P < 0.01$) from controls.

The decrease in cecal pH in the treatment groups provided dietary lactose was accompanied by a marked increase in the concentrations of undissociated acetic, propionic, and butyric acids (Table 3). The concentrations of undissociated acetic increased 5- to 15-fold; propionic acid, 6- to 24-fold; and butyric acid, 2- to 6-fold in the groups provided lactose compared with the control groups.

Effect of Dietary Lactose on Cecal Lactic Acid Concentrations. The concentration of lactic acid markedly increased in both Experiments #1 and #2 in those treatment groups provided lactose in the water or feed (Table 4). Increased lactic acid concentration was associated with decreased cecal pH and the lowest cecal pH values occurred in the treatment groups with the highest concentrations of lactic acid.

Table 3. Effect of dietary lactose on cecal pH and concentration of undissociated volatile fatty acids.

Treatment Group	pH	Undissociated Volatile Fatty Acids (μM/g) Acetic	Propionic	Butyric
	Experiment #1			
1. Controls	6.1	3.63	2.15	0.72
2. Anaerobic cultures	6.2	2.95	1.62	0.59
3. Lactose 2.5% in water	4.8	18.07 (5)[a]	12.47 (6)	0.55
4. Lactose 2.5% in water and anaerobic cultures	5.0	17.00 (5)	13.72 (6)	1.25 (2)
	Experiment #2			
1. Controls	6.5	1.55	0.75	0.30
2. Lactose 2.5% in water	5.0	23.67 (15)	18.08 (24)	1.78 (6)
3. Lactose 5% in feed	5.5	10.60 (7)	6.47 (9)	1.70 (6)
4. Lactose 10% in feed	4.9	19.22 (12)	16.57 (22)	1.48 (5)

[a]Values in parenthesis indicate the fold increase in undissociated volatile fatty acids compared to Control Group.

DISCUSSION

The results of the present study indicated that dietary lactose provided in the feed or water, combined with the establishment of normal cecal flora, decreased the number of *Salmonella* per gram of cecal content by more that 3.5 $\log_{10}$ units (99.9%) in ten-day-old broiler chicks. The number of *Salmonella* cecal-culture positive chicks in both Experiments #1 and #2 was reduced from 100% in the control groups to 53, 40, and 14% in the treatment groups provided lactose and inoculated with cecal flora. In comparison, providing lactose alone or inoculation with anaerobic cultures alone (Experiment #1) decreased the number of *Salmonella* in the ceca by 1.78 and 1.62 $\log_{10}$ units, respectively.

Table 4. Effect of dietary lactose on ceca pH and lactic acid concentration.

Treatment Group	Cecal pH	Lactic Acid (μM/g)
	Experiment #1	
1. Controls	6.1	5.38
2. Anaerobic Cultures	6.2	3.37
3. Lactose 2.5% in Water	4.8	28.10
4. Lactose 2.5% in Water and Anaerobic Cultures	5.0	25.86
	Experiment #2	
1. Controls	6.5	0.00
2. Lactose 2.5% in Water	5.0	17.30
3. Lactose 5% in Feed	5.5	4.59
4. Lactose 10% in Feed	4.9	21.39

The increased susceptibility of newly hatched chicks to *Salmonella* colonization has been attributed to the absence of normal cecal flora (2-5, 17, 21, 24-26, 33-37), and to low concentrations of bacteriostatic VFA produced by anaerobic cecal flora and the high pH present in the ceca of newly hatched chicks (2-5). It has been suggested that successful prevention of *Salmonella* colonization in broiler chicks by competitive exclusion may work by early establishment of normal flora which act to increase the concentrations of VFA and lower the cecal pH (3). Volatile fatty acids, particularly acetic, propionic, and butyric acids, produced by anaerobic bacteria in the cecum and colon, were reported to inhibit *Salmonella* growth *in vitro* and *in vivo* in mice (9, 10, 20). Other studies have reported that VFA's produced by anaerobic cultures of cecal flora from poultry, when coupled with low pH, inhibit *S. typhimurium* growth *in vitro* (3). In chickens two weeks of age and older, the normal levels of VFA's produced by anaerobic bacterial, combined with low cecal pH were reported to inhibit *Salmonella* multiplication (3).

Lactose provided in the feed or water in both Experiments #1 and #2 caused a decrease in cecal pH of 1.0 to 1.6 units. Inhibition of *Salmonella* growth *in vitro* (3) and decreased cecal colonization *in vivo* (2, 3) were reported to be associated with decreased pH. The bacteriostatic action of VFA on enteropathogens is exerted when they are present in the undissociated lipophic state (3, 5, 9, 10, 20). The amount of undissociated VFA progressively increases as the pH of the ceca decreases. The decrease in cecal pH of 1.0 to 1.6 units resulted in a 5- to 24-fold increase in the concentrations of undissociated acetic, propionic, or butyric acids.

Decreased cecal pH in the treatment groups provided lactose was associated with increased concentrations of lactic acid in the cecal contents. The data suggests that decreased cecal pH in the groups provided dietary lactose may have resulted from lactose fermentation and the production of lactic acid. Although the mechanism by which lactose in the diet causes low cecal pH is not confirmed, the results of the present study agree with early reports that lactose and mile sugar products markedly increase the acidity of the cecum (1, 6, 7).

The results of the present experiments indicate that providing lactose in the diet of broiler chicks causes a decrease in cecal pH, and when combined with early establishment of VFA-producing anaerobic flora, markedly increases the concentration of undissociated bacteriostatic VFA's and effectively prevents cecal colonization by *S. typhimurium*.

REFERENCES

1. **Ashcraft, D. W.** 1933. Effect of milk products on pH of intestinal contents of domestic fowl. Ohio Exp. Sta. Bull., Reynoldsburg, Ohio. **XII:**292-298.
2. **Barnes, E. M.** 1979. The intestinal microflora of poultry and game birds during life and after storage. J. Appl. Bacteriol. **46:**407-419.
3. **Barnes, E. M., C. S. Impey, and B. J. H. Stevens.** 1980. Factors affecting the incidence and anti-*Salmonella* activity of the anaerobic cecal flora of the young chick. J. Hyg. Camb. **82:**263-283.
4. **Barnes, E. M. and C. S. Impey.** 1980. Competitive exclusion of *Salmonellas* from the newly hatched chick. Vet. Rec. **106:**61-62.
5. **Barnes, E. M., C. S. Impey, and D. M. Cooper.** 1980. Manipulation of the crop and intestinal flora of the newly hatched chick. Am. J. Clin. Nutr. **33:**2426-2433.
6. **Beach, J. R.** 1925. The effect of feeding *Bacillus acidophilus*, lactose, dry skim milk or whole milk on the hydrogen ion concentration of contents of the ceca of chickens. Hilgardia **1:**145-166.
7. **Beach, J. R., and D. E. Davis.** 1925. The influence of feeding lactose or dry skim milk on artificial infection of chicks with *Eimeria avium*. Hilgardia **1:**167-181.
8. **Beach, J. R., and J. C. Card.** 1925. Studies on the control of avian coccidiosis. Poult. Sci. **4:**83-93.
9. **Bohnhoff, M., C. P. Miller, and W. R. Martin.** 1964. Resistance of the mouse's intestinal tract to experimental *Salmonella* infection. I. Factors which interfere with the initiation of infection by oral inoculation. J. Exp. Med. **120:**805-816.
10. **Bonhoff, M., C. P. Miller, and W. R. Martin.** 1964. Resistance of the mouse's intestinal tract to experimental *Salmonella* infection. II. Factors responsible for its loss following streptomycin treatment. J. Exp. Med. **120:**817-828.
11. **Freter, R.** 1956. Experimental enteric *Shigella* and *Vibrio* infections in mice and guinea pigs. J. Exp. Med. **104:**411-417.

12. **Freter, R.** 1962. *In vivo* and *in vitro* antagonism of intestinal bacteria against *Shigella flexneri*. II. The inhibitory mechanism. J. Infect. Dis. **110:**38-46.
13. **Freter, R.** 1974. Interactions between mechanisms controlling the intestinal microflora. Am. J. Clin. Nut. **27:**1409-1416.
14. **Fuller, R.** 1977. The importance of lactobacilli in maintaining normal microbial balance in the crop. Brit. Poult. Sci. **18:**85-94.
15. **Huttner, B., H. Landgraf, and F. Vielits.** 1981. Dontrolle der Salmonelleinfectionen in Masteltern tierbestanden durch verobreichhung von SPF-darmflora of eintagskuken. Dtsch. Tierarztl. Wschr. **88:**529-534.
16. **Impey, C. S., and G. C. Mead.** 1989. Fate of *Salmonellas* in the alimentary tract of chicks pre-treated with a mature caecal microflora to increase colonization resistance. J. Appl. Bacteriol. **66:**469-475.
17. **Loyd, A. B., R. B. Cumming, and R. D. Kent.** Prevention of *Salmonella typhimurium* infection in poultry by pretreatment of chickens and poults with intestinal extracts. Aust. Vet. J. **53:**82-87.
18. **Luginbuke, R. C., and S. D. Schlozhauer.** 1987. SAS/STAT guide for personal computers. 6th ed. SAS Institute, Cary, NC. pp. 555-573.
19. **Maier, B. R., A. B. Onderdonk, R. C. Baskett, and D. J. Hentges.** 1972. *Shigella*, indigenous flora interactions in mice. Am. J. Clin. Nutr. **25:**1433-1440.
20. **Meynell, G. G.** 1963. Antibacterial mechanisms of the mouse gut. II. The role of EH and volatile fatty acids in the normal gut. Brit. J. Exp. Pathol. **44:**209-211.
21. **Nurmi, E., and M. Rantala.** 1973. New aspects of *Salmonella* infection in broiler production. Nature **241:**210-211.
22. **Oyofo, B. A., J. R. DeLoach, D. E. Corrier, J. O. Norman, R. L. Ziprin, and H. H. Mollenhauer.** 1989. Prevention of *Salmonella typhimurium* colonization of broiler chickens with D-mannose. Poult. Sci. (In press).
23. **Oyofo, B. A., J. R. DeLoach, D. E. Corrier, J. O. Norman, R. L. Ziprin, and H. H. Mollenbauer.** 1989. Effect of carbohydrates on *Salmonella typhimurium* colonization of broiler chickens. Avian Dis. **33:**531-534.
24. **Pivnick, H., B. Blanchfield, and J. Y. D'Aoust.** 1981. Prevention of *Salmonella* infection in chicks by treatment with fecal culture from mature chickens. (Nurmi cultures). J. Food Protect. **44:**909-913.
25. **Pivnik, H. and E. Nurmi.** 1982. The Nurmi concept and its role in the control of salmonellae in poultry. *In* Developments in Food Microbiology, (R. Davis, ed.), Vol. 1, Chap. **2:**41-56.
26. **Rantala, M., and E. Nurmi.** 1973. Prevention of the growth of *Salmonella infantis* in chicks by the flora of alimentary tract of chickens. Brit. Poult. Sci. **14:**627-632.
27. **Rettger, L. F., W. F. Kirkpatrick, and F. H. Stoneburn.** 1912. Bacillary white diarrhea in young chicks. Storrs Agricl Exp. Sta. Bull., Storrs, Conn. **74:**153-162.
28. **Rettger, L. F., W. F. Kirkpatrick, and R. E. Jones.** 1914. Sour milk feeding and its influence on growth and mortality. Storrs Agric. Exp. Sta. Bull., Storrs, Conn. **77:**279-304.
29. **Rettger, L. F.** 1915. The influence of milk feeding on mortality and growth and on the character of the intestinal flora. J. Exp. Med. **21:**365-388.
30. **Rettger, L. F., W. F. Kirkpatrick, and L. E. Card.** 1915. Comparative study of sweet and sour milk. Storrs Agric. Sta. Bull., Storrs, Conn. **80:**1-28.
31. **Seuna, E. M., M. Raevuori, and E. Nurmi.** 1978. An epzootic of *Salmonella typhimurium* var *copenhagen* in broilers and the use of cultures chicken intestinal flora for its control. Brit. Poult, Sci. **19:**309-314.

32. **Snoeyenbos, G. H., O. M. Weinack, and C. F. Smyser.** 1978. Protecting chicks and poults from salmonellas by oral administration of "normal gut microflora." Avian Dis. **22**:273-287.

EVALUATION OF NATURAL *CAMPYLOBACTER JEJUNI* IN BROILER ENVIRONMENTS

W.L. Willis
T.L. Hanner
C. Murray

Department of Animal Science
North Carolina A&T State University
Greensboro, North Carolina 27411

ABSTRACT

Two studies were conducted to examine variations in natural *Campylobacter jejuni* colonization in broilers under different conditions of rearing. Daily high temperatures were obtained from the National Weather Service at the local airport for the implications to the studies. In the first 49-day study, 300 day-old broilers were placed 50 each in six large floor pens with new (3 pens) or used (3 pens) shavings. Weekly sampling revealed no broilers testing positive in this study conducted from November through December, with average monthly high temperatures of 16 ° C and 10 ° C, respectively. In the second study, two 49-day trials using a total of 400 broilers were conducted with battery and floor pen broilers fed no antibiotic, Virginiamycin (20 g/t), Flavomycin (2 g/t), and Bacitracin (50 g/t). Of the 25 broilers per treatment group, broilers tested positive only at seven weeks in the battery pens with Bacitracin (19.0%) and no-antibiotic floor broilers (6.3%) that had a six-week average high of 26 ° C in May. In the second 49-day trial, broilers were positive at four weeks in battery pens with Flavomycin (14.3%), and Bacitracin (4.2%). The average high was 30 ° C for the third week, and broilers appeared hotter in batteries due to stocking density. Only Virginiamycin (5.0%) was positive in the floor pens at the 5th week, with all groups in the batteries positive and the average weekly high being 29 ° C. By the sixth week, the floor pen broilers in all treatments had a substantial rise in positives as well as temperatures from 29 ° C to 31 ° C. These studies suggest that variations in environmental temperatures can influence initial colonization of *Campylobacter jejuni* in broilers. Additionally, antibiotics routinely used in feeds for poultry appear not to offer any preventive protection against colonization of *Campylobacter jejuni*.

Colonization Control of Human Bacterial Enteropathogens in Poultry

INTRODUCTION

Campylobacter jejuni is commonly found in the gastrointestinal tract of all classes of poultry. Many reports in the scientific literature have shown that this organism causes infection and disease in humans. Isolations have been made from poultry throughout the growing process, but many flocks escape infection. Although many important aspects of the transmission of *Campylobacter jejuni* remain unanswered, a great deal of progress has been made in understanding of reservoirs and prevalence of infection. The Centers for Disease Control (1977, 1984) have identified poultry products as a major source of contamination, leading to human illness. The organism has regularly been isolated from raw poultry carcasses, parts and edible viscera during processing and is present in chilled poultry products at the retail level (Acuff *et al.*, 1986; Stern *et al.*, 1985; Grant *et al.*, 1980; Lovett *et al.*, 1981). Much knowledge is to be gained on how we can commercially raise *Campylobacter*-free chickens. It is clear that much work is needed because the epidemiology of *Campylobacter jejuni* colonization in poultry at the farm is unclear. Many of the factors that support natural colonization of *Campylobacter jejuni* have not been adequately identified. The present study was undertaken to evaluate factors that influence the natural colonization of broiler chickens and to determine the effectiveness of three antibiotics routinely used in poultry rations fed to broilers in batteries and floor litter pens.

MATERIALS AND METHODS

In study 1, 300 day-old Ross X Arbor Acres broiler chicks were obtained from a local hatchery and divided equally (50/pen) in six large (6.1 X 6.1 m) pens. Three pens contained new wood shavings and the other three contained old woodshavings from a flock that was colonized with *Campylobacter jejuni* in an earlier study. Chicks were cloacal swabbed weekly using the culturette collection and transport system. Five chicks per pen were sacrificed weekly and intestinal swabs of the ceca and small intestine were taken for early detection of *Campylobacter jejuni* in the gastrointestinal tract. Samples were immediately transported to the lab and plated directly onto Campylobacter blood agar (BBL prepared media). All plates were incubated at 42 ° C for 72 hours in anaerobic jars containing 5% O_2, 10% CO_2, 85% N_2. Smears from colonies on the blood agar were examined by phase contrast microscopy for typical morphology and motility. Colonies were further confirmed as *Campylobacter jejuni* by the following biochemical tests: Oxidase and catalase production, hippurate hydrolysis, growth temperature, H_2S production, inhibition of 3.5% NaCl, and sensitivity to nalidixic acid.

In study 2, 400 day-old Ross X Arbor Acres broiler chicks were used in two 7-week batteries vs. litter rearing trials with medicated feeds. Chicks were divided into four groups (25 each) in a battery cage and in floor pens containing new wood shavings in trial 1 and old shavings in trial 2. Groups were given either no antibiotic, Virginiamycin (20g/t), Flavomycin (2g/t) or Bacitracin MD (50g/t). Chicks were transferred to finisher batteries at three weeks of age. Brooder heat was turned off at three weeks in the floor rearing broilers. All heating and cooling after three weeks came from mother nature.

Climatological data were obtained from the National Weather Service at the local airport. Sampling, isolation and identification of *Campylobacter jejuni* follow the same procedures as in experiment 1.

RESULTS AND DISCUSSION

Experiment 1. In this study, no broilers tested positive for *Campylobacter jejuni* for the entire seven-week rearing period. Weekly samples taken from the small intestine, ceca, and the cloaca did not reveal positive cultures. Apparently, broilers did not colonize during the period from November through December, probably because of lower average monthly high temperatures of 16 ° C and 10 ° C. In a prior study conducted in the same house during the months of September and October, birds became colonized with *Campylobacter jejuni* that had average monthly temperatures of 24.4 ° C and 17.8 ° C. Litter that was used from broilers previously colonized with *Campylobacter jejuni* did not aid in the colonization of broilers as suggested by a previous report by Genigeorgis *et al.* (1986). Similar findings are cited with other reports that indicate many flocks escape infection from *Campylobacter jejuni*. This may be due to inadequate environmental temperatures needed to warm the rearing environment to stimulate the growth of this organism in the gastrointestinal tract of the chicken.

Experiment 2. The percentage of broilers testing positive for *Campylobacter jejuni* in batteries and floor reared broilers on the same farm are shown in (Table 1). The results suggest that rearing environment is not a major factor influencing initial colonization in young broilers, except for temperatures. Only a few broilers became positive for *Campylobacter jejuni* at the seventh week in the first trial. Broilers tested positive in the battery with Bacitracin having (19.0%) and no antibiotic floor broilers (6.3%). The six-week average high temperature was 26 ° C in May. In the second trial, broilers were positive at four weeks in battery pens with Flavomycin (14.3%), and Bacitracin (4.2%). The average high for the third week was 30 ° C in June, and in both trials broilers appeared hotter in the batteries which may have been due to stocking density and building design. The remaining groups in the batteries tested positive at the 5th week as shown in (Table 1). Only Virginiamycin (5.0%) was positive in the floor pens at week 5, which had an average four-week high of 29 ° C. By the sixth week, the floor pen broilers in all treatments had a substantial rise in positive samples as well as temperature from 29 ° C to 31 ° C.

These results imply that seasonal and conditional house temperatures are major factors in *Campylobacter jejuni* colonization. This is in agreement with an earlier suggestion from a report of Blaser *et al.* (1983). In addition, Genigeorgis *et al.* (1986) observations seem to suggest a seasonal and conditional relationship in the colonization of broiler chickens. *Campylobacter jejuni* is very sensitive to ambient temperatures. It appears to colonize poorly under low temperatures, which may contribute epidemiologically in the colonization process with broiler chickens. The rearing environment must have a temperature high enough to give an increase in the body temperature of chickens of certain ages to stimulate *Campylobacter jejuni* incubation. Several conclusions can be drawn from these studies. Litter does not appear to aid broilers in the colonization or transmission of *Campylobacter jejuni*. Secondly, there appears to be a very strong relationship between the seasonal ambient environmental air temperature and *Campylobacter jejuni* initial colonization with warmer temperatures favoring colonization.

Table 1. Occurrence of *Campylobacter jejuni* in broiler chickens given medicated feeds in batteries and floor trials.[1]

Treatment	WK4	WK5	WK6	WK7
		TRIAL 1 [April (19.8 ° C) - May (23.6 ° C)][2]		
		CAGE		
No Antibiotic	0/24(0)	0/23(0)	0/21(0)	0/20(0)
Virginiamycin	0/20(0)	0/19(0)	0/19(0)	0/18(0)
Flavomycin	0/22(0)	0/22(0)	0/21(0)	0/20(0)
Bacitracin MD	0/23(0)	0/23(0)	0/22(0)	4/21(19.0)
		FLOOR		
No Antibitotic	0/17(0)	0/16(0)	0/16(0)	1/16(6.3)
Virginiamycin	0/22(0)	0/22(0)	0/22(0)	0/22(0)
Flavormycin	0/19(0)	0/19(0)	0/19(0)	0/18(0)
Bacitracin MD	0/18(0)	0/18(0)	0/18(0)	0/17(0)
		TRIAL 2 [June (29.4 ° C) - July (30.4 ° C)][2]		
		CAGE		
No antibiotic	0/24(0)	7/23(30.4)	17/23(73.9)	23/24(95.8)
Virginiamycin	0/22(0)	9/22(40.9)	17/21(80.9)	19/20(95.0)
Flavomycin	3.21(14.3)	7/21(33.3)	14/21(66/6)	18/20(90.0)
Bacitracin MD	1.24(4.2)	11/24(45.8)	21/24(87.5)	22/24(91.7)
		FLOOR		
No antibiotic	0/20(0)	0/21(0)	16/21(76.2)	19/20(95.0)
Virginiamycin	0/21(0)	1/20(5.0)	18/20(90.0)	18/20(90.0)
Flavomycin	0/25(0)	0/25(0)	22/23(95.6)	23/23(100)
Bacitracin MD	0/23(0)	0/23(0)	22/23(95.6)	21/22(95.4)

[1] No. Positive / No. Sampled (%).
[2] Represents average monthly maximum temperatures.

Finally, antibiotics used in this study did not prevent birds from becoming colonized with *Campylobacter jejuni* in batteries or floor-reared broilers. Further studies are needed to evaluate the influence of temperature in environmentally controlled rearing environments.

REFERENCES

1. **Acuff, G. R., C. Vanderzant, M. 0. Hanna, J. G. Ehlers, and F. A. Gardner.** 1986. Effects of handling and preparation of turkey products on survival of *Campylobacter jejuni*. J. Food Prot. **49:**627-631.
2. **Blaser, M. J., D. N. Taylor, and R. A. Feldman.** 1983. Epidemiology of *Campylobacter jejuni* infections. Epid. Rev. **5:**157-176.
3. **Centers for Disease Control.** 1977. Morbidity and mortality. Weekly Rep. Annu. Summ. as cited by Silliker. J. H., 1980. Status in Salmonellae ten years later. J. Food Prot. **43:**307-313.
4. **Centers for Disease Control.** 1984. Surveillance of the flow of *Salmonella* and *Campylobacter* in a community. Communicable Disease Control Section. Seattle-King County Department of Public Health. Communicable Disease Control Section Cent. Vet. Med. Fed Drug Admin., Arlington, VA.
5. **Genigeorgis, C., M. Hassuneh, and P. Collins.** 1986. *Campylobacter jejuni* infection on poultry farms and its effect on poultry meat contamination during slaughtering. J. Food Prot. **49:**895-903.
6. **Grant, I.H. N., N. J. Richardson, and N. D. Bokkenheuser.** 1980. Broiler chickens as potential sources of *Campylobacter* infections in humans, J. Cm. Microbiol. **11:**508-510.
7. **Lovett, J., J. Hunt, and C. L. Park.** 1981. Incidence of *Campylobacter fetus* ss. *jejuni* in southern Ohio fresh whole (eviscerated) market chickens. Page 11 m. Proc. 81st Ann. Meet., Am. Soc. Microbiol., Dallas, TX. March.
8. **Stern, N. J., M. P. Hernandez, L. C. Blankenship, K. E. Deibel, S. Doores, M. P. Doyle, H. Ng, M. D. Pierson, J. N. Sofos, W. H. Sveum, and D. C. Westhoff.** 1985. Prevalence and distribution of *Campylobacter jejuni* and *Campylobacter coli* in retail meats. J. Food Prot. **48:**595-599.

CHARACTERISTICS OF *SALMONELLA* ISOLATED IN 1989 FROM GEORGIA POULTRY

W. D. Waltman
A. M. Horne

Georgia Poultry Laboratory
Oakwood, Georgia 30566

ABSTRACT

An important part of any *Salmonella* or other bacterial control program is an understanding of the characteristics or types of microorganisms present. As part of our *Salmonella* monitoring program, we have isolated 209 *Salmonella* from Jan - July 1989. These have included ten somatic serogroups and 26 serotypes. The most frequently isolated serotypes were *johannesburg* (44) and *heidelberg* (39). Utilizing the API-20E system, the isolates were positive for ornithine decarboxylase, glucose, mannitol, rhamnose, and arabinose fermentation; were negative for ONPG, urea, tryptophane deaminase, indole, gelatinase, VP, and oxidase; and variable for arginine dehydrolase, lysine decarboxlase, citrate, H_2S, inositol, sorbitol, melibiose, salicin, and amygdalin fermentation. The antimicrobial susceptibility pattern showed that >90% of the isolates were susceptible to chloramphenicol, SXT, nalidixic acid, nitrofurantoin, gentamicin, neomycin, and ampicillin; 80 - 90% were susceptible to chlortetracycline, oxytetracycline, and triple sulfa; all were resistant to erythromycin and penicillin G.

Colonization Control of Human Bacterial Enteropathogens in Poultry

INTRODUCTION

Salmonella is of major importance to the poultry industry not only because of the direct economic losses due to morbidity and mortality, but also indirectly due to economic losses suffered by its association with human diseases and its perception by the public. The industry has successfully eliminated *S. pullorum* and *S. gallinarum* from commercial flocks and is aware of the need to control, and if possible, eliminate other *Salmonella* serotypes. For a control or eradication program to be successful, there must exist a means of monitoring the presence of *Salmonella* and the effectiveness of the control measures. As a state laboratory, which is also responsible for coordinating the NPIP, we have an ongoing monitoring program for *Salmonella*. We operate for the benefit of the poultry industry by notifying companies of potential problems with *Salmonella* and thus hopefully heading off major disease losses.

The data presented here summarizes the results of our *Salmonella* monitoring program for the first seven months of 1989.

MATERIALS AND METHODS

As part of our *Salmonella* monitoring program we routinely culture four types of samples: 1) diagnostic cases, 2) pullorum-typhoid reactors, 3) day-old chicks, and 4) environmental samples.

We have adapted the isolation procedure of Dr. Ed Mallinson. The samples are added to tubes of tetrathionate (TT) broth at a 1:10 ratio and incubated overnight at 37 ° C. They are plated onto Brilliant Green Agar (BGA) and BGA plus 20 μg/ml novobiocin (BGAN). The TT broth is left at room temperature for five days and then 1 ml transferred to a fresh tube of TT broth. This tube is incubated overnight at 37 ° C and then plated onto BGAN. The agar plates are incubated overnight at 37 ° C and observed for characteristic colonies . Routinely, three colonies are picked to TSI and then serogrouped using somatic antisera. The isolate (only one per case) is identified further using the API-20E system and an antimicrobial susceptibility pattern is determined using the standard Kirby-Bauer disk method. The isolate is sent to NVSL for serotyping.

RESULTS AND DISCUSSION

In the first seven months of 1989, we have made 209 independent *Salmonella* isolations. The sources of these isolations have been from the following: diagnostic cases (16.2%), pullorum-typhoid reactors (9.0%), day-old chicks (45.4%), fluff (7.1%), egg swabs (6.7%), drag swabs (7.2%), dead-in-shell embryos (5.3%), and other (2 . 7%).

The somatic serogroups of these isolates are shown in Table 1. Serogroups B, R, Dl, and Cl have been most prevalent. In the last ten years in our laboratory, the most frequently isolated *Salmonella* have been serogroups B (29.5%), Cl (28.5%), and R (17.8%).

The serotype of each *Salmonella* isolate is shown in Table 2 by the month it was isolated. The most commonly isolated serotypes were *johannesburg, heidelberg, enteritidis,*

Table 1. Serogroups of *Salmonella* isolated each month.

	% of Isolates*							
Serogroup	Jan (14)	Feb (21)	Mar (10)	Apr (25)	May (56)	Jun (30)	Jul (53)	Total (209)
B	35.7	38.1	20.0	20.0	35.7	43.3	13.2	28.7
C1	42.8	23.8	60.0	8.0	7.1	6.7	9.4	14.4
C2	21.4	4.8	10.0	4.0	3.6	0.0	1.9	4.3
C3	0.0	4.8	0.0	4.0	0.0	3.3	7.5	3.3
D1	0.0	0.0	0.0	8.0	35.7	6.7	20.8	16.7
E1	0.0	4.8	10.0	4.0	0.0	0.0	0.0	1.4
E2	0.0	0.0	0.0	4.0	0.0	0.0	0.0	0.5
E4	0.0	9.5	0.0	12.0	0.0	0.0	3.8	3.3
G2	0.0	0.0	0.0	0.0	0.0	0.0	5.7	1.4
R	0.0	9.5	0.0	36.0	14.3	33.3	28.3	21.0
?	0.0	4.8	0.0	0.0	3.6	6.7	9.4	4.8

* Number of *Salmonella* isolated each month is in parenthesis.

pullorum, and *typhimurium*. This compares to the most common serotypes over the last ten years in our laboratory; *johannesburg* (16.7%), *heidelberg* (14.3%), *infantis* (7.7%), and *typhimurium* (5.7%) .

Table 3 lists the biochemical characteristics of the *Salmonella* isolates using the API-20E system. Variability of 20% or greater was found for arginine, lysine, and inositol. Variability of 10 - 20% was found for citrate, H_2S, and melibiose. Of concern are the variabilities in lysine, citrate, and H S reactivity, since these are typically considered characteristic for *Salmonella*. In particular, lysine and H_2S reactivity has been involved in the development of differential isolation media for *Salmonella* such as xylose-lysine-desoxycholate agar and *Salmonella-Shigella* agar. These atypical strains of *Salmonella* may behave differently on these media and thereby be missed.

Table 2. Serotypes of *Salmonella* isolated each month.

Serotypes	% of Isolates*							
	Jan (14)	Feb (21)	Mar (10)	Apr (25)	May (56)	Jun (30)	Jul (53)	Total (209)
johann	0.0	9.5	0.0	36.0	14.3	33.3	28.3	21.1
heidel	7.1	28.6	20.0	8.0	32.1	26.7	3.8	18.7
enterit	0.0	0.0	0.0	0.0	32.1	0.0	0.0	8.6
pullorum	0.0	0.0	0.0	0.0	3.6	6.7	20.8	7.1
typhimur	14.3	0.0	0.0	12.0	0.0	13.3	9.4	6.7
tennessee	0.0	23.8	40.0	4.0	0.0	0.0	1.9	5.2
thompso	14.3	0.0	10.0	4.0	0.0	0.0	7.5	3.8
hadar	21.4	4.8	10.0	4.0	0.0	0.0	1.9	3.3
senften	0.0	9.5	0.0	12.0	0.0	0.0	3.8	3.3
montevid	0.0	0.0	0.0	0.0	7.1	6.7	0.0	2.9
kentucky	0.0	4.8	0.0	4.0	0.0	0.0	3.8	1.9
haardt	0.0	0.0	0.0	0.0	0.0	3.3	3.8	1.4
worthing	0.0	0.0	0.0	0.0	0.0	0.0	5.6	1.4
livings	14.2	0.0	0.0	0.0	0.0	0.0	0.0	0.9
califor	0.0	4.8	0.0	0.0	1.8	0.0	0.0	0.9
berta	0.0	0.0	0.0	8.0	0.0	0.0	0.0	0.9
blockley	0.0	0.0	0.0	0.0	3.6	0.0	0.0	0.9
oranien	14.3	0.0	0.0	0.0	0.0	0.0	0.0	0.9
typh (cop)	0.0	0.0	0.0	0.0	0.0	3.3	0.0	0.5
indiana	7.1	0.0	0.0	0.0	0.0	0.0	0.0	0.5
mbanda	0.0	0.0	10.0	0.0	0.0	0.0	0.0	0.5
give	0.0	0.0	10.0	0.0	0.0	0.0	0.0	0.5
new bruns	0.0	0.0	0.0	4.0	0.0	0.0	0.0	0.5
anatum	0.0	0.0	0.0	4.0	0.0	0.0	0.0	0.5
bredeney	7.1	0.0	0.0	0.0	0.0	0.0	0.0	0.5
london	0.0	4.8	0.0	0.0	0.0	0.0	0.0	0.5
unknown	0.0	9.5	0.0	0.0	5.4	6.7	9.4	5.7

*Number of *Salmonella* isolated each month is in parenthesis.

Table 3. Biochemical characteristics of *Salmonella* isolates using the API-20E system.

Substrate	% Positive* Jan (14)	Feb (21)	Mar (10)	Apr (25)	May (56)	Jun (30)	Jul (53)	Total (209)
ONPG	0.0	0.0	0.0	0.0	0.0	0.0	1.8	0.5
Arg	92.8	76.2	100.0	32.0	32.5	36.7	43.3	48.8
Lys	100.0	81.0	50.0	84.0	75.0	56.7	81.1	76.0
Orn	100.0	100.0	100.0	100.0	100.0	100.0	100.0	100.0
Citrate	92.8	85.7	80.0	88.0	85.7	90.0	83.0	86.1
H2S	100.0	95.2	100.0	100.0	69.6	73.3	83.0	83.2
Urea	0.0	0.0	0.0	0.0	0.0	0.0	0.0	0.0
TDA	0.0	0.0	0.0	0.0	0.0	0.0	0.0	0.0
Indole	0.0	0.0	0.0	0.0	0.0	0.0	0.0	0.0
VP	0.0	0.0	0.0	0.0	1.8	0.0	0.0	0.5
Gelatin	0.0	0.0	0.0	0.0	0.0	0.0	0.0	0.0
Glucose	100.0	100.0	100.0	100.0	100.0	100.0	100.0	100.0
Man'l	100.0	100.0	100.0	100.0	100.0	100.0	100.0	100.0
Inos'l	50.0	33.3	80.0	32.0	5.4	20.0	18.8	23.4
Sorb'l	100.0	100.0	100.0	100.0	96.4	93.3	81.1	93.3
Rham	100.0	100.0	100.0	100.0	100.0	100.0	96.2	99.0
Salicin	0.0	0.0	0.0	0.0	0.0	3.3	1.8	1.0
Melo	71.4	100.0	100.0	100.0	64.3	93.3	81.1	82.8
Amygd	0.0	0.0	0.0	0.0	1.8	3.3	0.0	1.0
Arab	100.0	100.0	100.0	100.0	100.0	100.0	100.0	100.0
Ox	0.0	0.0	0.0	0.0	0.0	0.0	0.0	0.0

* Number of *Salmonella* isolated each month is in parenthesis.

The antimicrobial susceptibility pattern of the *Salmonella* isolates are shown in Table 4. With exception of erythromycin and penicillin G, which generally have little activity against gram negative bacteria, resistance of 10% or greater was found only for chlortetracycline, oxytetracycline, and sulfa. All total, 7.7% of the isolates were resistant to one antimicrobial, 10.2% resistant to two, 2.5% were resistant to three, 1.3% were resistant to four, and 5.1% were resistant to five antimicrobials.

From 1979-1988, our laboratory isolated only 512 *Salmonella*. After the adoption of Dr. Mallinson's techniques earlier this year, our isolations increased significantly. For example, of 2325 cultures examined in one study, 128 were positive after 24 hours but 182 were positive after extended incubation. The 24 hour plating recovered 33 *Salmonella* not detected with the extended incubation, but 87 *Salmonella* were recovered only after extended incubation; therefore indicating the need for both culture.

Table 4. Antimicrobial susceptibility of *Salmonella* isolates by month of isolation.

Antibiotic	% Susceptibility*							
	Jan (14)	Feb (21)	Mar (10)	Apr (25)	May (56)	Jun (30)	Jul (53)	Total (209)
Chloram	100.0	100.0	100.0	96.0	100.0	100.0	100.0	99.5
Oxytet	28.6	76.2	80.0	72.0	98.2	100.0	94.3	86.6
SXT	92.8	100.0	100.0	100.0	100.0	100.0	100.0	99.5
Chlortet	14.3	76.2	70.0	60.0	96.4	90.0	88.6	80.4
Nalidix	100.0	100.0	90.0	100.0	100.0	100.0	100.0	99.5
Sulfa	71.4	76.2	80.0	76.0	96.4	100.0	94.3	89.5
Nitrofur	92.8	90.5	90.0	92.0	100.0	96.7	100.0	96.7
Gent	85.7	90.5	80.0	88.0	98.2	100.0	100.0	95.2
Neo	85.7	90.5	90.0	96.0	100.0	100.0	98.1	96.7
Erythro	0.0	0.0	0.0	0.0	0.0	0.0	0.0	0.0
Pen G	0.0	0.0	0.0	0.0	0.0	0.0	0.0	0.0
Amp	92.8	100.0	90.0	92.0	100.0	96.7	100.0	97.6

*Number of *Salmonella* isolated each month is in parenthesis.

SUMMARY

1. Paramount to a good *Salmonella* monitoring program is the development of sensitive methods for recovering the organism. The present procedures using novobiocin in the plating medium along with extended incubation has greatly increased our recovery of *Salmonella*.

2. With the exceptions of *enteritidis* and *pullorum* (serogroup DI) the common serogroups and serotypes have been fairly constant over the last ten years.

3. The increased isolation of *pullorum* has resulted from the importation of "backyard-type" birds from hatcheries in another state, which was picked up by our monitoring program.

4. The biochemical characteristics showed some variability which would be expected with this many isolates. However, the presence of lysine or H S negative strains (by API-20E) should be of concern to those who utilize these characteristics for recovery.

5. The antimicrobial susceptibility pattern indicates a high degree of susceptibility to the antimicrobials tested, especially in those strains isolated in late spring and summer.

EFFICACY OF UNDEFINED AND DEFINED BACTERIAL TREATMENT IN COMPETITIVE EXCLUSION OF *SALMONELLA* FROM CHICKS

S. Stavric
T.M. Gleeson
B. Blanchfield

Bureau of Microbial Hazards
Health Protection Branch
Health and Welfare Canada
Ottawa, Ontario KlA OL2

ABSTRACT

The intestinal microflora from adult birds (undefined treatments) protect newly hatched chicks against subsequent challenge with *Salmonella*. Such treatments, prepared from fecal and cecal contents, whole ceca or whole intestines have been administered to chicks by gavage, via drinking water or by spraying. In this study we evaluated the efficacy of various undefined treatments administered to chicks by five different methods. We also examined factors affecting the efficacy of defined bacterial mixtures. Treatments were given to one-day-old chicks and tested for their ability to protect chicks against 10^5 colony forming units of *Salmonella typhimurium*. Undefined treatments prepared from cecal homogenate were the most protective; those from subcultured fecal material, the least. Administration of treatment by gavage and by a newly developed agar-plate method were the most effective. In general, undefined treatments were always more effective than defined mixtures. Mixtures containing strains from a single genus were ineffective in comparison with mixtures containing strains from different genera. The efficacy of the mixtures, prepared from stored isolates, gradually decreased as the time of storage increased.

Colonization Control of Human Bacterial Enteropathogens in Poultry

INTRODUCTION

The protection of chicks against infection with *Salmonella,* offered by native gut microflora from *Salmonella*-free adult birds was first shown by Nurmi and Rantala (7) and later confirmed by others. Cecal or fecal contents from adult birds, cultured in bacteriological media, are common sources of protective microflora. Serial subculturing of this material should eliminate potential parasites and viruses (9). Whole ceca or whole intestines from specific-pathogen-free (SPF) birds have also been used (4). In most laboratory experiments these treatments have been administered to chicks orally (gavage), but in large trials they were provided via drinking water (14) or through spraying in hatcheries (4). Treatments were also administered to chicks in shipping boxes via agar plates (10). Comparative results on the efficacy of various administration methods, and on the protective activity of treatments from various sources have not been reported.

The bacterial composition of the above treatments is not defined. Therefore, attempts have been made in several countries to develop an effective treatment prepared from pure bacterial cultures (12). Although substantial progress has been achieved, defined culture(s) with a potency equivalent to that of undefined has not been developed (6).

The objectives of the present study were: l) to compare the efficacy of different undefined treatment materials, and of various methods for their administration, and 2) to evaluate the efficacy and stability of pure culture mixtures. These were tested in one-day-old chicks against a challenge of 10^5 colony forming units (CFU) of *Salmonella typhimurium*

METHODS

Undefined Treatments. Fecal suspension and fresh fecal culture: Fresh fecal material from four adult *Salmonella*-free birds was used to prepare 5% fecal suspension in VL medium (1). After gauze-filtration, it was either used directly as a treatment or 2 ml of this suspension, representing 10^{-1} fecal material, was inoculated into VL medium and grown at 37 ° C for three days.

Subcultured fecal culture: The fecal culture was subcultured in VL medium at 2% inoculum (each subculture was grown for one or two days at 37 ° C before the next subculture was prepared).

Cecal homogenate: Whole ceca (cecal content and cecal pouches) from donor birds were homogenized in VL medium (1:6) in a Sorval Omni-mixer for 2 min (16,000 rpm). Aliquots containing 10% glycerol were stored at -70 ° C. For treatment of chicks, the homogenates were quickly thawed and gauze-filtered. All undefined treatments were prepared and cultured under anaerobic conditions.

Defined Treatments. Intestinal material from donor birds was used as a source for the isolation of bacterial cultures (3, 13). Pure cultures in VL medium containing 10% glycerol were stored at -70 ° C. For each experiment, the isolates were revived by rapidly thawing frozen cultures. Each isolate was streaked on a separate Columbia blood agar (CBA) plate, which was incubated anaerobically for two days at 37 ° C. Approximately the same amount of growth of each isolate on CBA plates was swabbed into a single bottle of VL medium and incubated as a mixture for 24 h at 37 °C before administration to chicks (3).

Administration Methods. The five methods of administering treatments to newly hatched chicks were: gavage, drinking water (9), spraying in hatchers (4), via agar plates (10), and via feed slurry. In the plate methods, treatments were either spread on the surface of CBA plates (spread-plate) and grown for three days before administration, or treatments were incorporated into CBA (pour-plate) and administered without growth. These plates were inserted into shipping boxes which were covered for 2 - 4 hours, or until agar has been consumed by chicks. The feed slurry was prepared under anaerobic conditions by combining non-medicated feed with water (1:6) and adjusting to pH 6.6 before autoclaving. For administration to chicks, feed slurry was thoroughly mixed with treatments (20:1) and spread on the bottom of shipping boxes.

Laboratory Trials. Chicks (White Rock and Cornish) were from the same commercial supplier as described previously (13). Only for trials in which spraying method was used, chicks were hatched in our facilities. In general, 20 birds per treatment were housed, ten per cage, in wire-floored starter brooders where non-medicated chick-starter feed and water were always available.

To one-day-old chicks (day of hatch), defined treatments were administered by gavage, while for undefined treatments several other administration methods were also used. Chicks were challenged two days later in the drinking water with 10^5 CFU of nalidixic-acid-resistant *Salmonella typhimurium* per chick. Eight days after administration of treatment, chicks were sacrificed and their cecal contents examined for *Salmonella* (2). Results were expressed as Infection Factor (IF) and Protection Factor (PF) (8).

RESULTS AND DISCUSSION

Undefined Treatments. Multiple subculturing of fecal material resulted in a progressive decrease of protective activity. This effect was noticed only after an appropriate dilution of treatment (1/50) was administered to chicks. When compared with other undefined treatments, subcultured fecal material was the least, and cecal homogenate the most, protective.

Results on the effect of various administration methods on the protective activity of fecal culture and subculture, fecal suspension and cecal homogenate showed that the gavage method was the best for all treatments. The newly developed pour-plate method was almost as effective as gavage. It should have an advantage over the gavage method since plates can be prepared before the hatch, and the chicks treated in shipping boxes during transport from hatcheries to poultry farms. Spraying of chicks was the least effective; poor protection was obtained with all treatments regardless of whether the chicks were sprayed when 50% or 100% of them were hatched. In summary, cecal homogenate was the best treatment material, while gavage and pour-plates were the best administration methods (Table 1).

Defined Treatments. Our previous results showed that several mixtures of pure bacterial cultures reduced *Salmonella* infection in chicks (3, 11, 13). In this study new mixtures were prepared and evaluated for their protective activity and stability. In general, mixtures containing large number of strains from several genera were the most protective (Table 2). The mixture of 50 cultures which showed protection comparable to undefined fecal culture against 10^4 CFU of *Salmonella* (12), was also the most effective against the higher challenge. Mixtures containing 28 or ten bacterial strains were less protective, while mixtures of strains from only one genus, e.g., *Escherichia*, *Bacteroides*, *Lactobacillus*, or *Bifidobacterium* were not protective. In an attempt to enhance survival of bifidobacteria in chicken gut, Neosugar (5) was supplied in drinking water in some experiments. With respect to protection of chicks against *Salmonella*, no beneficial effect of Neosugar was evident on either chicks treated with bifidobacteria mixture or non-treated control chicks.

Table 1. Summary of results on protection of chicks against 10^5 CFU of *Salmonella* with four undefined treatments administered to chicks by various methods.

Administration Method	Protection Factor			
	Fecal Culture	Subculture FC (4x)[a]	Fecal Suspension	Cecal Homogenate
Gavage	>50.0	50.0	>50.0	>50.0
Drinking	4.6	2.2	16.3	34.0
Spraying	3.1	1.2	5.3	7.8
Pour-Plate	>50.0	ND[b]	22.5	>50.0
Spread-Plate	6.7	ND	10.5	19.7
Feed Slurry	5.1	ND	3.4	ND

[a] Fecal culture subcultured four times.
[b] = Not done.

Bacterial cultures that were effective as mixtures in protecting chicks against *Salmonella* were stored individually at -70 ° C. For each experiment, the cultures were recombined and tested in chicks. Results in Tables 2 and 3 showed that the protective activity of mixtures containing ten or 50 pure cultures progressively decreased as the storage time for isolates increased. A similar decrease in efficacy was reported for a mixture of 28 cultures (3). Storing the 50 cultures as a frozen mixture produced the same effect. Undefined treatments, on the other hand showed much less decrease in efficacy over the same storage period (Table 3). Similar results were obtained with lyophilized preparations.

Despite the progress in the development of defined treatments, factor(s) affecting the efficacy of such treatments upon storage remain unknown. Since the potential for using defined cultures as a colonization control method against human bacterial enteropathogens in poultry remains high, we feel that better understanding of factor(s) affecting their efficacy will be the key for designing more effective strategies to achieve this potential. Colonization control method, if successfully applied under field conditions, will not only reduce the incidence of *Salmonella* in poultry, but will also reduce this pathogen in the environment.

Table 2. Bacterial composition of defined treatments and protection of chicks against 10^5 CFU of *Salmonella*.

	Number of Strains in Mixture									
Genus	2	3	3	5	8	8	10	18	28	50
Escherichia			3				1			6
Streptococcus							1		7	8
Bacteroides	2						3		6	11
Bacillus										
Fusobacterium							1		2	2
Lactobacillus					8		3	18	4	10
Eubacterium									3	3
Propionibacterium									1	1
Clostridium										2
Bifidobacterium	3			5		8			1	1
Gram-Positive Anaerobic Rods							1		4	6
Protection Factors	0.84	1.03	0.89	1.09	0.97	1.01	8.70[a]	1.04	20.8[a]	26.5[a]

[a] The best protection achieved with these mixtures.

Table 3. Effect of storage on the protective activity of a mixture containing ten pure cultures[a].

Storage Time (Mo.)	Percent Infected		Infection Factor		Protection Factor
	Treat.	Non-Treat.	Treat.	Non-Treat.	Treat.
0	15	100	0.6	5.2	8.7
1	55	85	1.3	4.3	3.3
2	27	100	0.8	4.7	5.9
3	20	75	0.6	2.9	4.8
6	40	90	1.4	4.2	3.0
7	45	90	1.6	3.7	2.3
8	65	100	2.2	4.5	2.1
10	75	85	2.6	3.7	1.5
12	75	90	2.9	4.2	1.5
13	50	70	1.9	3.5	1.8
14	90	100	3.7	5.2	1.4

[a] Pure cultures were stored individually.

Table 4. Effect of storage on the protective activities of (a) a mixture of 50 pure cultures[a] and (b) cecal homogenate.

Storage Time (Mo.)	Percent Infected		Infection Factor		Protection Factor
	Treat.	Non-treat.	Treat.	Non-Treat.	Treat.
Pure Cultures					
0	10	100	0.2	5.3	26.5
1	10	95	0.4	5.8	14.5
2	20	100	0.5	5.3	10.6
6	25	100	0.6	4.6	7.7
38	80	100	3.3	5.7	1.7
Cecal Homogenate					
0	3	100	<0.05	5.0	>50.0
38	20	100	0.3	5.2	17.3

[a] Pure cultures were stored individually.

ACKNOWLEDGMENTS

We thank Barbara Buchanan for technical assistance on parts of this study.

REFERENCES

1. **Barnes, E.M., and C.S. Impey.** 1971. The isolation of the anaerobic bacteria from chicken caeca with particular reference to members of the family Bacteroidaceae. *In* Isolation of anaerobes. D.A. Shapton and R.G. Board, eds. S.A.B. Technical Series No. 5, Academic Press, London, pp. 115-123.
2. **Blanchfield, B., S. Stavric, T. Gleeson, and H. Pivnick.** 1984. Minimum intestinal inoculum for Nurmi cultures and a new method for determining competitive exclusion of *Salmonella* from chicks. J. Food Prot. **47:**542-545.
3. **Gleeson, T.M., S. Stavric, and B. Blanchfield.D** 1989. Protection of chicks against *Salmonella* infection with a mixture of pure cultures of intestinal bacteria. Avian Diseases. (In Press).
4. **Goren, E., W.A. de Jong, P. Doornenbal, N.M. Bolder, R.W.A.W. Mulder, and A. Jansen.** 1988. Reduction of salmonella infection of broilers by spray application of intestinal microflora: a longitudinal study.1988. Vet. Quart. **10(4):**249-255.
5. **Hidaka, H., T. Eida, T. Takizawa, T. Tokunaga, and Y. Tashiro.** 1986. Effects of fructooligosaccharides on intestinal flora and human health. Bifidobacteria Microflora **5(1):**37-50.
6. **Mead, G.C., and C.S. Impey.** 1986. Current progress in reducing salmonella colonization of poultry by "competitive exclusion". J. Appl. Bacteriol. Symp. Suppl. 67S-75S.
7. **Nurmi, E., and M. Rantala.** 1973. New aspects of *Salmonella* infection in broiler production. Nature **241:**210-211.
8. **Pivnick, H., D. Barnum, S. Stavric, T. Gleeson, and B. Blanchfield.** 1985. Investigations on the use of competitive exclusion to control *Salmonella* in poultry. *In* Proceedings of the International Symposium on Salmonella. G.H. Snoeyenbos, ed. American Association of Avian Pathologists, Inc. University of Pennsylvania, PA., pp. 80-87.
9. **Pivnick, H., and E. Nurmi.** 1982. The Nurmi concept and its role in the control of salmonellae in poultry. In: Developments in Food Microbiology-l. R. Davies, ed. Applied Science Pub., Ltd., Barking, England, pp. 41-70.
10. **Stavric, S., T.M. Gleeson, and B. Blanchfield.** 1989. Competitive exclusion of *Salmonella* from young chicks: new method for the administration of treatment. Abstract. 2nd Congress of French Society of Microbiology, Sept. 18-21, Strasbourg, France.
11. **Stavric, S., T.M. Gleeson, and B. Blanchfield.** 1988. Some factors influencing the efficacy of defined treatments in competitive exclusion of *Salmonella* from chicks. Abstract. IFT Annual Meeting, New Orleans, LA.
12. **Stavric, S.** 1987. Microbial colonization control of chicken intestine using defined cultures. Food Technol. **41(7):**93-98.
13. **Stavric, S., T.M. Gleeson, B. Blanchfield, and H. Pivnick.** 1985. Competitive exclusion of *Salmonella* from newly hatched chicks by mixtures of pure bacterial

cultures isolated from fecal and cecal contents of adult birds. J. Food Protect. **48:**778-782.

14. **Wierup, M., M. Wold-Troell, E. Nurmi, and M.M. Hakkinen.** 1988. Epidemiological evaluation of the *Salmonella*-controlling effect of a nationwide use of a competitive exclusion culture in poultry. Poultry Sci. **67:**1026-1033

MICROSCOPIC STUDY OF COLONIZING AND NONCOLONIZING *CAMPYLOBACTER JEJUNI*

R. J. Meinersmann
W. E. Rigsby
N. J. Stern

Poultry Microbiological Safety Research Unit
USDA, ARS, Russell Research Center
Athens, Georgia 30613

ABSTRACT

Campylobacter jejuni A74/O and A74/C are congenic strains. The latter, which can colonize chickens, was derived from the former, which can not colonize chicks at an oral dose of 10^5 organisms. In this study, the association of the two strains with the cecal epithelium was compared. Sections of cecal tissue from fasted day-of-hatch chicks were placed into organ cultures along with 5 X 10^7 bacteria per ml. After two hours of incubation at 37 ° C, the tissues were glutaraldehyde fixed and prepared for light and electron microscopy. By light microscopy, tissues cultured with A74/C were characterized by areas of degenerative and necrotic changes of epithelial cells and increased mitotic index of crypt cells. These changes were not prevalent in tissues cultured with A74/O. By electron microscopy, organisms were found deep in crypts in tissues cultured with A74/C, but not A74/O. Direct contact of bacteria with epithelial cells was not noted. The differences in epithelial association of the two strains can not be accounted for by differences in motility.

METHODS

Day-of-hatch chicks were sacrificed and ceca were removed, cut into sections and placed in cultures with RPMI 1640 and no additives. *Campylobacter jejuni* were grown on brucella FBP plates, harvested with buffered peptone and a standard number were added to the cecal tissues. Two hours later the tissues were removed from culture and for light microscopy and transmission electron microscopy the tissues were fixed with glutaraldehyde in cacodylate buffer or for scanning EM, the tissues were fixed with Parducz's solution. Tissues for light microscopy were embedded in paraffin, sectioned and stained with either hematoxylin and eosin or Warthin Starry's silver stain. Sections were made for fluorescent antibody staining from the same paraffin blocks. The sections were rehydrated and treated with antiserum developed against *C. jejuni* outer membrane proteins and absorbed with normal chick intestines. Affinity purified rhodamine labeled anti-rabbit IgG was used to stain the tissue. Tissues for transmission electron microscopy were post fixed in 1% osmium, dehydrated, embedded in Spurs' or LR White's embedding medium, sectioned and stained with uranyl acetate and lead citrate. Tissues for scanning electron microscopy were ethanol dehydrated, critical point dried, and sputter coated. *C. jejuni* grown on brucella FBP was also prepared for negative contrast electron microscopy by washing with PBS, pelleting and resuspending in 1% phosphotungstic acid, and drying onto grids.

RESULTS AND DISCUSSION

Cecal tissues kept in culture were found to have sub-epithelial edema, regardless of bacterial treatment. In the presence of both strains of *Campylobacter jejuni* necrosis and exfoliation of epithelial cells was prominent. Early studies suggested that the colonizing strain caused more extensive damage but this was not verified. Also, an increase in the mitotic index of crypt cells was not verified. No pathology of the epithelial cells in the crypts was noted.

Tissue sections cultured with *C. jejuni* strain A74/O or strain A74/C were stained with fluorescent antibody. Positively fluorescent bacteria were readily found in five of eight tissues cultured with strain A74/C. The organism was found along the lumenal surface of the epithelium and deep in crypts. Positive staining bacteria were found in only one of eight tissues cultured with strain A74/O. These bacteria were only found in an area mixed with exfoliated (presumably necrotic) cells and never in crypts.

Tissues which were silver stained showed the same results as the fluorescent antibody stained tissues. Bacteria of strain A74/C could readily be found in crypts whereas strain A74/O never was. Bacteria were only ever found in the lumen, never inter- or intracellularly.

Campylobacter jejuni strain A74/O and strain A74/C were placed on grids and stained with phosphotungstic acid. Three morphologic types of organism were seen in equal numbers in both strains. These types were: 1) long form - length about six times width, with bi-polar singular flagella, 2) short form - length about three times width, with single flagellum at one end, and 3) short form - length about three times width with bi-polar singular flagella. The body of the bacteria were characteristically gently 'S' curved with the ends narrower than the center. The base of the flagellum was usually projected from a flattened end of the organism.

Bacteria were only found in tissues inoculated with the colonizing strain (A74/C). The bacteria were deep in crypts with no apparent contact (adherence) to the epithelial cells.

The A74/C strain of *Campylobacter jejuni* was easily found by scanning electron microscopy. Only one organism of the A74/O strain was found. The bacteria were invariably found associated with mucin layers. A few endogenous organisms, mostly cocci and some assorted rods and fungi-like organisms were seen. If tissues were treated with acetylcysteine (Mucomyst) before addition of bacteria, the amount of mucin was greatly reduced and no bacteria could be found.

CONCLUSIONS

1. *Campylobacter jejuni* strain A74/O and strain A74/C differ in their ability to colonize chick ceca. No difference in morphology was seen in these strains by negative contrast electron microscopy.
2. Both *C. jejuni* strain A74/O and strain A74/C appear to be toxigenic to mature cecal epithelial cells. Therefore, toxigenicity probably does not account for differences in colonization ability. It was noted that crypt epithelial cells were not affected even though, in the case of strain A74/C, bacteria could be immediately adjacent to these cells.
3. The colonizing strain (A74/C) could be found in the crypts but the noncolonizing strain never was. Colonization is apparently dependent on the presence of mucin. This suggests that the colonizing strain is chemo-attracted into the crypts by mucin and the non-colonizing strain (A74/O) is not.

A COMPARISON OF CONVENTIONAL VS. A DNA HYBRIDIZATION METHOD FOR THE DETECTION OF *SALMONELLA* IN HENS AND EGGS

J.E. Murphy
J.D. Klinger

GENE-TRAK Systems, Inc.
Framingham, Massachusetts 01701

R.L. Taylor, Jr.

Department of Animal and Nutritional Sciences
University of New Hampshire
Durham, New Hampshire 03824

P. F. Cotter

Biology Department
Framingham State College
Framingham, Massachusetts 01701

ABSTRACT

Experiments were designed to compare the efficiency of the GENE-TRAK Systems Inc. DNA hybridization method (GT-DNAH) for *Salmonella* detection with a procedure using conventional bacteriological methods. Samples obtained from commercial laying hens and other sources were tested in parallel, giving the following results. No *Salmonella* were detected by either method in a carriage survey comprising a 2.5% sample from over 5,000 commercial laying hens representing five major breeders. Conversely, *Salmonella* were recovered more efficiently by the GT-DNAH procedure from hens given a direct cloacal exposure, whereas hens exposed by mouth did not become infected. Moreover, cloacal exposure resulted in recovery of *Salmonella* from 5/117 (4.3%) eggs produced by such hens. *Salmonella*-positive eggs were sometimes defective, misshapen, or thin-shelled, but *Salmonella* were recovered from normal eggs as well. We concluded that the GT-DNAH *Salmonella* detection method is useful in poultry and egg samples, and we have shown that cloacal exposure may result in ovarian infection accompanied by the production of infected eggs.

Colonization Control of Human Bacterial Enteropathogens in Poultry

INTRODUCTION

Recent epidemiological investigations of human *Salmonella* infections have implicated Grade A shell eggs as the source (1, 2). A reduction of *Salmonella* carriage in laying flocks would therefore potentially benefit both consumers as well as producers. Such a reduction will depend in part on the availability of sensitive detection methods that can be applied to specimens of poultry origin.

While the GENE-TRAK DNA hybridization (GT-DNAH) method has been successfully applied to the detection of *Salmonella* in food samples, its utility in screening poultry for *Salmonella* colonization has not yet been demonstrated. Thus, it was the purpose of our experiments to determine if the GT-DNAH method could be applied to various samples derived from commercial laying hens, eggs, and experimental lines of poultry. The results obtained suggest that this method is useful for poultry research and it was shown to be superior to a detection method based on conventional bacteriology.

METHODS

Bacteria. A clinical isolate of *Salmonella enteritidis* (CDC stk 1411-82 NH 107-F) from human stool or *S. typhimurium* (ATCC 23566) were used throughout.

Chickens. Mature (20-30 weeks) commercial White Leghorns weighing 3.5 to 4.0 pounds or similar brown egg layers were used. All hens used were negative for *S. pullorum* antibody. They were housed in standard laying cages in an egg test facility or in individual cages at the experimental farm located at the University of New Hampshire, Durham. These hens were obtained from a nationwide distribution of breeders and were fed a commercial diet and water *ad libidum*.

Exposure. Eighteen hour trypticase soy broth (Difco Laboratories, Detroit) cultures were incubated at 37 ° C. One ml of broth culture representing either isolate was inoculated *per os* using an 18 g gavage cannula, or dripped onto the vent, to stimulate a sucking reflex, until the entire 1 ml was consumed. Contact exposure refers to the potential for colonization of hens housed in the same room as those exposed directly.

Sample Collection. Cloacal samples were obtained using sterile cotton swabs which were immediately dropped into tubes containing 10 ml 1% peptone broth (Difco). Egg samples were collected, marked with the hen number, dated, and stored at room temperature until testing. Unwashed eggs were either broken out separately to provide 1 ml yolk samples, or were pooled in groups of three, placed in plastic sandwich bags, crushed and stomached manually for 30 seconds. Such homogenized whole egg preparations were sampled in aliquots of 1 ml.

Conventional Bacteriology. Ten ml peptone broth cultures were incubated 18 hours at 39-40 ° C. One ml aliquots and the swab (where applicable) were transferred to 10 ml tetrathionate broth containing 0.01% brilliant green and sulfathiazole (1.25 *v*g/ml) and incubated for an additional 24 hours. Brilliant green agar (BGA) and Hektoen Enteric agar (HEA; Difco) plates were streaked from the tetrathionate broths and incubated for 24 hours. Suspect *Salmonella* colonies were picked and inoculated in triple sugar iron agar (TSI) and lysine iron agar (LIA; Difco) slants. Slants showing reactions consistent with the positive controls (either *S.enteritidis* or *S. typhimurium*) were tested for carbohydrate fermentation reactions using the Minitek disk system (BBL; Baltimore) (3).

Those isolates having fermentation patterns consistent with the positive controls were tested by agglutination using somatic group-specific antisera for *S. enteritidis* (factor 9) and for *S. typhimurium* (factors 4, 5; Difco).

DNA Hybridization (DNAH). GN broths were inoculated from the same tetrathionate cultures (18 hours, 35 ° C) used in conventional bacteriology. Cellular DNA was obtained by chemical lysis and vaccum filtration onto nylon membranes provided in the assay kit. The DNA samples were fixed on the filters and air-dried for two minutes and then hybridized with a specific P^{32} labeled probe mixture according to the manufacturers protocol (4). Bound probe was detected in a scintillation counter (without scintillant), and the counts per minute were compared to those of positive and negative controls run with the samples.

RESULTS

Carriage Survey. Commercial type laying hens representing the entries of five breeders in a competitive egg production test were sampled for *Salmonella* carriage using cloacal swabs. In the first trial, 90 samples were obtained; in a second trial an additional 48 samples were taken. The combined 138 samples represented approximately 2.5% of the total number of hens in the house. All samples were negative for *Salmonella* by conventional bacteriology and the DNAH test. The second trial included three additional kinds of samples: a) cloacal swabs from 24 hens of an experimental strain housed at the University of New Hampshire, b) cecal specimens obtained aseptically at necropsy from the same 24 hens as above, and c) cloacal swabs (n = 24) from experimental racing pigeons housed in a coop located about five miles from the poultry buildings. All samples obtained from these sources were negative by conventional and DNAH methods.

Direct Challenge. Commercial type White Leghorns were exposed to 1 ml of an 18-hour broth culture of *S. typhimurium* (ATCC 23566) as described in methods. Contact exposed hens were caged directly below these in such a manner as to be partially bombarded by feces produced by hens given the direct exposure. All hens used in the direct challenge studies had been pre-tested for antibody to *S. pullorum* by the whole blood method (5). These hens were also pre-tested for *Salmonella* carriage by both conventional bacteriology and the GENE-TRAK DNAH method. All such pre-challenge tests were negative. Two weeks post-challenge, cloacal swabs were obtained.

During these two post-challenge weeks, eggs were sampled on three occasions, i.e., those first layed after the challenge, those collected one week later, and those layed closest to the date of the cloacal swab samples. Fecal specimens were also collected two weeks post-challenge from accumulations below those hens given direct exposure. The results are given in Table 1. In a second experiment an overnight broth culture of *S. enteritidis* was dropped onto the cloacae of ten mature hens. In the two week period following challenge three eggs per hen were collected and stored at 10 ° C. The hens were then killed by cervical dislocation, and samples of eggs, ovaries, cecas, and a one inch segment of the large bowel were obtained aseptically at necropsy. The hybridization assay and conventional bacteriology results are given in Table 2. *Salmonella* were recovered from six of ten hens exposed via the cloacal route, the large bowel being source in three cases, the cecum in three cases, and an ovary in one case. Although no *Salmonella* were recovered from 30 egg samples (three per hen), two hens produced misshapen eggs, with conspicuously wrinkled shells. *Salmonella* were recovered from the ovary of one such hen but not the other. In no instance were *Salmonella* detected by the conventional method missed by the DNAH method. Conversely, *Salmonella* were found by the DNAH method in three cecal samples and from one large bowel specimen which were missed by theconventional bacteriology method.

Table 1. Recovery of *Salmonella* from hens given an oral or contact exposure.

		Route	
Challenge Strain	Specimen	P.O.	Contact
S. typhimurium	cloaca	0/10[a]	0/10
"	eggs	0/30	0/30
"	feces	0/10	n.t.[b]
S. enteritidis	cloaca	0/10	0/10
"	eggs	0/10	0/10
"	feces	n.t.	n.t.

[a] # positive by conventional bacteriology and/or GENE-TRAK DNAH over # tested.

Because eggs having gross defects were observed in hens exposed to *S. enteritidis* via the cloacal route, it was decided that a more careful study of egg characteristics should be made. To this end, 20 additional commercial type White Leghorn hens in lay were exposed to *S. enteritidis* via the cloaca. All hens were pullorum negative and free of *Salmonella* prior to the challenge as determined by whole blood tests, conventional bacteriology and DNAH testing. All pre-challenge eggs from such hens were also free from *Salmonella*. Three weeks post-challenge, cloacal swabs and eggs were obtained. Two hens had *Salmonella* positive cloacal swabs as determined by conventional bacteriology. Interestingly, however, 11 out of 20 hens were producing eggs with conspicuous defects ranging from slight to severe wrinkles, and many had thin shells. A total of 20 out of 50 eggs (40%) were defective. Two of the 20 defective eggs tested bacteriologically were positive for *Proteus* spp. *Salmonella* were detected in one of the *Proteus*-positive eggs, but only by the DNAH test. An additional sample of eggs from the same 20 hens was collected and tested during the 4th and 5th week post-challenge. The results are given in

Table 2. Recovery of *S. enteritidis* from hens given a cloacal exposure.

Hen No.	Ovary	Ceca	Bowel	Eggs (3 per hen)
1.	-/-[a]	-/+	-/-	-/-
2.	-/-	-/+	-/-	-/-
3.	-/-	-/-	-/-	-/-
4.	-/-	-/-	-/-	-/-
5.	-/-	-/-	-/-	-/-
6.	+/+	-/-	+/+	-/-[b]
7.	-/-	-/-	-/+	-/-
8.	-/-	-/-	+/+	-/-
9.	-/-	-/-	-/-	-/-[b]
10.	-/-	-/+	-/-	-/-

[a] Conventional bacteriology/GENE TRAK DNAH.
[b] Misshapen eggs recovered.

Table 3. Five of 117 (4.3%) of these eggs (Table 3) were positive for *Salmonella* by either method; one positive egg showed no gross defects, all others were conspicuously misshapen and/or thin-shelled. The majority (55 of 59; 93%) of the defective eggs were *Salmonella* negative by both methods. While the majority of the hens produced six or seven recoverable eggs during the seven day collection period, one hen produced only one recoverable egg, which was extremely thin shelled. Additional eggs were apparently broken before collection, and were found in the litter below that hens' cage.

DISCUSSION

One of the difficulties encountered during the detection of *Salmonella* in mixed-flora samples is masking by competitive organisms. In the present study, *Proteus* spp. were most frequently isolated from all types of specimens. This happened in spite of the inclusion of sulfathiazole in the selective enrichment broths (tetrathionate-brilliant

Table 3. Recovery of *Salmonella* from eggs produced by hens given a cloacal challenge.

Hen No.	No. Eggs Recovered	No. Defective	No. + Bacteriology	No. + GT-DNAH
151	7	7	2	2
1684	6	2	1	1
116	6	1	1	1
553	6	0	1	1
16 Others (Pooled)	92	49	0	0

green) (6). Visual inspection is the linchpin for presumptive identification in conventional procedures and small numbers of *Salmonella* colonies may be overlooked among many similar appearing *Proteus* colonies. The GENE-TRAK DNAH procedure depends on the presence of *Salmonella* target DNA sequences capable of hybridizing with specific labeled probes (7). Our results have demonstrated the superiority of this method because of a better detection rate (13/556 DNAH vs. 10/556 conventional) and specifically in the detection of *Salmonella* in the presence of *Proteus*. No false positives were obtained using the DNAH method during these trials. Superiority of the DNAH method is also realized by a shorter time-to-result (optimally 48 hours) and the ease with which larger sample sizes can be handled. Potentially interfering nonspecific DNA sequences from competitor organisms did not contribute to false positive results suggesting this method should be applicable to a wide variety of poultry specimens.

The results of this initial study also provide some additional insight into the occurrance of *Salmonella* in poultry. We found no naturally occurring carriage in the survey population. We surveyed 2.5% of the 5000+ hens in an egg laying contest which consisted of entries from five major commercial breeders. These results may be more representative than larger samples drawn from more homogeneous populations, though such a conclusion requires statistical analysis beyond the scope of this study. By direct oral challenge with the strain used in this study *S. typhimurium* (ATCC 23566) we failed to establish chronic colonization. Earlier sampling may have revealed transient carriage following these high level inocula. Conversely, cloacal challenge (*S. enteritidis*) resulted in established colonization in six of ten, and five of 20 cases (trials 2 and 3, respectively). The oral route represents the cheif way humans contract salmonellosis and gastroenteritis, and occasionally asymptomatic low level carriage and excretion are the most frequent clinical outcomes. In future studies, individual challenge strain characteristics must also be considered as potential variables in the ability of *Salmonella* to establish colonization in poultry. Such characteristics may be highly host specific. Finally, the high rate of production of defective eggs and the recovery of *Salmonella* from such eggs suggests that the cloacal exposure method should be of use in studies of transovarian transmission of this pathogen. In conclusion, we have demonstrated the utility of the GENE-TRAK DNAH method for *Salmonella* detection in poultry specimens. We have shown that *Salmonella* infection of the ovary established after cloacal exposure can be accompanied by production of both defective and normal appearing eggs, some of which may contain *Salmonella*.

REFERENCES

1. Increasing rate of *Salmonella enteritidis* infections in the Northeastern United States. MMWR **36:**10-11,1987.
2. **St. Louis, M. E.,** ***et al.*** 1988. The emergence of Grade A eggs as a major source of *Salmonella enteritidis* infection. J. Am. Med. Assoc. **259:**2103-2107.
3. **Cox, N. A., and J. E. Williams.** 1976. A simplified biochemical system to screen *Salmonella* isolates from poultry for serotyping. Poultry Sci. **55:**1968-1971.
4. **Flowers, R. S., M. A. Mazola, M. S. Curiale, D. A. Gabis, and J. H. Silliker.** 1987. Comparative study of a DNA hybridization method and the conventional culture procedure for the detection of *Salmonella* in foods. J. Food Sci. **52:**781-785.
5. **Williams, J. E., E. T. Mallinson, and G. H. Snoeyenbos.** 1980. Salmonellosis and Arizonosis, Ch 1. *In* Isolation and Identification of Avian Pathogens, 2nd ed. S.B. Hatchner *et al.* (eds) Am. Assoc. Avian Pathologists (publishers).
6. ***ibid.***
7. **Fitts, R.** 1985. Development of a DNA-DNA hybridization test for the presence of *Salmonella* in foods. Food Technol. **39:**95-102.

TREATMENT OF *CAMPYLOBACTER JEJUNI* WITH ANTIBODY INCREASES THE DOSE REQUIRED TO COLONIZE CHICKS

N. J. Stern
R. J. Meinersmann

Poultry Microbiological Safety Research Unit
USDA, ARS, Russell Research Center
Athens, Georgia 30613

H. W. Dickerson

Department of Medical Microbiology
College of Veterinary Medicine
University of Georgia
Athens, Georgia 30602

ABSTRACT

This study was designed to clarify the role of antibodies in controlling chicken colonization by *Campylobacter jejuni*. Cecal colonization by *C. jejuni* was compared after the organism was exposed either to phosphate buffered saline, normal rabbit serum, rabbit hyperimmune anti-*C. jejuni* serum, or anti-*C. jejuni* antibodies extracted from chicken bile. Antibodies from bile were extracted by affinity absorption against outer membrane proteins from the challenge organism. Sera were heated one hour at 56 ° C to abrogate Complement activity. Bacterial inoculum levels were enumerated after one hour exposure at 4 ° C to the various treatments. The heated sera and the bile antibodies were not bactericidal nor was bacterial agglutination evident. Serial dilutions of the antibody treated *C. jejuni* were gavaged into day-old chicks. Six days later the ceca were removed from the chicks and samples were cultured on *Campylobacter*-Charcoal Differential Agar. The colonization dose-50% was increased by 2- to 160-fold when the organism was preincubated with hyperimmune antiserum or the bile antibodies as compared to preincubation with PBS. We conclude that antibodies inhibit chicken cecal colonization by *C. jejuni*.

Colonization Control of Human Bacterial Enteropathogens in Poultry

INTRODUCTION

Tauxe *et al.* (1987) and Deming *et al.* (1988) provide strong evidence indicating the important role which chicken plays as a vehicle in human *Campylobacter jejuni* enteritis. Substantial efforts in the processing of poultry carcasses has not resolved this human health issue. The goal of our research is to diminish the presence of *C. jejuni* in the GI tract of chickens.

Intestinal immunity to enteric bacteria is often accompanied by the host producing secretory immunoblobulin A which interferes in the colonization of the bacterial agent. Immunity to the disease produced by *C. jejuni* could be attributed to the presence of either maternal antibody passed through milk or by active immunity developing as a response to the constant exposure of *Campylobacter* (Abimuku and Dolby, 1988; Blaser *et al.*, 1985). We assessed the role of antibody in the commensal colonization of *C. jejuni* in the chicken. Colonizing doses of *C. jejuni* that were or were not treated with specific antibodies were compared.

MATERIALS AND METHODS

Outer Membrane Proteins (OMP). The OMPs were prepared by the method of Logan and Trust (1982), using *C. jejuni* isolate A74/C (described in Stern *et al.*, 1988).

Rabbit Immune Sera: Two rabbits, free of *C. jejuni*, were immunized IM with 1 mg of OMP in Freund's, and twice boosted with 1 mg OMP in water at two week intervals.

Chicken Handling. Chickens were housed in isolation units at six to ten per unit. Cecal/fecal droppings were assayed for presence of *C. jejuni* to ensure absence of the organism.

Challenge Protocol. *C. jejuni* A74/C was serially diluted and incubated at 4 ° C with the antibody containing samples. We enumerated these bacterial suspensions after one hour, and 0.2 ml of the dilutions were gavaged per 24-hour-old chick.

Sampling Protocol. Seven-day-old chicks were killed by cervical dislocation and ceca were dissected. Cecal materials were quantitatively assessed for *C. jejuni*.

Data Analysis. Colonization dose-50% (CD-50) were assessed by Probit analysis. Colonization quotient (CQ), the average log of the cfu *C. jejuni* per gm cecal material within a group, was determined.

RESULTS

1. Pre-incubating *C. jejuni* with anti-*C. jejuni* antiserum or with bile antibody for one hour increased the colonization dose-50% in one-day-old chicks.
2. Heat inactivated, normal rabbit antiserum increased the CD-50% of *C. jejuni* in one-day-old chicks as compared with the PBS controls.
3. Neither complement inactivated serum or bile antibody showed evidence of bactericidal activity or bacterial agglutination.
4. CD-50% analysis is the most sensitive means to assess *C. jejuni* colonization of chickens, but inconsistencies in the data were observed.

5. The CD-50% for the PBS treated *C. jejuni*-challenged chicks was *ca.* 1.3 X 10^3 cfu; for the normal serum treated *C. jejuni*-challenged chicks was *ca.* 6.0 X 10^3 cfu; and for the hyperimmune antiserum treated *C. jejuni*-challenged chicks was *ca.* 3.7 X 10^4.

6. The CD-50% of the organism treated with bile antibodies was, in Experiment 4, approximately 50-fold greater, and in Experiment 5, approximately 3-fold greater than the CD-50% observed with the PBS treated *C. jejuni* used to challenge the chicks.

DISCUSSION AND CONCLUSIONS

Specific anti-*Campylobacter jejuni* antibody preincubated with *C. jejuni* before challenge of one-day-old chicks dimninishes the ability of the organism to colonize the animals. In each of the five experiments, the CD-50% was increased in the antibody treated group when compared with the PBS-control treated groups. Antibody from rabbit sera or from chicken bile were both effective in diminishing the colonization. The activity of these antibody containing solutions are being assessed.

The presence of these neutralizing antibodies on the organism does not kill the organism nor does it agglutinate *C. jejuni*. As serum is bactericidal for the organism, complement inactivation by heating the serum, attenuated this lethal property.

Specific antigens involved in the colonization of the organism in chick ceca are expressed on the organism and are neutralized by the antibody. The antibodies used for neutralizing colonization by the organism were obtained by either immunizing against the outer membrane proteins of the homologous organism or by affinity purifying chicken bile against the same outer membrane proteins. Presumably, all these outer membrane proteins are not involved in cecal colonization in the chick. By identifying and immunizing the host chicken against these specific colonization factor antigens it should be possible to diminish colonization by the organism. Likewise, by enhancing the sIgA response to these colonization factor antigens, stronger and directed immunological response against *C. jejuni* colonization may be accomplished.

ACKNOWLEDGEMENTS

Funds for this research were provided in part by the Southeastern Poultry and Egg Association (Grant No. M#5O) and the USDA--Agricultural Research Service. Technical help by Douglas Cosby, Margaret Myszewski, Debbie Posey, and Lalla Tanner is acknowledged with appreciation.

REFERENCES

1. **Abimiku, A. G., and J. M. Dolby.** 1987. The mechanism of protection of infant mice from intestinal colonisation with *Campylobacter jejuni*. J. Med. Microbiol. **23**:339-344.
2. **Blaser, M. J., P. F. Smith, and P. F. Kohler.** 1985. Susceptibility of Campylobacter isolates to the bactericidal activity of human serum. J. Infectious Dis. **151**:227-235.

3. **Deming, M. S., R. V. Tauxe, P. A. Blake, S. E. Dixon, B. S. Fowler, T. S. Jones, E. A. Lockamy, C. M. Patton, and R. O. Sikes.** 1987. Campylobacter enteritis at a university: transmission from eating chicken and from cats. Am. J. Epidemiol. **126:**526-534.
4. **Logan, S. M., and T. J. Trust.** 1982. Outer membrane characteristics of *Campylobacter jejuni*. Infect. Immun. **38:**898-906.
5. **Stern, N. J., J. S. Bailey, L. C. Blankenship, N. A. Cox, and F. McHan.** 1988. Colonization characteristics of *Campylobacter jejuni* in chick ceca. Avian Dis. **32:**330-334.
6. **Tauxe, R. V., D. A. Peues, and N. H. Bean.** 1987. Campylobacter infections: the emerging national pattern. Am. J. Public Health **77:**1219-1221.

A COMPARISON OF AN ENZYME IMMUNOASSAY, DNA HYBRIDIZATION, ANTIBODY IMMOBILIZATION, AND CONVENTIONAL METHODS FOR RECOVERY OF NATURALLY OCCURRING SALMONELLAE FROM PROCESSED BROILER CARCASSES

J. S. Bailey
N. A. Cox
L.C. Blankenship

Poultry Microbiological Safety Research Unit
USDA, ARS, Russell Research Center
Athens, Georgia 30613

ABSTRACT

Three hundred and ninety processed broiler carcasses were obtained from four processing plants and evaluated for the presence of salmonellae by the whole bird rinse procedure. Enzyme immunoassay (Salmonella-Tek™), colorimetric DNA hybridization (GENE-TRAK®), and antibody immobilization (1-2 Test™) methods were compared to the Food Safety Inspection Service (FSIS) culture method for detection of salmonellae. Additionally, any discrepancies among these procedures were reanalyzed using a culture confirmation procedure. All samples were preenriched in buffered peptone water incubated at 37 ° C then transferred to and enriched in TT broth incubated at 42 ° C for 24 h. Following enrichment, instructions of the manufacturers were followed for 'rapid' procedures and the conventional recovery and confirmation scheme was that of the FSIS. If any discrepant results were observed between procedures, selective plates were restreaked from TT broths and from GN broths in the interest of identifying every positive sample and explaining differences among the four methodologies. These additional positive samples were called confirmed conventional positives.

Salmonellae were detected on 71% of carcasses using confirmed conventional procedures, 65% with regular conventional, 66% with the 1-2 Test, 71% with GENE-TRAK, and 76% with Salmonella-Tek. As compared to confirmed conventional, only 1.4% false positives were observed with the 1-2 Test and GENE-TRAK while 6% false positives were found with Salmonella-Tek. However, with the important false negatives were analyzed with the confirmed conventional, Salmonella-Tek, GENE-TRAK, and 1-2 Test had 0.4, 2.5% and 7.2% respectively. When the rapid methods were compared to the FSIS culture method, false negative rates for Salmonella-Tek, GENE-TRAK and the 1-2 Test were 0.4%, 2.3% and 10.6% respectively. The close correlation of these rapid procedures to the conventional procedure warrants consideration of these procedures for detection of salmonellae from broiler chickens.

INTRODUCTION

Traditional recovery methods for foodborne *Salmonella* involve 5 basic steps: 1) Preenrichment - The initial step in which the food sample is enriched in a nonselective medium to restore injured *Salmonella* cells to a stable physiological condition, 2) Selective enrichment - A step in which the sample is further enriched in a growth-promoting medium containing selectively inhibitory reagents. This medium allows a continued increase of *Salmonella* while simultaneously restricting proliferation of most other bacteria, 3) Selective plating - A step using solid selective media that restrict growth of bacteria other than *Salmonella* and provide visual recognition of pure, discrete colonies suspected to be *Salmonella*, 4) Biochemical screening - An elimination of most organisms other than *Salmonella* that also provides a tentative generic identification of *Salmonella* cultures, and 5) Serotyping - A serological technique which provides a specific identification of the cultures. These procedures take from four to seven days to complete.

Acceptance and use of any new method will probably require that the method be faster, cheaper or easier than current conventional methods. In addition, any new method for salmonellae should be at least as sensitive as the conventional method and should not have excessive false positives reactions. Three different technologies are being used to speed up the detection of salmonellae from foods. These include enzyme immunoassay, DNA hybridization and antibody immobilization. These methods take 48 to 53 hours to complete. They use conventional preenrichment and/or selective enrichment procedures with time saved by eliminating the selective plating, biochemical screening and serology. In this study, commercially available kits representing each of these technologies were evaluated for their efficacy in relation to conventional methods for detection of salmonellae from processed broiler carcasses.

MATERIALS AND METHODS

Three hundred and ninety processed broiler carcasses were obtained from four processing plants, individually bagged, frozen, transferred to our laboratory within 48 h, and sampled within three weeks. Carcasses were thawed in the refrigerator for 48 h and sampled by the whole carcass rinse method (2) using buffered peptone (BP) as the preenrichment broth. After overnight incubation at 37 ° C, a 1 ml aliquot of the BP was transferred to a 10 ml tube of TT (3) enrichment broth which was incubated for 24 h at 42 ° C. Selective plating media was that used by the Food Safety Inspection Service (FSIS), BG Sulfa (Difco) and MLIA (1). Typical isolates were screened on Lysine Iron Agar slants and confirmed with 'H' serology. This conventional scheme we called FSIS conventional. Also, the presence of salmonellae was determined from TT enrichment broths using the post-enrichment Salmonella-Tek™ enzyme immunoassay (Organon Teknika, Durham, NC), the colorimetric DNA hybridization (GENE-TRAK® Systems,

Framingham, MA), and the 1-2 Test™ (BioControl, Bothell, WA). The manufacuturer's instructions for post-enrichment and method performance were followed for each of the 'rapid methods'. If there were any discrepant results between the conventional and/or any of the 'rapid methods', additional BG sulfa and MLIA plates were streaked from the TT broth which had been held at 4 ° C and the GN broth used in the DNA hybridization assay. Any additional salmonellae detections were called confirmed conventionals. This confirmation method was employed to ensure identification of every positive sample and to explain differences among the four methodologies compared in this study. Streaking from the GN broth has the effect of providing another nutrient enrichment following selective enrichment. It should be noted that this confirmation method is used for purposes of this study only.

RESULTS AND DISCUSSION

Salmonellae were detected on 277 (71%) of carcasses using confirmed conventional procedures, 254 (65%) with FSIS conventional, 258 (66%) with the 1-2 Test, 278 (71%) with GENE-TRAK, and 295 (76%) with Salmonella-Tek. GENE-TRAK, the 1-2 Test, and Salmonella-Tek had 4 (1.4%), 4 (1.4%), and 16 (6%) false positives, respectively. However, with the more important false negatives, Salmonella-Tek had 1 (0.4%) while GENE-TRAK and 1-2 had 7 (2.5%) and 20 (7.2%), respectively when compared to the confirmed culture method (Table 1).

Table 1. Comparison of regular conventional and three rapid systems for recovery of *Salmonella* from processed broiler carcasses.

Test	#(+)/#(Tested) (%)	False (+)	False (-)
Conventional	254/390 (65%)	-	-
Salmonella-Tek	295/390 (76%)	42/254 (16%)	1/254 (0.4%)
Gene-Trak	278/390 (71%)	29/254 (11%)	6/254 (2%)
BioControl	258/390 (66%)	23/254 (9%)	27/254 (11%)

The fewest number of salmonellae were detected using the FSIS conventional procedure. Therefore, when compared to the FSIS conventional, there were 23 (9%), 29 (11%), and 42 (16%) apparent false positives with the 1-2 Test, GENE-TRAK, and Salmonella-Tek, respectively (Table 2). The number of false negative samples was similar to the true conventionals.

Table 2. Comparison of true conventional and three rapid systems for recovery of *Salmonella* from processed broiler carcasses.

Test	#(+)/#(Tested) (%)	False (+)	False (-)
Conventional	277/390 (71%)	-	-
Salmonella-Tek	295/390 (76%)	16/277 (6%)	1/277 (0.4%)
Gene-Trak	278/390 (71%)	4/277 (1.4%)	7/277 (2.5%)
BioControl	258/390 (66%)	4/277 (1.4%)	20/277 (7.2%)

These data demonstrate the difficulty in accurately evaluating alternative methods to standard conventional methods. The procedure used by the FSIS for recovery of salmonellae from poultry and meat products is widely accepted as being as good as any regulatory approved method. If only the FSIS conventional method had been used, it would seem that the 'rapid' methods were producing excessively large numbers of false positives when in fact, the 'rapid'procedures were more sensitive than the FSIS conventional method.

Each of these procedures takes from 48 to 53 hours from time of initial sample to time of a negative or presumptive positive result. In addition to savings in time, total technician time for preparation and transfer of media is reduced, incubator space is reduced and the number of samples which can be simultaneously evaluated is increased. Specifically, the 1-2 Test was by far the easiest system to inoculate and read the results, but gave the highest number of false negative results. The GENE-TRAK system had low false positives (1.4%) and low false negatives (2.5%) but was probably the most labor intensive to perform. The Salmonella-Tek had the fewest false negatives (0.4%), slightly higher false positives (6%), but was less labor intensive than GENE-TRAK. These systems have all received AOAC approval and while each has their advantages and disadvantages, they all detected more salmonellae than did the FSIS conventional procedure, and the close correlation of these rapid procedures to the confirmed conventional procedure warrants consideration of these procedures for the routine detection of salmonellae from broiler chickens.

ACKNOWLEDGMENTS

We gratefully acknowledge the technical assistance of Mark Berrang, Mike Musgrove, Debbie Posey, and Lalla Tanner.

Mention of specific brand names does not imply endorsement by the authors or institutions at which they are employed to the exclusion of others not mentioned.

REFERENCES

1. **Bailey, J. S., J. Y. Chiu, N. A. Cox, and R. W. Johnston.** 1988. Improved selective procedure for detection of salmonellae from poultry and sausage products. J. Food Prot. **51**:391-396.
2. **Cox, N. A., J. E. Thomson, and J. S. Bailey.** 1981. Sampling of broiler carcasses for *Salmonella* with low volume water rinse. **60**:768-770.
3. **Hajna, A. A., and S. R. Damon.** 1956. New enrichment and plating media for the isolation of *Salmonella* and *Shigella* organisms. Appl. Microbiol. **4**:341-345.

MINIMUM INFECTIVE NUMBER OF *CAMPYLOBACTER* BACTERIA FOR BROILERS

N. M. Bolder
R. W. A. W. Mulder

Spelderholt Centre for Poultry Research and Information Services
Agricultural Research Service
7361 DA, Beekbergen
The Netherlands

ABSTRACT

In the literature poultry is often accused of being one of the main vectors in spreading *Campylobacter*. The route of infection in live birds is not clear. *Campylobacter* isolations from feed, hatchery debris, or from young broilers are seldom reported. Therefore, we studied the effect of the number of *Campylobacter* bacteria needed to infect day-old chicks. In one experiment, two different genetic types of broilers were infected with approx. 0, 10, 100, 1000, 10,000 cfu of *Campylobacter* bacteria, and in another experiment eight groups of broilers, all from a different genetic background, were infected with the same doses. Samples of the faeces and caecal contents were examined. Minimum infective numbers varied from 100 to 1000 cfu, with small differences between the different genetic strains used.

Colonization Control of Human Bacterial Enteropathogens in Poultry

INTRODUCTION

Campylobacter jejuni/coli can cause acute enteritis in man. In many countries all over the world *Campylobacter* is considered to be one of the primary causes of human foodborne bacterial disease. Some years ago, *Salmonella* was considered more important, but advanced techniques and more attention for *Campylobacter* has shown that this bacterium causes at least as much trouble.

Foods of animal origin are the main carrier for *Campylobacter*, and *Campylobacter jejuni* is isolated frequently from poultry. Human and poultry isolates appear to belong to the same serotype, indicating that poultry is one of the main vectors.

There are several studies about the spread of *Campylobacter* in poultry rearing farms. Rosef and Kapperud (1983) found more that 50% of the flies captured in a chicken farm to be carriers of the organism. Lindblom *et al.* (1986) also consider flies and other insects as possible transmittants, but they stress the need for research on the colonization mechanism. Smitherman *et al.* (1984) surveyed four broiler farms and found birds colonized after 12 days, when raised on old litter in a non-cleaned shed. Stern and Meinersmann (1989) infected different genetic lines of broiler chickens with *Campylobacter jejuni* bacteria and saw differences in colonization between the lines. The colonization dose 50% (CD-50) of a mixed culture of *Campylobacter jejuni* strains appeared to be 35 colony forming units (cfu) per bird. At a level of 3500 cfu per bird, complete colonization was established.

At the Spelderholt Centre for Poultry Research and Information Services, some broiler lines were bred (Leenstra, 1986) with which two tests were carried out to estimate the number of *Campylobacter jejuni* bacteria needed for colonization of the gut. As a control, several commercial lines with different genetic backgrounds were tested.

The second experiment was designed so that some information could be gathered on cross contamination between flocks and about the speed of flock infection.

MATERIALS AND METHODS

In the first test, an experimental Feed Conversion line (FC) was compared with a commercial one. Eggs of both lines were hatched at the experimental hatchery at Spelderholt, and the chicks wre placed in isolators of ten birds each for two weeks. After three days the broilers were orally infected with 0, 10, 100, 1000, and 10,000 cfu per bird of *Campylobacter jejuni*, cultured for 48 hours in Nutrient Broth (Oxoid CM 1).

For the second experiment, eggs of both the experimental FC-line and of the Growth Restriced line (GR) plus four commercial lines (1 - 4) were also hatched at the experimental hatchery at Spelderholt. The one-day-old chicks were placed in a shed equipped for experiments with pathogenic microorganisms, and with the possibility of cross contamination between the groups. From every line, 20 broilers out of 24 were orally infected at day 1 with 0, 10, 100, 1000, and 10,000 cfu per bird of *Campylobacter jejuni*, cultured for 48 hours in Nutrient Broth (Oxoid CM 1). The groups were randomly mixed so that cross contamination between groups was possible.

During the rearing period of two to three weeks, samples of faecal material and dead chicks were taken and cultured for *Campylobacter*. Culture conditions were as follows: 1) enrichment in Nutrient Broth with growth supplement (Oxoid SR84) for 48 hours at 43 ° C, followed by, 2) selective plating on Brucella agar (Difco 0964) with Skirrow supplement (Oxoid SR69) for 48 hours at 43 ° C under microaerophilic conditions.

RESULTS AND DISCUSSION

Table 1 shows the numbers of *Campylobacter* bacteria that were given to the one-day-old chicks. In the first experiment no chicken belonging to groups 1 and 2 became *Campylobacter*-positive. From the third group in the experimental FC-line,samples of faeces and caeca were negative, whereas the commercial line showed the first infected samples at day 5 (Table 2). It took more that ten days to infect all birds of this group. Groups 4 and 5 of both lines were heavily infected at day 8, although the commercial line showed *Campylobacter*-positive birds already at day 2 after infection.

This experiment demonstrates that the minimum infective dose of the commercial line is between 100 and 1000 cfu/chick. According to Stern and Meinersmann (1989), the CD-50 dose appeared to be 150 cfu per bird. More than 1000 cfu/chick are needed to colonize the experimental FC-line.

Table 1. Infective dose of *Campylobacter jejuni* given to one-day-old chicks.

Group	CFU/Chick	Experiment 1	Experiment 2
1	0	0	0
2	10	2.0×10^1	8.0×10^1
3	100	1.5×10^2	7.8×10^2
4	1,000	1.5×10^3	9.5×10^3
5	10,000	7.3×10^3	7.2×10^4

In the second experiment, four commercial lines were mixed groupwise with two experimental lines. The number of *Campylobacter* bacteria isolated was slightly higher than that in the first experiment (Table 1). In Table 3 the results of the same lines as those presented from the first experiment are shown. There were no significant differences between the four groups of the commercial lines or between the two experimental lines from Spelderholt.

For both lines, given in Table 3, results were similar to those observed in the first experiment. Colonization of the commercial line occurred with a lower dose of *Campylobacter* bacteria compared to the experimental line. The design of the experiment made it possible to get some information on the cross-contamination between broilers. After two weeks, chicks that were not infected orally (the control groups) appeared to be *Campylobacter*-positive.

Table 2. Colonization percentage of *Campylobacter jejuni* in two genetically different lines of broilers.

	Group (Table 1)									
	1		2		3		4		5	
Day / Line	FC	1	FC	1	FC	1	FC	1	FC	1
1	0	0	0	0	0	0	0	0	10	0
2	0	0	0	0	0	0	0	10	10	0
3	0	0	0	0	0	0	0	30	50	100
5	0	0	0	0	0	50	20	75	100	100
8	0	0	0	0	0	30	90	80	100	70
10	0	0	0	0	0	40	100	70	100	100

Table 3. Colonization percentage of *Campylobacter jejuni* in two lines of broilers.

	Group (Table 1)									
	1		2		3		4		5	
Day / Line	FC	1	FC	1	FC	1	FC	1	FC	1
1	0		0	0	0	0	0	10	0	0
3	0	0	0	10	0	0	0	40	0	40
6	0	0	0	0	0	0	60	90	80	50
8	0	0	0	30	10	40	60	60	60	40
13	0	0	0	*	*	*	*	*	100	100
15	*	*	90	100	100	100	100	100	100	100
17	50	40	70	*	90	100	100	*	100	*
20	90	90	90	100	100	*	*	*	*	*

*not determined

CONCLUSION

There was a difference in resistance to *Campylobacter* bacteria between commercial and experimental lines of broilers. Commercial lines show colonization characteristics similar to those of Stern and Meinersmann (1989). When broilers are exposed to natural environmental contamination, it may take two weeks before the entire flock is infected with *Campylobacter*.

REFERENCES

1. **Leenstra, F. R., P. F. G. Vereijken, and R. Pit.** 1986. Phenotype and genetic variation in, and correlations between, abdominal fat, body weight and feed conversion. Poultry Sci. **65:**1225-1235.
2. **Lindblom, G. B., E. Sjögren, and B. Kaijser.** 1986. Natural Campylobacter colonization in chickens raised under different conditions. J. Hyg. Cambridge **96:**385-391.
3. **Rosef, O., and G. Kapperud.** 1983. House flies as possible vectors of *Campylobacter fetus* subsp. *jejuni*. Appl. and Environ. Microbiol. **45:**381-383.
4. **Smitherman, R. E., C. A. Genigeorgis, and T. B. Farver.** 1984. Prelimenary observations on the occurrence of *Campylobacter jejuni* at four California chicken ranches. J. Food Prot. **47:**293-298.
5. **Stern, N. J., and R. J. Meinersmann.** 1989. Potentials for colonization of *Campylobacter jejuni* in the chicken. J. Food Prot. **52:**427-430.

REDUCTION OF *CAMPYLOBACTER* INFECTION OF BROILERS BY COMPETITIVE EXCLUSION TREATMENT OF DAY-OLD BROILER CHICKS--A FIELD STUDY

R.W.A.W. Mulder
N.M. Bolder

Spelderholt Centre for Poultry Research and Information Services
Agricultural Research Service
7361 DA Beekbergen, The Netherlands

ABSTRACT

In a previous field study the positive effect of competitive exclusion treatment of day-old broiler chicks in reducing the *Salmonella* infection of broilers was demonstrated. No attention was paid towards reduction of *Campylobacter* infection by this treatment. Therefore, a small field trial including approx. 200,000 broilers divided over 58 flocks was executed under the same experimental conditions (Goren *et al.*, 1988). Caeca, liver, and gall were examined for the presence of *Campylobacter* bacteria. *Campylobacter* was isolated after enrichment and selective plating under microaerophilic conditions. The treatment proved to be successful. The number of *Campylobacter* infected flocks was reduced from 62% to 41%, whereas the number of contaminated caeca was reduced from 40% positive in the non-treated group to 21% in the treated group. At the same time the incidence in *Campylobacter*-positive flocks was lowered significantly by the treatment.

Colonization Control of Human Bacterial Enteropathogens in Poultry

INTRODUCTION

The results of a large field trial about the effect of competitive exclusion (CE) treatment of day-old broiler chicks on colonization resistance of *Salmonella* organisms were reported by Goren *et al.* (1988). During an 18 month period, 284 flocks (with more than 8 million broilers) housed on 46 farms were tested in this study. Half of the flocks were CE-treated, the other half were not. The CE treatment proved very effective in prevention of *Salmonella* colonization and also reduced the incidence of *Salmonella* organisms in infected flocks. At that time *Campylobacter* was not as important as today. The National Health Council (Anon., 1988) reported on *Campylobacter jejuni* infections in the Netherlands and highlighted the role of poultry in the spreading of the organisms. Therefore, a smaller field trial was executed to study the effect of CE treatment on *Campylobacter* colonization in the young bird. Results of the trial are reported here.

MATERIALS AND METHODS

Experiment. The study was conducted at 58 broiler farms, which received chicks from one hatchery. Half of the participating farms received a CE-treated flock, the other half received non-treated flocks. The broilers were all slaughtered in the same slaughterhouse. In total, 58 flocks involving some 200,000 broilers participated in the trial.

Per flock, ten caecal samples were taken which were combined per five, resulting in two mixed samples. Per flock, ten liver + gall samples were also taken and examined individually for the presence of *Campylobacter* organisms.

***Campylobacter* Isolation Procedure.** Enrichment (1:10) was in a *Campylobacter* enrichment medium containing per liter: 13 g nutrient broth (OXOID-CMl), 2 g yeast extract, 3 g Bile Salts no. 3 (OXOID L56), 1.5 g agar (Merck 1614). After sterilizing the medium, four flasks of OXOID SR84 growth supplement were added. Incubation was at 43 ° C for 48 hours under microaerophilic conditions by using BBL Campypak plus in anaerobic jars. Thereafter, the samples were inoculated on a *Campylobacter* blood medium containing per liter: 28 g Brucella broth (Difco), 17 g agar (Merck 1614), antibiotic supplement (OXOID SR69), 0.15 ml cephaloridin (10%), 100 ml defribinated sheep blood (Biotrading Benelux). Incubation was at 43 ° C for 48 hours under microaerophilic conditions. After biochemical identification *Campylobacter* isolates were serotyped at the Dutch National Health Institute (RIVM, Bilthoven).

Method of CE Treatment. The method of application of CE flora was as described by Goren *et al.* (1988) by spray treatment in the hatchery.

RESULTS

Table 1 summarizes the results of the CE treatment of the 29 flocks compared with the 29 non-treated flocks. As can be seen, the number of *Campylobacter jejuni* infected flocks (as measured by *Campylobacter* positive results in liver + gall or caecal samples) was reduced from 62% to 41%. When all other isolated *Campylobacter* species (*coli*, *fetus*) were taken into account the infection was reduced from 69% to 55%.

Table 2 gives the results from the individual samples. The incidence of *Campylobacter* organisms was reduced significantly by the CE treatment, both in liver + gall and in caecal samples.

Table 1. Effect of CE treatment on *Campylobacter* infection of flocks.

Campylobacter jejuni	Non-Treated	CE-Treated	Total
Number of Flocks	29	29	58
Liver + Gall Samples	16 (55%)	9 (31%)	25 (43%)
Caecal Samples	13 (45%)	9 (31%)	22 (38%)
Total of Samples	18 (62%)	12 (41%)	30 (52%)
Including *Campylobacter coli* and *C. fetus*			
Total of Samples	20 (69%)	16 (55%)	36 (62%)

Table 2. Effect of CE treatments on *Campylobacter jejuni* contamination of liver + gall and caecal samples.

	Non-Treated		CE-Treated		Total	
	No. of Samples	Positive	No. of Samples	Positive	No. of Samples	Positive
Liver + Gall Samples	286	79 (28%)	290	35 (12%)	576	114 (20%)
Caecal Samples	58	23 (40%)	58	12 (21%)	116	35 (30%)

Table 3 gives the number of colony forming units (cfu)of *Campylobacter* per gram of liver + gall or caecal contents. This number is reduced by the CE treatment. The effect is very clear in the range of 10^3 to 10^5, the range where the usual counts of non-treated birds can be found. For caecal samples the effect is very clear in the range 10^6 to 10^7 cfu/g material, *Campylobacter* counts that usually are found in *Campylobacter* contaminated flocks.

Table 3. Number of *Campylobacter* cfu in liver + gall and caecal samples as reduced by CE treatment of broilers.

Liver + Gall Samples											
Range cfu/g	10^2		10^3		10^4		10^5		10^6		10^7
Number of Samples Within this Range:											
Non-Treated (n = 79)		24		39		11		5			
CE-Treated (n = 35)		20		10		5		-			
Caecal Samples											
Non-Treated (n = 23)				1		3		4		15	
CE-Treated (n = 12)				2		5		4		1	

CONCLUSION

In a small field trial, it was shown that a CE treatment applied by spraying day-old broiler chicks in the hatchery reduces the number of *Campylobacter*-infected flocks considerably. The other most striking result is that the CE treatment also reduced the number of *Campylobacter* cfu's in liver + gall and caecal samples. Especially with respect to possible spreading of the organisms during processing, this is an important finding.

REFERENCES

1. **Anon.** 1988. *Campylobacter jejuni* infections in the Netherlands. Report by a committee of the Health Council of the Netherlands. The Hague, October 10, 1988.
2. **Goren, E., W.A. de Jong, P. Doornenbal, N.M. Bolder, R.W.A.W. Mulder, and A. Jansen.** 1988. Reduction of Salmonella infection of broilers by spray application of intestinal microflora: a longitudinal study. The Veterinary Quarterly **10:** 249-255.

EFFECT OF GENETICS AND PRIOR *SALMONELLA ENTERITIDIS* INFECTION ON THE ABILITY OF CHICKENS TO BE INFECTED WITH *S. ENTERITIDIS*

W. H. Benjamin, Jr.

Department of Microbiology
University of Alabama at Birmingham
Birmingham, Alabama 35294

W. E. Briles

Department of Biologic Sciences
Northern Illinois University
DeKalb, Illinois 60115

W. D. Waltman

Georgia Poultry Laboratory
Oakwood, Georgia 30566

D. E. Briles

Department of Microbiology
University of Alabama at Birmingham
Birmingham, Alabama 35294

ABSTRACT

In mice it has been clearly shown that prior sublethal *Salmonella typhimurium* infection is very protective against subsequent *Salmonella* infection. As a first step in the development of a live attenuated vaccine for *Salmonella enteritidis*, we have begun studies to determine the extent to which prior sublethal infection with *S. enteritidis* reduces the infectivity of *S. enteritidis* strain Y8P2. In our studies to date, the interval between immunizing and challenge infections in baby chicks has been seven days. At seven

Colonization Control of Human Bacterial Enteropathogens in Poultry

days post challenge with 8 x 10^7 Y8P2 *S. enteritidis*, we observed at least two logs more *Salmonella* in the spleen and liver of the unimmunized than immunized chicks. In the course of these studies we detected a large variation in the resistance and susceptibility within the closed Ancona population we were using. This variation appeared to have a genetic basis since examination of progeny from specific families within the population demonstrated that chicks from some families were much more resistant to *Salmonella* infection than others. These differences were not observed to be related to the blood group B alleles segregating within these families.

INTRODUCTION

One means to prevent the spread of *Salmonella enteritidis* through eggs would be to immunize hens so that they could resist invasive colonization with this organism. In mammalian species it has been shown that the best immunogen against invasive *Salmonella* infection is a live vaccine that elicits protective cellular immunity (1, 2, 3). The development of live *Salmonella* vaccines for mammals was proceeded by studies demonstrating that prior or active invasive infection with *Salmonella* prevented reinfection with the same or related *Salmonella* isolates (3). Specific antibodies have been found to be relatively inefficient at preventing invasive disease (4). In the studies described here we have begun to investigate whether this fundamental property of immunity to invasive *Salmonella* infections also holds true for chickens.

Our initial studies have been directed at the immunization of chickens by an initial infection with the Y8P2 strain of *S. enteritidis* followed by challenging immunized and non-immunized chickens with the same strain of *S. enteritidis*. Strain Y8P2 is a field isolate that has been implicated in outbreaks of egg-borne *S. enteritidis* infection in humans (5). In many of these studies we have utilized strains of Y8P2 for immunization and challenge that carry different antibiotic resistance markers. This allows us to separately enumerate the immunizing and challenge *Salmonella* in the tissues of the chick at different times after infection.

Our results are still very preliminary and constitute initial studies examining the doses of the Y8P2 *Salmonella* strain appropriate for immunization and challenge and the effect of the genetic background of the chickens on their ability to be immunized and infected with *S. enteritidis* strain Y8P2.

METHODS

Immunization of Leghorn Chicks. One-week-old male Leghorn chicks were immunized iv with Y8P2 carrying a transposon which codes for tetracycline resistance (tc) (insert *zbi421*::Tn*10* near *aroA* at 19 units on the *Salmonella* chromosome). Seven chicks were killed on day 6 and liver and spleen plated to determine the number of the immunizing dose remaining at six days post infection. On days 7 and 14 immunized and non-immunized controls were challenged with a kanamycin resistant derivative of Y8P2 (*hisT*::Tn*5*). The chicks were killed three days later and the number of *Salmonella* recovered from the liver and spleen of each bird was determined by plating.

Immunization of Ancona Chicks. Chicks for this study were from a closed population derived from the Wisconsin Ancona inbred line 3. The population is maintained at Northern Illinois University and carries a number of different blood group B alleles that have been backcrossed into the inbred line for six to eight generations. One-week-old Ancona (Wisconsin inbred line 3) chicks were infected iv with 6 x 10^7 CFU of Y8P2 and one week later these and an uninfected control group were challenged with 8 x 10^7 CFU of Y8P2. One week later all chicks were killed and the number of CFU of *Salmonella* determined. Preliminary evidence indicates that over 90% of the *Salmonella* recovered from the immiunized-challenged chicks originated from the immunizing inoculum rather than the challenge inoculum.

Inheritance of Salmonella Resistance. One-week-old chicks from matings of Wisconsin inbred line 3 chickens segregating for different B blood group alleles were infected iv with about 5 x 10^6 CFU of Y8P2. They were challenged with the same dose one week later. They were killed six days after the last injection. The 97 chicks for the study came from families derived from five sires (66, 67, 68, 69, and 71) each mated to two to four dams.

Test for Association with *Salmonella* Resistance and B Blood Group. The B blood group of all Wisconsin line 3 chicks were determined by agglutination with specific iso-antisera prepared to type the alleles segregating in this closed population.

RESULTS AND DISCUSSION

Immunization Study. We have immunized one-week-old Leghorn chicks with two doses of *S. enteritidis* strain Y8P2, 3 x 10^6 and 6 x 10^7 CFU/chick. We observed that in most chicks the immunizing *Salmonella* were rapidly cleared such that after six, ten, and 17 days post infection the number of organisms in the liver and spleen represented only about 0.01%, 0.001% and $<$0.0001%, respectively, of the initial inoculum.

Analysis of the protective effects of the prior *Salmonella* injections was complicated by the observation that in almost every group of chicks we observed some highly infected individuals. We believe that this variation is the result of genetic differences in *Salmonella* susceptibility. If this is assumed to be the case, then analysis of the data from chicks with intermediate levels of Salmonella provides an evaluation of the protective effects of prior immunization. The amount of protection against subsequent challenge appeared to be dependent on the size of the initial immunization dose. The larger of the two doses (given to Ancona birds) appeared to give at least two logs of protection against subsequent *S. enteritidis* infection, as judged by numbers of *Salmonella* post challenge. The smaller dose (given to Leghorns) provided only about one log of protection.

We anticipate that the initial immunization would have probably been more successful if: 1) larger doses of salmonella had been injected, 2) immunizations had included more than one injection of live salmonella, or 3) a more invasive strain of salmonella had been used. We suspect that the ideal *S. enteritidis* vaccine will be a highly virulent invasive strain of *S. enteritidis* carrying mutations that will prevent its continued growth *in vivo*. Studies to investigate these possibilities are underway.

Genetic Studies. During analysis of the immunization and challenge data it became apparent that there was a large variation in the number of *Salmonella* recovered from similarly treated chickens. One explanation for this spread in the data is that some of the chicks were genetically much more susceptible than others. If large genetic differences in resistance of chicks to *S. enteritidis* infection exists, they could be important in immunization studies for several reasons. The availability of resistant and susceptible strains of chickens would allow comparison of the efficacy of the vaccination for chicks genetically resistant and susceptible to *Salmonella*. They would permit optimization of immunization procedures for resistant as well as susceptible chicks. Furthermore an understanding of the inheritance of *Salmonella* resistance in chickens would allow the production of commercial chickens with increased *Salmonella* resistance.

1. Inheritance of *Salmonella* Susceptibility. For this study we made matings among members of a closed colony of Wisconsin Inbred Line 3 (Ancona) chickens. Matings were carried out by artificial insemination, and chicks were raised as two sequential hatches. At one week of age the chicks in each hatch were subjected to the same immunization and challenge scheme used for the immunization studies already described. When we examined the numbers of total Y8P2 *S. enteritidis* in the progeny of each sire-dam family, it is apparent that some families, such as those from sire 67, produced all resistant progeny, whereas others, such as the families of sire 68, produced a significant number of highly susceptible chicks. It is also apparent that the same general results were observed for each of the two hatches of chicks. These findings provide evidence for a strong genetic component in the resistance of chicks to infection with *S. enteritidis* strain Y8P2.

2. Association of the B Locus with *Salmonella* Resistance? Inheritance at the major histocompatibility (MHC) loci of birds and mammals has been shown to have a major impact on immune responsiveness and resistance to certain infections. In the mouse, the MHC locus has been reported to have an effect on *S. typhimurium* resistance. The MHC of the chicken is called the B blood group. We have examined the relationship of inheritance of alleles at the B locus of the chicken to their resistance to infection with *S. enteritidis*.

The offspring of sire 68 showed the highest frequency of susceptible chicks. The sire and both dams of this family were 8/11 heterozygotes at the B locus. Although the two chicks with the highest number of *Salmonella* were the only 8/8 homozygotes, when the data from all 17 of the chicks of this sire are examined there is no statistically significant evidence for a linkage between *Salmonella* resistance and the B locus. There were no statistically significant associations between the blood group B genotype and susceptibility to *Salmonella* infection when the levels of infection were compared for chicks from the other familes of the 8/8, 8/11, and 11/11 genotypes .

It should be noted, however, that although the data provide no evidence for linkage of these B locus alleles to *Salmonella* susceptibility, these data do not rule out an effect of other MHC alleles on *Salmonella* susceptibility. In mice it has been shown that a major non-MHC locus, *Ity*, controls the resistance to *S. typhimurium* early in the infection. We hope to continue our genetic studies with chickens to determine if a comparable major *Salmonella* resistance locus can be identified in chickens.

REFERENCES

1. **Hobson, D.** 1957. Resistance to reinfection in experimental mouse typhoid. J. Hyg. (London) **55:**334-343
2. **Mitsuhashi, S., M. Kawakami, Y. Yamaguchi, and M. Nagai.** 1958. Jpn. J. Exp. Med. **28:**249-258

3. **Collins, F. M., G. B. Mackaness, and R. B. Blanden.** 1966. Infection-immunity in experimental salmonellosis. J. Exp. Med. **124:**601-619.
4, **Eisenstein, T. K., L. M. Killar, and B. M. Sultzer.** 1984. Immunity to infection with *Salmonella typhimurium*: Mouse-strain differences in vaccine- and serum-mediated protection. J. Infect. Dis. **150(3):**425-435.
5. **St. Louis, M. E., D. L. Morse, M. E. Potter, T. M. DeMelfi, J. J. Guzewich, R. V. Tauxe, P. A. Blake,** ***et al.*** 1988. The emergence of grade A eggs as a major source of *Salmonella enteritidis* infections. JAMA **259:**2103-2107.

EVALUATION OF ANIMAL AND AGAR PASSAGE FOR MODIFYING THE POTENTIAL OF *SALMONELLA* STRAINS TO COLONIZE THE CECA OF BROILER CHICKS

S. E. Craven
N. A. Cox
J. S. Bailey
N. J. Stern
L.C. Blankenship

Poultry Microbiological Safety Research Unit
USDA, ARS, Russell Research Center
Athens, Georgia 30613

ABSTRACT

Variants of *Salmonella* strains with altered colonization potential were isolated for use in experiments to identify the colonization factors of *Salmonella* in the chick intestinal tract. Three strains (*S. california*, *djakarta*, and *wentworth*) were characterized as poor cecal colonizers after an initial screening of 12 strains. In experiments designed to enhance colonization potential, strains were passaged seven times through the intestinal tracts of chicks. To assess colonization, a nalidixic acid-streptomycin-resistant ($NAL^r STR^r$) strain of each serotype was given orally to chicks simultaneously with its nalidixic acid-resistant (NAL^r) animal-passaged derivative strain, or each of the two strains was given individually to chicks. The numbers of *Salmonella* in the cecal contents of sacrificed chicks were determined at selected intervals by differential recovery on antibiotic-containing plating medium. Results indicated that cecal colonization was similar for each strain and its derivative strain obtained through animal passage. A *S. california* $NAL^r STR^r$ strain passaged 40 times on recovery agar was deficient in colonization as compared to its NAL^r parent strain or the NAL^r animal-passaged derivative of the parent strain. Deficiency in colonization for this strain was observed with oral or intracloacal challenge of chicks. Motility, mannose-sensitive hemagglutination, and growth rates of the parent and agar-passaged *S. california* strains were similar. Investigations of other properties of *Salmonella* strains and their colonization-deficient variants will be made to identify colonization factors.

INTRODUCTION

The colonization of the intestinal tract of poultry by *Salmonella* is important since the carcass may become contaminated with intestinal contents containing this organism and serve as a vehicle of transmission to people who handle or consume the meat. The mechanisms involved in the colonization of the alimentary canal of the chicken by *Salmonella* are not well understood (2) but are likely complex. More knowledge about colonization factors is needed to develop new strategies to intervene in colonization and to reduce *Salmonella* contamination of poultry.

MATERIALS AND METHODS

Bacterial Strains. The strains used in this study were spontaneous nalidixic acid-resistant (NAL^r) mutants of strains obtained from the culture collection of Dr. N. A. Cox, Russell Research Center. Selected strains were passaged seven times by fecal-oral transfers in chicks or passaged forty times by streaking for colony isolation on BGS + NAL agar. For some strains, spontaneous streptomycin-resistant (STR^r) mutants were isolated.

Broiler Chicks. Peterson-Arbor Acres' chicks obtained from a single commercial flock were housed in isolation units.

Challenge Protocol. Cultures from BHI agar slants (20 h at 37 ° C) were diluted, and 0.1 to 0.2 ml of the dilutions were introduced by oral gavage or intracloacal challenge to give 10^8 CFU/strain/day-of-hatch chick (unless stated otherwise).

Sampling Protocol. At selected intervals after challenge, chicks were killed by cervical dislocation, ceca removed, and serial dilutions of cecal contents plated on BGS agar + 100 ppm NAL and BGS agar + 100 ppm NAL and 100 ppm STR. CFU were enumerated after 24 h at 37 °C.

Test for Mannose-Sensitive Hemagglutination (MSHA). The experimental strains, and fimbriate (Fim^+) and Fim^- control strains, were taken through two serial, 48-h cultures in BHI broth at 37 ° C. Cultures were washed in saline and concentrated to 1.5×10^{10} CFU/ml. Twenty-five μl of two-fold dilutions of bacteria in saline was added to 50 μl of a 1.5 % saline suspension of guinea pig erythrocytes in wells. After shaking at 22 ° C for 30 min, the highest dilution with observable hemagglutination was recorded, and its reciprocal taken as the MSHA titre. Hemagglutination was also tested in the presence of 0.1 % D-mannose.

Motility Determination. Motility after growth in BHI broth was determined in BHI motility agar (3.5 mg agar and 50 μg triphenyltetrazolium chloride/ml BHI broth) according to the procedure of McCormick *et al.* (9).

Generation Times. BHI broth was inoculated with about 10^4 CFU of each strain and incubated at 37 ° C. At intervals serial dilutions were plated on BHI agar, and plates were incubated at 37 ° C for 24 h prior to counting.

RESULTS

1. The colonization potential in day-of-hatch chicks of three strains of *Salmonella* characterized as weak colonizers was not improved by passage seven times through the intestinal tracts of young chicks.
2. A derivative strain of *S. california* isolated after 40 X passage on agar was deficient in colonizing ability as compared to the parent strain.
3. The deficiency in colonization of the agar-passaged *S. california* strain was observed when it was competing directly with its parent strain after simultaneous challenge of chicks with the two strains.
4. The *S. california* agar-passaged strain colonized the chicks to a lesser extent than its parent strain after intracloacal as well as after oral challenge.
5. Growth rates, motility, and mannose-sensitive hemagglutination activities were similar for the *S. california* colonization-deficient and *S. california* parent strains.

DISCUSSION

The conversion of virulent to avirulent form by passage on agar and the recovery of virulence by passage of the avirulent form through animals is known to occur for several bacterial pathogens. In this study, the change in the colonization potential was determined for several *Salmonella* strains after passage in animals or on agar. For the three strains of *Salmonella* passaged through chicks, none were altered in their ability to colonize day-of-hatch chicks. A *S. california* strain that was passaged on agar was converted to a deficient colonizer. An isolate of this agar-passaged, colonization-deficient *S. california* strain, which was recovered from ceca ten days post-challenge, also was deficient in colonization after subsequent challenge, demonstrating stability for the trait of colonization deficiency. Catrenich and Johnson (3) reported that the virulence conversion of *Legionella pneumophila* was a one-way phenomenon, virulent to avirulent form by agar passage but not avirulent to virulent form by animal passage.

Phenotypic factors associated with virulence interconversions include polysaccharide capsules (1, 8), pili, and outer membrane proteins (5, 7, 13, 14). The role of mannose-sensitive pili of *Salmonella* in the invasion of mucosa, colonization and/or infection of animals has beem controversial with some studies supporting a role for mannose-sensitive adhesions (10, 11, 12) and others not (4, 6). The results of our study are not consistent with a role for mannose-sensitive adhesion in accounting for the differences in colonization between the agar-passaged and parent *S. california* strains, because their MSHA titers were similar. Also motility and growth rates for the two strains were similar. Other phenotypic factors of these strains will be examined in order to identify those that may be important to their colonization of the intestinal tract of chicks.The difference in the extent of colonization when comparing the parent and agar-passaged strains was observed after intracloacal or after oral challenge suggesting that a host factor restricted to the upper portion of the intestinal tract (e.g., acidity) was not responsible for these differences.

REFERENCES

1. **Austrian, R.** 1953. Morphologic variation in pneumococcus. J. Exp. Med. **98**:21-34.
2. **Barrow, P. A., J. M. Simpson, and M. A. Lovell.** 1988. Intestinal colonisation in the chicken by food-poisoning *Salmonella* serotypes; microbial characteristics associated with faecal excretion. Avian Pathol. **17**:571-588.
3. **Catrenich, C. E., and W. Johnson.** 1988. Virulence conversion of *Legionella pneumophila*: a one-way phenomenon. Infect. Immun. **56**:3121-3125.
4. **Dugoid, J. P., M. R. Darekar, and D. W. F. Wheator.** 1976. Fimbriae and infectivity in *Salmonella typhimurium*. J. Med. Microbiol. **9**:459-473.
5. **Hultgren, S. J., T. N. Porter, A. J. Schaeffer, and J. L. Duncan.** 1985. Role of type 1 pili and effects of phase variation on lower urinary tract infections produced by *Escherichia coli*. Infect. Immun. **50**:370-377.
6. **Jones, G. W., and L. A. Richardson.** 1981. The attachment to and invasion of HeLa cells by *Salmonella typhimurium*: the contribution of mannose-sensitive and mannose-resistant haemagglutinating activities. J. Gen. Microbiol. **127**:361-370.
7. **Maayan, M. C., I. Ofek, O. Medalia, and M. Aronson.** 1985. Population shift in mannose-specific fimbriated phase of infection in mice. Infect. Immun. **49**:785-789.
8. **Masson, L., B. E. Holbein, and F. E. Ashton.** 1982. Virulence linked to polysaccharide production in serogroup B *Neisseria meningitidis*. FEMS Microbiol. Lett. **13**:187-190.
9. **McCormick, B. A., B. A. D. Stocker, D. C. Laux, and P. C. Cohen.** 1988. Roles of motility, chemotaxis, and penetration through and growth in intestinal mucus in the ability of an avirulent strain of *Salmonella typhimurium* to colonize the large intestine of streptomycin-treated mice. Infect. Immun. **56**:2209-2217.
10. **Oyofo, B. A., R. E. Drolesky, J. O. Norman, H. H. Mollenhauer, R. L. Ziprin, D. E. Corrier, and J. R. Deloach.** 1989. Inhibition by mannose of *in vitro* colonization of chicken small intestine by *Salmonella typhimurium.* Poultry Sci. **68**:1351-1356.
11. **Tanaka, Y., and Y. Katsube.** 1978. Infectivity of *Salmonella typhimurium* for mice in relation to fimbriae. Jpn. J. Vet. Sci. **40**:671-681.
12. **Tavendale, A., C. K. H. Jardine, D. C. Old, and J. P. Dugoid.** 1983. Haemagglutinins and adhesion of *Salmonella typhimurium* to HEp-2 and HeLa cells. J. Gen. Microbiol. **16**:371-380.
13. **Wardlaw, A. C., and R. Parton.** 1979. Changes in envelope proteins and correlation with biological activities in *B. pertussis*, pp. 94-98. *In* C.R. Manclark and J.C. Hill (ed.), International Symposium on Pertussis, U.S. Department of Health, Education, and Welfare. Washington, DC.
14. **Weiss, A. A., and S. Falkow.** 1984. Genetic analysis of phase change in *Bordetella pertussis*. Infect. Immun. **43**:263-269.

SUMMARY

Leroy C. Blankenship

USDA, ARS, Russell Research Center
Athens, Georgia

Food safety is an international responsibility that encompasses all nations and societies of the world. Each country and region has its own unique food safety problems that relate to culture, climates, economic status, and many other factors. However, the same human bacterial enteropathogens, *Salmonella, Campylobacter*, and *Listeria* are common to all fresh poultry and meat.

The intestinal tract of poultry is a complex multi-compartmented organ serving primarily a digestive function providing nutrients for survival and growth, as well as other functions. It also provides a unique ecological niche for a broad variety of microorganisms, most of which exist in a friendly commensal relationship with the host and many of which provide benefit to the host. In contrast, pathogenic microorganisms occasionally find their way into the intestinal tract and produce disease. Pathogenic colonization elicits defense responses (immunological, febrile, etc.) from the host that limit the damage caused by the pathogen and speed its elimination from the host. Generally, the defense responses are not called-up as a consequence of commensal colonization. Some microorganisms which commensally colonize the intestinal tract of poultry are pathogenic colonizers of humans, such as *Salmonella, Campylobacter*, and *Listeria*.

Many factors influence which microbial genera and species colonize the poultry intestinal tract. These include: (1) fortuitous ingestion as a result of food and water contamination, (2) coprophagy, (3) survival through gastric barrier passage, (4) locating an hospitable colonization site, (5) favorable symbiotic interactions with other microorganisms, (6) effectively competing with other bacteria, (7) nature of the host's diet, (8) physiological status of the host, (9) health and disease status of the host, (10) host age, (11) environmental stresses, (12) medication effects, and (13) host genetic background. Thus, microbes that successfully colonize the poultry intestine do so in a highly complex, dynamic, competitive, and interactive environment.

Dr. Bill Hubbert, Food Safety and Inspection Service, USDA, in opening remarks, pointed out that food safety has been a big issue of the 1980's and is likely to continue to be well in to the 1990's. He emphasized that the most effective control of human enteropathogens in our meat and poultry supply is to prevent colonization of animals during production so enteropathogen-free animals can be presented for slaughter and processing. Dr. Hubbert defined food safety risk as the sum of the hazard plus outrage and observed that food safety is not a health issue only, but an economic and trade issue as well.

Dr. Glen Snoeyenbos, University of Massachusetts, who has devoted a large portion of his research career to the study of the control of paratyphoid salmonella in poultry, presented the introduction. He was the convener of the 1984 salmonella symposium held in New Orleans. He noted that the observations and recommendations made in that symposium are still valid. He considers that *Salmonella enteritidis* infection of layer chickens and its linkage to egg-borne salmonellosis is a major new problem. He pointed out that an acceptable cost of producing enteropathogen-free poultry could be elevated by public perceptions and demands.

Session I was devoted to presentations concerned with environmental factors and sources associated with colonization control of human bacterial enteropathogens in poultry. The session was convened by Dr. Stan Bailey, Agricultural Research Service, USDA.

Dr. Frank Jones, North Carolina State University, in discussing environmental factors that contribute to salmonella colonization of chickens, described ovarian infection of breeder/multiplier flocks by *S. typhimurium* in which management practices, disease status, house contamination, and T-2 toxin were contributing factors. Reduced litter moisture and water chlorination were found to be important for reducing the contamination rate on farms.

Dr. Damien Gabis, Silliker Laboratories, in discussing environmental factors affecting enteropathogens in feed and feed mills, noted that 50% of meat and bone meal samples were found to be positive for salmonellae and that post-cooking contamination was of major importance. He indicated that improved plant design could help control recontamination. The finished product should be kept dry and well separated from areas of the rendering plant where raw product is processed.

Dr. Simon Shane, Louisiana State University, lectured on environmental factors associated with *Campylobacter jejuni* colonization in chickens. He noted that high rates of intestinal carriage have been confirmed among breeder hens and pre-slaughter broiler and turkey flocks. Horizontal spread occurs rapidly via water receptacles and coprophagy; however, vertical transmission is not involved in commercial production and colonization is not linked to any disease state in chickens. Thorough chemical disinfection followed by a one-week flock replacement interval will eliminate *C. jejuni* in the grow out environment, but flies, fomites, wild birds, and vermin can reintroduce the bacterium. Thus, effective biosecurity can help limit reintroduction into the grow out house.

Dr. David Rollins, Naval Medical Research Institute, reported on a farm study in England in which 86% of all chicken samples were *Campylobacter*-positive while air, water, feed, litter, and other samples were culture-negative. However, the presence of viable, non-culture, spiral bacteria (*Campylobacter*) could be demonstrated by fluorescent antibody microscopy. Cleaning and sanitizing the water system significantly reduced the incidence of chicken colonization in subsequent flocks in the same house. When intervention measures were withdrawn, the broiler flocks gradually became recolonized with the same endemic serotype.

Progress in development of competitive exclusion as a treatment to prevent colonization of poultry by human bacterial enteropathogens was the subject of Session II. Dr. Esko Nurmi, Director, National Veterinary Institute, Helsinki, who was first to report successful competitive exclusion for protecting chickens from *Salmonella* colonization was the convener.

Competitive exclusion (CE) has probably existed as long as complex life has existed. It likely has played an important role in the evolutionary selective survival process of the various species in nature. CE is a natural protective process. CE is a complex interaction of microbes, nutrients, and host factors that selectively excludes specific groups or genera/species/strains of microorganisms from colonizing the intestinal tract. The mechanisms of CE are complex and poorly understood at this time. Mechanisms of exclusion vary for the particular bacterium of concern as illustrated by the inability of *Salmonella* excluding CE cultures to provide protection against *Campylobacter* colonization. Although CE had been previously studied in other species, it was Dr. Nurmi in 1973 who first reported convincing data that CE treatment of day-of-hatch chicks could be used as a means of limiting or preventing colonization by challenge with *Salmonella* cells. This was accomplished by simply gavaging chicks with aqueous slurries of *Salmonella*-free fecal matter from healthy adult chickens or anaerobic bacteriological cultures of the fecal material.

Following some preliminary characterization of the process by Dr. Nurmi, several other laboratories in different countries initiated in-depth studies attempting to enhance the protective capacity of CE cultures.

Dr. Rial Rolfe, Texas Tech University, presented a discussion entitled "Population Dynamics of the Intestinal Tract." He emphasized the complex microbial composition of the intestinal tract and the even more complex interactions of microorganisms present in the gut ecosystem that are involved in preventing both commensal and pathogenic bacteria from colonizing the intestinal tract. Hundreds of species may be involved along with host factors, in producing an environment that is antagonistic to human bacterial enteropathogen colonization of chickens. Considerable additional research will be necessary in order to understand fully the mechanisms of competitive exclusion and to learn how to exploit them to the fullest extent possible.

Dr. Roel Mulder, Spelderholdt Institute, The Netherlands, described the experience of scientists in the Netherlands with competitive exclusion. They demonstrated in both laboratory studies and large scale, commercial field trials that competitive exclusion can reduce the incidence of intestinal colonization of chickens by *Salmonella*. However, the beneficial reduction of intestinal colonization by competitive exclusion is offset by cross-contamination during transportation of chickens to slaughter plants and by processing procedures. He stressed the value of effective transport crate washing. The method of competitive exclusion culture treatment involved the spraying of culture into hatching cabinets when the hatching was 30% complete. He also noted some success had been achieved during a field trial study in reducing *Campylobacter* colonization in positive flocks. Dr. Mulder pointed out the value determining the human bacterial enteropathogen status of flocks prior to shipment to processing plants so that non-colonized flocks can be processed ahead of contaminated flocks, thereby reducing the incidence of contaminated carcasses.

Dr. Geoff Mead, AFRC Institute of Food Research, Bristol, England, spoke about developments in competitive exclusion to control *Salmonella* carriage in chickens in his country. After considerable research effort to develop competitive exclusion cultures with identified isolates obtained from cecal contents and/or anaerobic cultures of cecal material, it was concluded that defined cultures are of limited effectiveness in field trials. The complexity of defined mixtures, as well as the loss of protective capability during *in vitro* culture and storage, led to these conclusions. Undefined competitive exclusion cultures were successfully used to prevent *Salmonella* following antibiotic treatment of breeder hens previously known to be *Salmonella*- positive. In laboratory experiments, competitive exclusion treatment was shown to reduce the spread of *S. enteritidis* phage type 4 among chicks in the presence of infected seeder chicks. Experimental treatment of feed with acids in conjunction with cecal culture administration was found to be protective against laboratory *Salmonella* challenge. Dr. Mead emphasized the need for strict attention to biological safety requirements for production and application of undefined competitive exclusion cultures.

Dr. Nelson Cox, Agricultural Research Service, USDA, reported recent research on alternative methods of administrating competitive exclusion treatment to chickens. Surveys of hatcheries revealed a high incidence of *Salmonella* contamination of egg fragments, transport belts, paper pads, and fluff. It is well documented that the effectiveness of competitive exclusion treatment is diminished when chicks are exposed to *Salmonella* before treatment. Consequently, successful experiments were conducted in which competitive exclusion cultures were administered *in ovo* three days before chicks hatched and were exposed to *Salmonella*. Chicks treated in this manner were protected against colonization by a challenge dose of 10^5 *S. typhimurium* on the day-of-hatch. Much remains to be determined before this technology can be transferred to commercial hatchery operations.

Dr. Robert Tauxe, Centers for Disease Control, delivered an informative and entertaining after dinner lecture about the transmission of human bacterial pathogens through poultry. He described the various epidemiological methods that were used to investigate disease outbreaks. Current trends in incidence of campylobacteriosis and salmonellosis indicate that increases continue to be observed. Unique characteristics associated with *Campylobacter* and *Salmonella* were discussed. The spread of *S. enteritidis* from the Northeastern states to the Atlantic coast and Midwestern states indicated that the current control measures are not adequate to contain the spread of this organism among egg laying flocks. Dr. Tauxe emphasized the importance of and a need for educational programs to teach children, young adults, and the public in general, safe raw and cooked food handling practice.

Session III was devoted to presentations concerning the mechanisms of colonization of chickens by *Salmonella* and *Campylobacter*. The session was convened by Dr. Norman Stern, Agricultural Research Service, USDA. Although much research has been devoted to the study of pathogenic colonization of poultry and other animals, we still have major gaps in our knowledge in this area. In contrast, our understanding of commensal colonization is in its infancy. Significant benefit can be expected to be derived from a clear understanding of commensal colonization in terms of improved health, performance, and resistance to pathogens among domestic animals as well as humans.

Dr. Michael Doyle, University of Wisconsin, discussed colonization characteristics of chicks by *Campylobacter jejuni*. Dr. Doyle pointed out that poultry is perhaps the most important reservoir of *C. jejuni* among domestic animals and is likely the leading meat source of human campylobacteriosis. The bacterium mainly colonizes the ceca, large intestine, and cloaca of chickens in a commensal manner. *C. jejuni* preferentially resides in cecal crypts without a requirement for attachments to epithelial cells. Chemoattraction to mucin and L-fructose, which are imported constituents of mucus, may explain why *C. jejuni* is found most often in cecal crypts that are naturally filled with mucus. The bacterium has been shown to be capable of surviving and growing using mucin as its sole substrate. Dr. Doyle indicated that control and prevention of colonization of chickens by *C. jejuni* would be a major step toward reduction of the incidence of human campylobacteriosis.

Dr. Carlton Gyles, University of Guelph, Ontario, Canada, addressed *Salmonella* virulence factors and intestinal colonization of chickens and calves. He reported that among a collection of avian isolates, six serovars possessed typical large virulence plasmids while two serovars harbored small virulence plasmids. Curing the serovars of plasmids significantly diminished virulence. No correlation was found between serovar or virulence and serum resistance. It was established by means of DNA probes that plasmid-borne virulence genes were related or identical in plasmids from different serovars. Dr. Gyles found that *S. enteritidis* was relatively more resistant to chicken serum compared to other serovars tested, but this finding could not be correlated with plasmid profiles. A 36 Mdal plasmid in an *S. enteritidis* strain was demonstrated to relate to virulence. In studies of wild-type and mutant strains of *S. typhimurium* administered orally or injected into ligated ileal loops of calves, Dr. Gyles reported that the ileum was severely affected, followed by the jejunum, colon, and cecum. The intestinal mucosa invasion observed in calves would probably not be observed among chickens. Much remains to be learned from research on virulence and commensal colonization factors.

Dr. Charles Benson, University of Pennsylvania, discussed his observations on virulence properties of *Salmonella enteritidis* isolates obtained from poultry layer flocks. Dr. Benson and associates were the first to isolate *S. enteritidis* from the ovaries of laying hens. Strikingly, they detected the organism in only 15 of 6,000 eggs obtained from serologically positive flocks. One of the isolates was found to be highly infectious for the HeLa cells. Using this *in vitro* method for comparative studies it was found that the lethal dose-50% differed among isolates. None of 34 isolates obtained from various epidemic outbreaks were capable of enterotoxin production. It was further determined that roosters probably do not transmit *S. enteritidis* via their semen. Dr. Benson expects that a considerable research effort will be required to fully understand all the mysteries of *S. enteritidis*.

Dr. Charles Beard, Agricultural Research Service, USDA, described ongoing research in his laboratory involving experimental *Salmonella enteritidis* in chickens. He initially noted that the emergence of grade A table eggs as a major source of *S. enteritidis* infections in humans is closely linked to product mishandling and abuse, primarily in institutional settings. A sensitive microslide agglutination test was described in which antigens form homologous challenge strains were used in studying the progress of experimental infections and contamination of eggs. Cloacal swabs were also useful in showing that about 18% of experimentally infected hens remained culture positive for at least 18 weeks. It was observed that infected hens laid substantially fewer eggs and culture positive eggs were most prevalent during the first two weeks after *S. enteritidis* challenge. Only about 25% of challenged hens were found to be ovarially infected. The bacterium was never found in yolk contents, but rather was found on the yolk membrane and the albumin.

Session IV was devoted to immunization aspects of controlling *Salmonella* commensal colonization of chickens. The session was convened by Dr. Richard Meinersmann, Agricultural Research Service, USDA. Potential exists to exploit the chicken immune system as a means of controlling intestinal colonization by human bacterial enteropathogens; however, the commensal nature of the colonization by almost all paratyphoid *Salmonella* presents many complex and interesting research problems relating to manipulating and stimulating the developing immune system of the chicken to respond to commensal colonizers as though they were pathogenic colonizers.

Dr. Roy Curtiss, III, presented his recent research on properties and uses of live recombinant and nonrecombinant avirulent *Salmonella* for poultry. Studies on invasive strains of *S. typhimurium* and *S. enteritidis* that are capable of causing death in day-of-hatch chicks after oral challenge showed that the large virulence plasmid contributes to virulence. Further, the invasive (*inv*) genes and the gene necessary for survival in macrophages (*pho*P) are required for complete virulence. Some experiments suggested that heterogenicity may exist among chicks to fend off infection from an avirulent mutant. The immune response of chicks was found to be strongest to avirulent strains possessing the *inv* genes. Mutant strains of *S. typhimurium* and *S. enteritidis* with paired deletions of adenylate cyclase and cyclic AMP receptor protein were found to be totally avirulent with respect to lethality, but were capable of colonizing and invading the gut while having limited ability to invade liver and spleen. These mutant strains appear to produce a strong immunity not only to deep infection, but also to colonization by wild-type strains. Cross-protection was also produced against *Escherichia coli* air-saculitis. These mutants have served as effective vectors for delivering other recombinant antigens for vaccination purposes. More research is needed to assess the capacity of the mutant strains to diminish or prevent chicken colonization, persistence and shedding of various *Salmonella* species by immunized birds.

Dr. Paul Barrow, Institute of Animal Health, Houghton Laboratory, England, prepared a lecture on the subject of immunological control of *Salmonella* in chickens. Because Dr. Barrow was unable to attend the symposium, the presentation was skillfully made by the session chairman, Dr. Richard Meinersmann. It was initially pointed out that much knowledge concerning immunizing animals against *Salmonella* has come from experiments with mice in which a systemic disease is produced. This knowledge is likely to be irrelevant with respect to chickens in which most *Salmonella* intestinal colonizations produced no pathology or disease state. Although humoral and cellular immune responses to *Salmonella* colonizations can be detected, their role in intestinal clearance is unknown. Live attentuated *Salmonella* vaccines are largely ineffective in protecting chickens from the intestinal colonization by host non-specific serotypes. Killed cell vaccines produce either poor or inconsistent response. Laboratory experiments have shown that chickens which have cleared themselves of an invasive avian *S. typhimurium* infection are highly resistent to subsequent challenge. More basic knowledge on the microbial and the molecular basis of colonization, the relationship between colonization and immunity, and the identification of effective protective immunogens must be forthcoming before immunological protection of chickens to *Salmonella* can be achieved.

Dr. Hyun Lillehoj, Agricultural Research Service, USDA, spoke on the subject of intestinal immunity and heritable factors affecting the potentials of colonization. She reviewed the immune system as it relates to antimicrobial response of animals to human bacterial enteropathogen colonization. Dr. Lillehoj emphasized that the gut associated lymphoid tissues (GALT) are at the forefront of resistance to intestinal colonization. The influence of GALT is dependent upon age, immune status and genetic background of the host. The various cell types of the immune system can now be isolated, including macrophages, natural killer cells, B-cells and several subpopulations of T-cells in the gut, in the epithelium, and lamina propria. Major histo-compatibility complexes (MHC) are linked to resistance to enteric bacterial and coccidial colonization and infections. A clear association of several MHC genes and non-MHC genes has been determined in mice. Much new information will be required for development of immunological-based control of commensal colonization of chickens by human bacterial enteropathogens.

Recent technological advances in molecular biology and gene manipulation involving cloning of MHC genes and transgenic insetion into the various germlines of chickens is expected to lead to improvements in poultry genetic stocks.

Dr. Kakambi Nagaraja, University of Minnesota, discussed the question, "Is vaccination a feasible approach for the control of *Salmonella*?" He reported on studies using formalin-killed whole cell antigens and outer membrane protein (OMP) antigens for immunization of chickens. Autogenous killed-cell vaccines are useful. Laboratory tests showed that OMP vaccines protected turkeys from hetero and homologous challenge. OMP vaccines and whole killed-cell vaccines greatly decreased *S. enteritidis* incidence in organ infection and fecal excretion among challenged chickens.

Many excellent presentations were made during this symposium which represent a reasonably complete statement of our current knowledge about commensal colonization of chickens by *Salmonella* and *Campylobacter* and how little we know about how to control those colonizations. Also, many valuable poster presentations were given which have not been summarized here. The reader is referred to the extended abstracts of these posters which appear elsewhere in this book.

A number of conclusions and observations about this fine symposium seem appropriate. First, it is obvious that we are only beginning to scratch the surface with regard to our understanding of the characteristics and mechanisms that control the commensal colonization of chickens by human bacterial enteropathogens. It is safe to say that our knowledge in this area is developing rapidly as a consequence of the driving force of increased public concern and awareness of enteropathogen contamination of our fresh poultry and meat supply. It has become crystal clear that only an integrated enteropathogen colonization control program involving all phases of the integrated poultry industry will be successful in reducing or elimination carcass and product contamination. And lastly, much valuable interdisciplinary knowledge will result from research directed toward control of commensal colonization of human bacterial enteropathogens in poultry. This knowledge will have spin-off benefit for improvement of the health status, performance, and general well-being of humans and animals.

Lastly, we believe that our primary line of defense against enteropathogen contamination of the fresh poultry meat supply is in the production area. It is here that prevention of commensal colonization and subsequent delivery of *Salmonella*- and *Campylobacter*-free chickens to processing plants will lead to enteropathogen-free carcasses and products for consumers. The second line of defense will require improvements in the processing technology to prevent cross-contamination to the greatest extent possible while reducing the microbial load on carcasses to the lowest possible level. And the third line of defense must be an effective public school educational program that will provide students with the knowledge they need to safely handle, process, cook, and store fresh poultry and meat products.

INDEX

D

FOOD SCIENCE AND TECHNOLOGY

A Series of Monographs

Maynard A. Amerine, Rose Marie Pangborn, and Edward B. Roessler, PRINCIPLES OF SENSORY EVALUATION OF FOOD. 1965.

Martin Glicksman, GUM TECHNOLOGY IN THE FOOD INDUSTRY. 1970.

L. A. Goldblatt, AFLATOXIN. 1970.

Maynard A. Joslyn, METHODS IN FOOD ANALYSIS, second edition. 1970.

A. C. Hulme (ed.), THE BIOCHEMISTRY OF FRUITS AND THEIR PRODUCTS. Volume 1—1970. Volume 2—1971.

G. Ohloff and A. F. Thomas, GUSTATION AND OLFACTION. 1971.

C. R. Stumbo, THERMOBACTERIOLOGY IN FOOD PROCESSING, second edition. 1973.

Irvin E. Liener (ed.), TOXIC CONSTITUENTS OF ANIMAL FOODSTUFFS. 1974.

Aaron M. Altschul (ed.), NEW PROTEIN FOODS: Volume 1, TECHNOLOGY, PART A—1974. Volume 2, TECHNOLOGY, Part B—1976. Volume 3, ANIMAL PROTEIN SUPPLIES, PART A—1978. Volume 4, ANIMAL PROTEIN SUPPLIES, PART B—1981. Volume 5, SEED STORAGE PROTEINS—1985.

S. A. Goldblith, L. Rey, and W. W. Rothmayr, FREEZE DRYING AND ADVANCED FOOD TECHNOLOGY. 1975.

R. B. Duckworth (ed.), WATER RELATIONS OF FOOD. 1975.

Gerald Reed (ed.), ENZYMES IN FOOD PROCESSING, second edition. 1975.

A. G. Ward and A. Courts (eds.), THE SCIENCE AND TECHNOLOGY OF GELATIN. 1976.

John A. Troller and J. H. B. Christian, WATER ACTIVITY AND FOOD. 1978.

A. E. Bender, FOOD PROCESSING AND NUTRITION. 1978.

D. R. Osborne and P. Voogt, THE ANALYSIS OF NUTRIENTS IN FOODS. 1978.

Marcel Loncin and R. L. Merson, FOOD ENGINEERING: PRINCIPLES AND SELECTED APPLICATIONS. 1979.

Hans Riemann and Frank L. Bryan (eds.), FOOD-BORNE INFECTIONS AND INTOXICATIONS, second edition. 1979.

N. A. Michael Eskin, PLANT PIGMENTS, FLAVORS AND TEXTURES: THE CHEMISTRY AND BIOCHEMISTRY OF SELECTED COMPOUNDS. 1979.

J. G. Vaughan (ed.), FOOD MICROSCOPY. 1979.

J. R. A. Pollock (ed.), BREWING SCIENCE, Volume 1—1979. Volume 2—1980.

Irvin E. Liener (ed.), TOXIC CONSTITUENTS OF PLANT FOODSTUFFS, second edition. 1980.

J. Christopher Bauernfeind (ed.), CAROTENOIDS AS COLORANTS AND VITAMIN A PRECURSORS: TECHNOLOGICAL AND NUTRITIONAL APPLICATIONS. 1981.

Pericles Markakis (ed.), ANTHOCYANINS AS FOOD COLORS. 1982.

Vernal S. Packard, HUMAN MILK AND INFANT FORMULA. 1982.

George F. Stewart and Maynard A. Amerine, INTRODUCTION TO FOOD SCIENCE AND TECHNOLOGY, second edition. 1982.

Malcolm C. Bourne, FOOD TEXTURE AND VISCOSITY: CONCEPT AND MEASUREMENT. 1982.

R. Macrae (ed.), HPLC IN FOOD ANALYSIS. 1982.

Héctor A. Iglesias and Jorge Chirife, HANDBOOK OF FOOD ISOTHERMS: WATER SORPTION PARAMETERS FOR FOOD AND FOOD COMPONENTS. 1982.

John A. Troller, SANITATION IN FOOD PROCESSING. 1983.

Colin Dennis (ed.), POST-HARVEST PATHOLOGY OF FRUITS AND VEGETABLES. 1983.

P. J. Barnes (ed.), LIPIDS IN CEREAL TECHNOLOGY. 1983.

George Charalambous (ed.), ANALYSIS OF FOODS AND BEVERAGES: MODERN TECHNIQUES. 1984.

David Pimentel and Carl W. Hall, FOOD AND ENERGY RESOURCES. 1984.

Joe M. Regenstein and Carrie E. Regenstein, FOOD PROTEIN CHEMISTRY: AN INTRODUCTION FOR FOOD SCIENTISTS. 1984.

R. Paul Singh and Dennis R. Heldman, INTRODUCTION TO FOOD ENGINEERING. 1984.

Maximo C. Gacula, Jr., and Jagbir Singh, STATISTICAL METHODS IN FOOD AND CONSUMER RESEARCH, 1984.

S. M. Herschdoerfer (ed.), QUALITY CONTROL IN THE FOOD INDUSTRY, second edition. Volume 1 — 1984. Volume 2 (first edition) — 1968. Volume 3 (first edition) — 1972.

Yeshajahu Pomeranz, FUNCTIONAL PROPERTIES OF FOOD COMPONENTS. 1985.

Herbert Stone and Joel L. Sidel, SENSORY EVALUATION PRACTICES. 1985.

Fergus M. Clydesdale and Kathryn L. Wiemer (eds.), IRON FORTIFICATION OF FOODS. 1985.

John I. Pitt and Ailsa D. Hocking, FUNGI AND FOOD SPOILAGE. 1985.

Robert V. Decareau, MICROWAVES IN THE FOOD PROCESSING INDUSTRY. 1985.

S. M. Herschdoerfer (ed.), QUALITY CONTROL IN THE FOOD INDUSTRY, second edition. Volume 2 — 1985. Volume 3 — 1986. Volume 4 — 1987.

F. E. Cunningham and N. A. Cox (eds.), MICROBIOLOGY OF POULTRY MEAT PRODUCTS. 1986.

Walter M. Urbain, FOOD IRRADIATION. 1986.

Peter J. Bechtel, MUSCLE AS FOOD. 1986.

H. W.-S. Chan, AUTOXIDATION OF UNSATURATED LIPIDS. 1986.

Chester O. McCorkle, Jr., ECONOMICS OF FOOD PROCESSING IN THE UNITED STATES. 1987.

Jethro Jagtiani, Harvey T. Chan, Jr., and William S. Sakai, TROPICAL FRUIT PROCESSING. 1987.

J. Solms, D. A. Booth, R. M. Pangborn, and O. Raunhardt, FOOD ACCEPTANCE AND NUTRITION. 1987.

R. Macrae, HPLC IN FOOD ANALYSIS, second edition. 1988.

A. M. Pearson and R. B. Young, MUSCLE AND MEAT BIOCHEMISTRY. 1989.

Dean O. Cliver (ed.), FOODBORNE DISEASES. 1990.

Marjorie P. Penfield and Ada Marie Campbell, EXPERIMENTAL FOOD SCIENCE, third edition. 1990.

Leroy C. Blankenship, COLONIZATION CONTROL OF HUMAN BACTERIAL ENTEROPATHOGENS IN POULTRY. 1991.

Yeshajahu Pomeranz, FUNCTIONAL PROPERTIES OF FOOD COMPONENTS, second edition. 1991.